POLYMER YEARBOOK 13

POLYMER YEARBOOK

An annual desk reference featuring mini-reviews and useful information for polymer scientists.

Edited by Richard A. Pethrick

This book is part of a series. The publisher will accept continuation orders which may be cancelled at any time and which provide for automatic billing and shipping of each title in the series upon publication. Please write for details.

POLYMER YEARBOOK 13

Edited by
RICHARD A. PETHRICK
University of Strathclyde, UK

Associate Editors
GENNADI E. ZAIKOV
Institute of Chemical Physics
Russian Academy of Sciences, Moscow

TEIJI TSURUTA
NAOYUKI KOIDE
Science University of Tokyo, Japan

CRC Press
Taylor & Francis Group
Boca Raton London New York

CRC Press is an imprint of the
Taylor & Francis Group, an **informa** business

First published 1996 by Harwood Academic Publishers

Published 2019 by CRC Press
Taylor & Francis Group
6000 Broken Sound Parkway NW, Suite 300
Boca Raton, FL 33487-2742

© 1996 by Taylor & Francis Group, LLC
CRC Press is an imprint of Taylor & Francis Group, an Informa business

First issued in paperback 2019

No claim to original U.S. Government works

ISBN 13: 978-0-367-45591-0 (pbk)
ISBN 13: 978-3-7186-5914-2 (hbk)

Visit the Taylor & Francis Web site at
http://www.taylorandfrancis.com

and the CRC Press Web site at
http://www.crcpress.com

Library of Congress Cataloging-in-Publication Data

Polymer yearbook. – 1st ed. – Chur, New York, N.Y.:
Harwood Academic Publishers, 1984–
v.: ill.; 24 cm.
Annual.
ISSN 0738-1743 = Polymer yearbook.

1. Polymers and polymerization-Collected works.
QD380.P654   547.7dc19   84–648226
                                    AACR2   MARC-S
Library of Congress   [8508]

Contents

PROGRESS IN POLYMER SCIENCE IN JAPAN

CURRENT AWARENESS

Preface

I am very pleased this year to be able to include in Polymer Yearbook a brief bibliography recognising the contribution which Professor G.E. Zaikov of the Institute of Chemical Physics in Moscow has made to polymer science. This year is his 60th birthday and it was felt that it would be appropriate to include a short bibliography to mark this event.

Professor Zaikov has been active over the last four years in identifying authors and topics for specialist reviews and providing information on various conferences in Russia and allied states. The Japanese editors, Professors Teiji Tsuruta and Naoyuki Koide of the Science University of Tokyo, Japan, have helped organise review papers presented at the spring and fall meetings of the Japanese Polymer Science Society. Plans put in place a few years ago have led to a decrease in the delay in the production of these reviews, which contain reference to material which will not have as yet appeared in the English literature. We believe this will be recognized as a very important resource of relevant information on current topics.

Because of the large volume of material available for inclusion in Polymer Yearbook, this year there has been a reduction in the section on Data on Polymeric Materials and also Current Awareness, allowing an increase in the section devoted to reviews on specific topics. This year we have omitted the sections on abbreviations and trade names and these will be included in future editions.

Contributions from conference organizers should be sent to the Chief Editor, Professor R.A. Pethrick for consideration for publication in future volumes of Polymer Yearbook. The editors welcome suggestions from readers concerning ways in which Polymer Yearbook might be developed and improved to cater better for polymer scientists in the future.

The editors gratefully acknowledge the continuing help and meticulous devotion to Polymer Yearbook by Helen Paton who sustains the project at the University of Strathclyde.

Contributors

Nikolai N. Bakhman	Russian Academy of Sciences, Moscow, Russia
G.P. Gladyshev	Saratov State University, Saratov, Russia
M.D. Goldfein	Saratov State University, Saratov, Russia
Tamotsu Hashimoto	Fukui University, Fukui, Japan
Kenichi Hatanaka	Tokyo Institute of Technology, Yokohama, Japan
Masahiro Irie	Kyushu University, Fukuoka, Japan
Tisato Kajiyama	Kyushu University, Fukuoka, Japan
V.V. Kharitonov	Institute of Chemical Physics, Chernogolovka, Russia
A.I. Kokorin	Russian Academy of Sciences, Moscow, Russia
S.V. Kolesov	Bashkir State University, Ufa, Russia
A.L. Kovarski	Russian Academy of Sciences, Moscow, Russia
Kohei Kugo	Konan University, Kobe, Japan
E.I. Kulish	Bashkir State University, Ufa, Russia
A.B. Mazaletskii	Russian Academy of Sciences, Moscow, Russia
K.S. Minsker	Bashkir State University, Ufa, Russia
Norio Nemoto	Kyushu University, Fukuoka, Japan
B.L. Psikha	Institute of Chemical Physics, Chernogolovka, Russia
O.F. Shlensky	Russian Academy of Sciences, Moscow, Russia
Gabriel O. Shonaike	Kyoto Institute of Technology, Kyoto, Japan
Atsushi Takahara	Kyushu University, Fukuoka, Japan
D.A. Topchiev	Russian Academy of Sciences, Moscow, Russia
A.V. Trubnikov	Saratov State University, Saratov, Russia
M. Tsunooka	University of Osaka Prefecture, Osaka, Japan
E.F. Vaynsteyn	Russian Academy of Sciences, Moscow, Russia
V.G. Vinogradova	Russian Academy of Sciences, Moscow, Russia
Hideki Yamane	Kyoto Institute of Technology, Kyoto, Japan
Yu. G. Yanovsky	Russian Academy of Sciences, Moscow, Russia
Gennadi E. Zaikov	Russian Academy of Sciences, Moscow, Russia

Review Section

GENNADI EFREMOVICH ZAIKOV
60th Birthday, 40 Years of Scientific Activity

Professor Gennadi E. Zaikov, DSc, was born on January 7, 1935 in Omsk in the family of a military officer and school teacher. He finished secondary school in Omsk in 1952 and entered the Chemistry Department of Moscow State University (MSU), as an undergraduate in December 1957 and gained a first class BSc degree.

Professor Zaikov started his scientific research in the Isotope Separation Unit of the Moscow State University (MSU), of which Professor Georgii M. Panchenkov was Head. This research was concerned with the separation of Li^6 and Li^7 isotopes for the nuclear industry.

When an undergraduate at MSU, Professor Nikolai M. Emanuel invited him to work in the Institute of Chemical Physics (ICP) of the USSR Academy of Sciences. Professor Emanuel liked jazz very much and requested G.E. Zaikov to find a jazz-band.

Since February 1958 G.E. Zaikov has worked in Institute of Chemical Physics. His postgraduate studies were concerned with the kinetics and mechanism of oxidation of a number of organic compounds in the liquid and gas phase nearly their critical conditions of temperature and pressure. The thesis, for which the PhD degree was awarded in January, 1964, was the basis of an industrial method for the production of acetic acid and methylethylketone by means of liquid-phase oxidation of n-butane. During the period of 1964–1968, G.E. Zaikov prepared his DSc thesis entitled "The role of medium in radical-chain processes of oxidation". He was awarded his DSc degree in May, 1968, and went to the laboratory of Professor Keith Ingold (National Research Council, Ottawa, Canada) for his first Postdoctoral Fellowship.

In 1966 G.E. Zaikov extend his research interest to Polymer Ageing and Stabilisation. Since 1967 he has held the position of Head of the Laboratory of Chemical Stability of

Polymers. He became the Deputy of Professor Emanuel in the Division of Polymer Ageing in the Institute of Chemical Physics. He has been the Chairman of a number of local and international conferences on Degradation and Stabilisation of Polymers, Flammability of Polymeric Materials.

Since 1970 G.E. Zaikov has been a Professor. He has been the supervisor of 25 PhD thesis. He is an expert in many fields: chemical physics, chemical kinetics, ageing and stabilisation of polymers, biological kinetics. His main research interests are oxidation of low-molecular and macromolecular compounds, ozonolysis, hydrolysis, biodegradation, mechanical decomposition of polymers, flammability of polymeric materials. He developed kinetic fundamentals in these areas, and enabled substantial progress in the reliable exploitation and storage of polymeric materials and products as well as in prediction of "life-times".

He wrote his first paper in 1958, and has now published about 1000 papers, including 70 reviews, 15 booklets and 44 monographs. Half of his books are published in English in the USA and Europe.

Zaikov is the Head of the Division and Leader of cooperation of Academies of Sciences of Socialistic Countries on Ageing and Stabilisation of Polymers.

G.E. Zaikov is the Member of the Editorial Board of 8 international journals, publishing in USA, UK, Bulgaria and Poland, as well as of 2 Russian journals.

He teaches a course on Ageing and Stabilisation of Polymers for undergraduate students of Lomonosov State Academy of Fine Chemical Technology, Moscow. Professor Zaikov is well-known in international scientific society and for 14 years he has been a member of several committees of the International Union on Pure and Applied Chemistry (IUPAC). He has contributed to many international conferences with plenary lectures.

Professor Zaikov is very active in the development of cooperation with Western countries (USA, Western Europe). His activity have helped to keep together scientists of the Department of Chemical and Biological Kinetics of the Institute of Chemical Physics, when Russian Government has reduced and even almost suspended financial support of science. Professor Gennadi E. Zaikov meets his 60th birthday full of energy and new scientific ideas.

PVC Degradation in Blends with Other Polymers

K.S. MINSKER,[1] S.V. KOLESOV,[2] E.I. KULISH[3] and G.E. ZAIKOV[4]

[1]Bashkir State University, 1–19, Frunze Str., Ufa, 450074; Tel. (office): 8-3472-22-47-85; Tel. (private): 8-3472-23-66-08
[2]Bashkir State University, 25/2–92, Perovskaya Str., Ufa, 450092
[3]Bashkir State University, 25/1–18, Gafuri Str., Ufa, 450076; Tel. (private): 8-3472-22-50-83
[4]Institute of Chemical Physics Russian Academy of Sciences, 64–188, Leninski Pr., Moscow, 117296; Tel. (office): 939-71-91; Tel. (private): 137-17-93

The PVC thermal stability is very dependent on the presence of chemical additives.[1] The problem of their effect on PVC degradation has not been extensively investigated despite their extensive use in blends obtained by milling and melt formation. This review considers the current understanding of the problem of PVC degradation.

SOME PECULARITIES OF PVC DEGRADATION

Some critical phenomena associated with degradation of PVC are the sharp variation of the reaction rate, the elimination of HCl (V_{HCl}), the dependence on the processing temperature and the sample size (thickness etc).[2] One should take into account that PVC is not indifferent to eliminating HCl. Therefore, if the PVC degradation is conducted in a closed space and HCl is not trapped by an inert additive — HCl acceptor, the process will develop an autoacceleration mode.[3] If, however the HCl is quickly trapped by the chemical additive, the HCl elimination rate from PVC is determined mainly by the rate of PVC and HCl solubility in the polymer. For thin films, when HCl diffusion is not a controlling process, V_{HCl} is determined by the rate of the uncatalyzed formation of HCl. If the sample size exceeds some critical dimension the HCl diffusion rate will be lower than the rate of HCl formation and a continuous increase of the PVC dehydrochlorination rate due to its catalysis under the influence of HCl is observed. When PVC is blended with another polymer the change in the HCl solubility can increase the rate of degradation. For filmed specimens from PVC-thermoplastic polyurethane (TPU) the critical film thickness ($l_{cr.}$) at 448 K is approximately 3 times lower than that obtained for PVC;[1] $l_{cr.}$ decreases as polyurethane content in the blend increases.

5

THERMAL DEGRADATION OF PVC BLENDS

PVC-TPU Blend

Polyurethane, polymers are thermodynamically compatible with PVC,[4] and degradation accelerates with elimination of HCl[5,6] due to the presence of nitrogen-containing groups and their ability to build electrophile complexes with HCl. The general process is as follows:

$$PVC \xrightarrow{K_1} HCl \tag{1}$$

$$PVC \xrightarrow[TPU]{K_3'} HCl \tag{2}$$

$$HCl + TPU \xrightarrow[K']{K_2} TPU\ HCl \tag{3}$$

$$PVC \xrightarrow[TPU \cdot HCl]{K_3''} HCl \tag{4}$$

The PVC degradation rate is considerably increased with increasing TPU content in the blend (Fig. 1, Curve 1). With the absence of the HCl stabilizer-acceptor the HCl forming in the PVC degradation reacts with urethane groups within the TPU molecules to produce the corresponding nitrile salt (Reaction (3)). The process of TPU interaction with HCl is reversible. A dynamic equilibrium is established between the reactions of hydrochlorination of urethane groups and dehydrochlorination of the complex. The kinetic parameters of HCl complex degradation are determined by the HCl elimination curves from TPU nitrile salts:

$$Z_{eq.} = [HCl]_{in.} \cdot (1 - e^{-K' \cdot \tau}) \tag{5}$$

here $Z_{eq.}$ is the equilibrium content of urethane groups within TPU, (mol/TPU mol); τ is the process duration time, (s); $[HCl]_{in.}$ is the instant mean concentration of HCl in the reaction volume, determined by the rate of HCl elimination from PVC; K' is the rate constant of the hydrochlorinated complex degradation within the TPU macromolecule (s^{-1}), Table (1).

The equilibrium conditions are expressed by:

$$K_2 \cdot [HCl]_{in.} \cdot (Z_0 - Z_{eq.}) = K' \cdot Z_{eq.} \tag{6}$$

where Z_0 is the initial content of urethane groups taking part in HCl linking, (mol/TPU mol); K_2 is the constant rate of HCl linked with the urethane group, $(HCl\ mol \cdot s)^{-1}$. From the known value of Z_0 and experimentally established $Z_{eq.}$ and K', the constant rate of formation of hydrochlorinated urethane groups K_2 is determined on the basis of (6) (Table 1). The value of the equilibrium constant for the given process (K_R) is estimated from the value of K' and K_2 (Table 1).

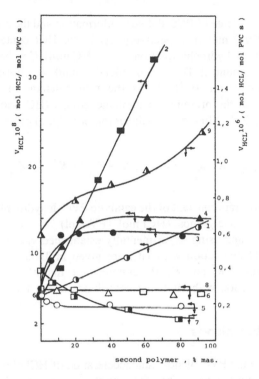

Figure 1 The dependence of the PVC thermal degradation rate on the second polymer content in the blend (1–8 for T = 423 K; 9 for T = 448 K): 1 – PVC-TPU blend; 2 – PVC-SNR-40 blend; 3 – PVC-SNR-18 blend; 4 – PVC-SNR 26 blend; 5 – PVC-PMMA blend; 6 – PVC-poly-α-olefins blend; 7 – PVC-MBS blend; 8 – PVC-ABS blend and 9 – elastically deformation dispersion PVC-PE blend.

In conditions when the HCl elimination is fast and quantatively linked to the effective stabilizer — acceptor (barium stearate), the HCl acceptance by the TPU urethane groups is premature. The kinetics of PVC dehydrochloration in the blends with TPU in this case corresponds to a linear dependence — "HCl output — time" and is described by the equation, characteristic of the PVC degradation with the additive catalyzing the HCl elimination from PVC:[1]

$$V_{HCl} = V_{HCl}^{PVC} + V_3 + K_1 \cdot a_0 + K_3 \cdot a_0 \cdot C \qquad (7)$$

TABLE 1
The rate constants of PVC catalytic dehydrochlorination in blends with TPU and nitrile rubbers (N_2, 423 K)

	TPU	SNR-18	SNR-26	SNR-40
$K'10^4$, s^{-1}	15,1±0,3	0,84±0,04	2,23±0.11	3,23±0.1
$K_2 10^{-2}$, (HCl mol·s)$^{-1}$	80,2±3,0	2,01±0,10	4,79±0,23	5,49±0,2
$K_R.10^2$, (HCl mol)$^{-1}$	5,24±0,26	2,38±0,11	2,05±0,10	1,69±0,0
$K'_3 10^6$, s^{-1} (HCl mol/PVC mol)	14,3±2,12	0,25±0,12	0,36±0,27	0,40±0,2
$K'''_3 10^2$, s^{-1} (HCl mol/second polymer mol)$^{-1}$	46,1±2,11	2,03±0,10	0,49±0,02	0,03±0,00

where V_{HCl}^{PVC} and V_3 are the rate of PVC dehydrochlorination and the rate of PVC catalyzed by TPU, (mol PVC/PVC mol·s) respectively; a_0 is the HCl content within the PVC monomer link before the dehydrochlorination, (mol/PVC mol); C is the content of urethane groups related to the amount of TPU in the blend, (mol/PVC mol); K_1 is the constant of PVC dehydrochlorination, (s^{-1}); K_3 is the rate constant of the catalytic PVC dehydrochlorination with the presence of urethane groups (HCl mol/PVC mol)$^{-1}$ s^{-1}.

The stationary value $V_{HCl}^{eq.}$, after the non-linear (autocatalytic) period of HCl elimination is:

$$V_{HCl}^{eq.} = K_1 \cdot a_0 + K_3' \cdot a_0 \cdot (Z_0 - Z_{eq.}) + K_3'' \cdot a_0 \cdot Z_{eq.} \qquad (8)$$

where K_3'' is the effective rate constant of the catalytic PVC dehydrochlorinative influenced by hydrochlorinated urethane groups, (HCl mol/TPV mol)$^{-1}$ s^{-1}. The K_3'' values obtained by solving equation (8) and using experimentally established values of $V_{HCl}^{eq.}$ and kinetic parameters of the TPU interaction with HCl are given in Table 1.

Comparing the values of K_3' and K_3'' the conclusion is that it's not the urethane groups directly but their interaction with HCl that contributes to the PVC degradation.

PVC — nitrile rubber blends

Using nitrile rubbers in the PVC blends, autoacceleration of HCl elimination is observed. The HCl elimination rate increases until the equilibrium content of N-substituted nitrile salts, is achieved determining the rate constant for the HCl elimination from PVC ($V_{PVC}^{eq.}$).[7]

For PVC-SNR-40 blends (full thermodynamic compatible[8]) in the film state, obtained from solution in 1,2-dichlorethane, an increase of $V_{HCl}^{eq.}$ is observed as the SNR-40 content in the blend is increased (Fig. 1, Curve 2). Using PVC blends with nitrile rubbers with CN-groups lower content (SNR-18 and SNR-26), the rate $V_{HCl}^{eq.}$ increases occurs only up to 10 and 20 mass.%, respectively (Fig. 1, Curves 3,4). The maximum volume being determined by the limited thermodynamic compatibility of PVC with the nitrile rubbers.[9] As the content of nitrile rubbers of SNR-18 and SNR-26 grades exceeds the value corresponding to thermodynamic compatibility, SNR is isolated into a separate phase. Consequently, further increase of SNR-18 and SNR-26 content and the rubber phase volume in the PVC blend do not change $V_{HCl}^{eq.}$. Therefore the nitrile groups content, capable of interaction with PVC remains constant with growth in SNR content.

An analogus procedure of kinetic data processing[10] may be applied to this system. The numerical values of the rate constants of PVC catalytic dehydrochlorination in the presence of the nitrile group and hydrochloric complexes [CN·HCl] effects the kinetic parameters, Table 1.[10]

Comparison of constants K_3' and K_3'' in the degradation of PVC blend with SNR and TPU in PVC blends indicates that the TPU-HCl increases the rate of urethane hydrochloric complexes dehydrochlorination by two orders of magnitude with respect to nitrile rubbers alone. This is explained by the higher basicity of TPU compared with SNR.

PVC-PMMA blend

The degradation of PVC blends with polymethylmethacrylate (PMMA) is characterized by a diminishing rate of PVC thermal dehydrochlorination in comparison with the pure polymer.[11,12] As the PMMA content in the PVC blend increases a reduction of the process is observed (Fig. 1, Curve 5). In the experimental conditions (448–463 K) PMMA is depolymerized to MMA monomer,[13] one can conclude that the inhibition of PVC dehydrochlorination is due to methylmethacrylate.

PVC-ABC and PVC-MBS blends

Methylmethacrylate-butadiene-styrene copolymer (MBS) within the range 5–50 mass.% in the blends of PVC and PMMA effectively reduces V_{HCl} in PVC thermal degradation.[14] The higher the MBS content and processing temperature, the greater the observed reduction of HCl elimination (Fig. 1, Curve 7).

In thermal degradation of PVC blends with acrylonitrile-butadiene-styrene copolymer (ABS) the rate of PVC thermal dehydrochlorination V_{HCl} remains the same as for pure PVC irrespective of the amount of polymer additive (Fig. 1, Curve 8).

PVC-poly-α-olefins blends

Poly-α-olefins, contain no functional groups are thermodynamically incompatible with PVC[4] and have no influence on PVC. In PVC blends with polyethylene (PE), polypropylene (PP), polyisobutylene (PIB) obtained as film, the PVC thermal stability remains constant (Fig. 1, Curve 6). However, under the conditions of dispersion from the melt,[15] a considerable increase of PVC dehydrochlorination rate is observed as the PE content in the blend is increased[15] (Fig. 1, Curve 9). Chemiluminescence methods did not identify oxidation products that could have accelerated PVC thermal degradation.

Thus, despite the chemical unreactivity of poly-α-olefins may change the PVC degradation rate in cases when polymers are well dispersed, unexpected as this effect may be.

THERMOOXIDATIVE DEGRADATION OF PVC BLENDS

In the presence of oxygen, HCl elimination from PVC is markedly increased except for polymethylmethacrylate (Fig. 2, Curve 1).[16] Degradation of PVC in thermooxidative conditions is increased by the products giving off during oxidation.

PVC-TPU blends

Degradation of PVC in blends with polyurethane is considerably accelerated in the presence of oxygen. Even a small TPU content (5–7 mass.%) in the PVC blend causes the rate

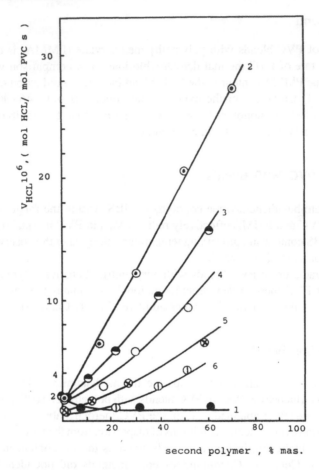

Figure 2   The dependence of the PVC thermooxidative degradation rate on the second polymer content in the blend (1–3,5 for T = 448 K; 4,6 for T = 443 K): 1 – PVC-PMMA blend; 2 – PVC-TPU blend; 3,5 – PVC-MBS blend and 4,6 – PVC-ABS blend.

of PVC thermooxidative degradation to go up (Fig. 2, Curve 2). Oxygen activates TPU oxidation and the products formed promote degradation indicated by a sharp increase in the chemiluminescence intensity (I_{chl}) compared to the values for the pure polymer (Fig. 3, Curves 1–4).

A linear dependence of the maximum chemiluminescence intensity I_{chl} on the oxidation-dehydrochlorination rate (V_{HCl}) of PVC blends with TPU is observed (Fig. 3, Curve 5):

$$\Delta I_{chl}^{max} = I_{chl}^{max} - I^* \quad \text{and} \quad \Delta V_{HCl} = V_3^* - V_{HCl}^*$$

where I_{chl}^{max} and V_3^* are the values of maximum chemiluminescence intensity and HCl elimination rate respectively. The formation of hydroperoxides in the thermooxidation of TPU may arise from both urethane and polyester groups. Thus, the kinetics of degradation of PVC blends with TPU in oxygen medium depends mainly on the oxidation stability of TPU.

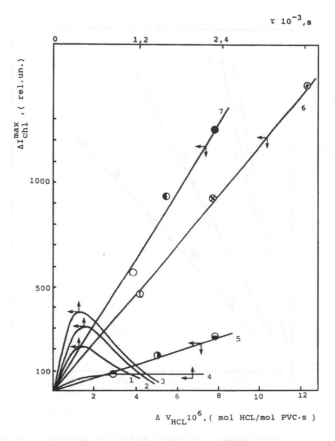

Figure 3   The variation of chemiluminescence intensity at the PVC thermooxidative destruction in the blend with TPU. TPU content in the 100% mas. PVC-TPU blend: 1 – 10; 2 – 25; 3 – 50; 4 – 0: and the dependence of chemiluminescence intensity variation at the PVC thermooxidative destruction in the blend with TPU (curve 5) and ABS (curve 6), MBS (curve 7) on the variation of oxidative dehydrochlorination rate of PVC (10–50% mas.second polymer content in the 100% mas. PVC-TPU blend (T = 433 K, O_2).

The chains generated on reactions are:[17]

$$RH + O_2 \xrightarrow{\quad V_{in} \quad} R^{\cdot} + HO_2^{\cdot}$$

$$R^{\cdot} + O_2 \longrightarrow ROO^{\cdot}$$

with an initiation rate of V_{in}. Hydroperoxides accumulation proceeds according to the reaction:

$$ROO\cdot + RH \xrightarrow{\quad k_2 \quad} ROOH + R\cdot$$

where k_2 is the rate constant of oxidative chains destruction, the former are decomposed into free radicals with the rate constant k_3:

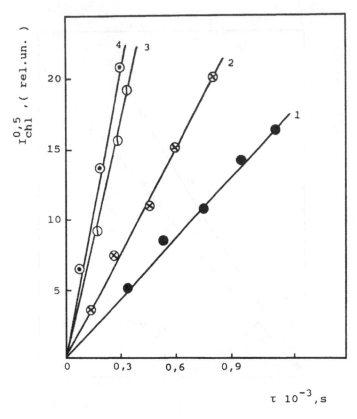

Figure 4   The dependence of chemiluminescence intensity on the TPU oxidation time at the temperature, K: 1 – 413; 2 – 418; 3 – 423; 4 – 433 K (atm. O_2).

$$ROOH \xrightarrow{\ k_3\ } RO^\bullet + OH^\bullet$$

This fact contributes to autoacceleration of the reaction with time.

In case of quadratic termination of the reaction chains in free radical oxidation of TPU:

$$RO_2^\bullet + RO_2^\bullet \xrightarrow{\ k_6\ } \text{molecular product}$$

The TPU oxidation kinetics is described by the famous parabolic dependence:

$$[ROOH]^{0,5} = 0,5 \cdot k_2 \cdot k_6^{-0,6} \cdot k_3^{-0,5} \cdot [RH] \cdot \tau$$

Kinetic chemiluminescence curves exhibit a linear dependence in the coordinates "$I_{chl}^{0,5}$ – time" (Fig. 4).

The angle of inclination of the line is proportional to the oxidation parameter: $K_{ef} = k_2 \cdot k_6^{-0,5} \cdot k_3^{0,5}$, Table 2. The above parameter, characterizes the autocatalytic kinetics of TPU oxidation and include the ratio of the constants k_2/k_6, which determine the rate

TABLE 2

The values of parameter $k_2 \cdot k_6^{-0,5} \cdot k_3^{0,5}$ for chemiluminescence analysis (atm O_2).

$k_2 \cdot k_6^{-0,5} \cdot k_3^{0,5}$, (rel. units)			Temperature,
ABS	MBS	TPU	(K)
0,8±0,1	1,5±0,1	0,2±0,02	413
2,0±0,2	2,6±0,2	0,4±0,04	418
2,9±0,3	3,3±0,3	0,6±0,07	423
4,2±0,3	4,8±0,4	0,9±0,10	433

of chain oxidation at a constant rate of initiation and the effective rate of initiation k_3 of TPU oxidation. The value of the TPU oxidability parameter:

$$K_{ef} = k_2 \cdot k_6^{-0,5} \cdot k_3^{0,5}$$

increases with the temperature and according to experiment, the dependence of the oxidability parameter on time is approximated by a straight line (Fig. 5, Curve 1).

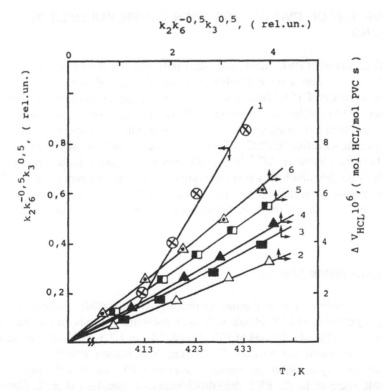

Figure 5   The dependence of oxidability parameter on the TPU oxidation temperature (the temperature interval — 413–433 K, O_2) — curve 1 and the dependence of dehydrochlorination rate variation at the PVC thermooxidative destruction in the blend with ABS (2,4,6) and MBS (3,5) on the oxidability parameter of copolymer. Copolymer content in the 100% mas. blend with PVC: 2,3–5; 4,5–10; 6–20 (413–433 K, O_2).

PVC-ABS and PVC-MBS blends

Both ABS and MBS copolymers accelerate the thermooxidative degradation of PVC blends. MBS has a greater effect than ABS (Fig. 2, Curves 3–6). The acceleration of thermooxidative degradation of PVC by ABS and MBS is connected to TPU, with catalytic formation of hydroperoxides. The chemiluminescence intensity variation is linearly dependent on the variation of HCl elimination rate (Fig. 3, Curves 6,7). The chemiluminescence intensity is dependent on the process temperature and the oxidability of the second polymer. In MBS, oxidation by atmospheric oxygen being, connected with the presence of the polybutadiene block is twice as great as in ABS. Table 2 gives values of the oxidability parameter $K_{ef} = k_2 \cdot k_6^{-0,5} \cdot k_3^{0,5}$ for ABS and MBS indicating the higher oxidability of MBS than ABS. For acrylonitrile butadiene styrene copolymer (ABS) the values of the parameter are considerably higher than in the case of TPU in PVC blends.

The thermooxidative degradation of PVC blends with other polymers is determined by the oxidation stability of the second polymer and is associated with the accumulation and decomposition of hydroperoxides formed.

DEGRADATION OF PVC BLENDS WITH OTHER POLYMER IN SOLUTIONS

Thermal degradation of PVC blends with other polymers in solutions is essentially different from that for PVC and a second polymer in the absence of solvent.[18]

Thermal stability of PVC blends with other polymers in solution are determined by two factors: (1) the influence of solvent; (2) the influence of the second polymer. The PVC degradation rate is influenced by two types of polymer — solvent interaction; specific and non-specific solvation of PVC.[19] Specific interaction of PVC in solutions is correlated with the basicity parameter "B".[3] In the solvents with basicity higher than that of PVC ($B > 50$ cm^{-1}), the dehydrochlorination rate is greater than in the absence of solvent. In solvents with basicity lower than that of PVC ($B < 50$ cm^{-1}), the degradation rate of PVC without solvent exceeds the value of the degradation rate in solution — solvation stabilization of PVC (Fig. 6).

PVC-poly-α-olefins blends

Even in the simplest case of thermodynamically incompatible, chemically unreactive solutions, degradation of PVC blends with poly-α-olefins depends on the chemical nature of the solvent. In diluted solutions of PVC (mass. 2%) in 1,2,3-trichloropropane and 1,2,4-trichlorobenzene (with the basicity $B = 20$ cm^{-1}), solvation stabilization is observed. Introduction of poly-α-olefins in amounts exceeding 10% of the PVC mass results in a considerable increase in the PVC dehydrochlorination rate V_{HCl} (Fig. 7, Curve 1). One of the causes could be oxidation of poly-α-olefins in the process of solution preparation. The accelerating effect of oxidation products is well known.[14,22] However, the chemi-

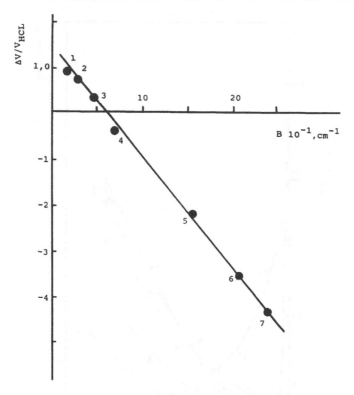

Figure 6   The dependence of relation variation of the PVC thermal dehydrochlorination rate on the basicity parameter B (N_2, T = 423 K): 1 – p-dichlorobenzene; 2 – o-dichlorobenzene; 3 – naphthalene; 4 – nitrobenzene; 5 – benzonitrile; 6 – acetophenone and 7 – cyclohexanone.

luminescence analysis of solutions, prepared at higher temperature in an inert gas flow proves that neither poly-α-olefin (PE) nor PVC are oxidated in these conditions.

In degradation of PVC blends with poly-α-olefins in cyclohexanol (with basicity $B = 242$ cm^{-1}), the PVC dehydrochlorination rate reduces as compared to V_{HCl} of PVC, with increase of poly-α-olefin content in the blend (Fig. 7, Curve 2). The similar dependence are observed with related systems.

The change in conditions, lowering the temperature to 383 K or increase of the concentration to 10 mass.% leads to phase separation. Poly-α-olefins under these conditions cease to influence thermal degradation.

The use of dodecane as a low-molecular analogue for PE results in unusual characteristics for the dehydrochlorination rate variation of PVC in solution (Fig. 7, Curves 5,6). Hydrocarbons play the part of "poor" and "good" solvents. Successive addition of the solvent leads to precipitation of PVC solution, whilst with thermodynamically "good" solvents the variation of V_{HCl} similar to that for polymer hydrocarbons, the PVC dehydrochlorination rate reducing with solvent basicity $B > 50$ cm^{-1} (cyclohexanone) and increasing in solutions with basicity of $B < 50$ cm^{-1} (o-dichlorobenzene). The effect of poly-α-olefins and low-molecular hydrocarbons is the result of precipitant.

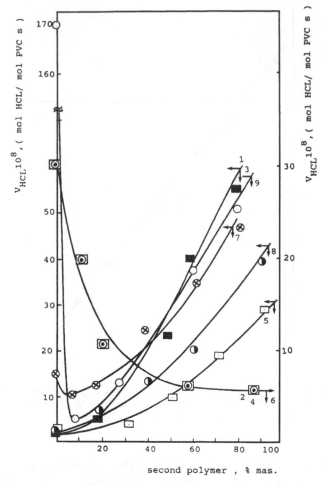

Figure 7   The dependence of the thermal degradation rate of the PVC in solution on the second polymer content in the blend (N_2, 2% polymer solution, T = 423 K); (1,3,5,7 — in the 1,2,3-trichloropropane solution; 2,4,6 — in the cyclohexanol solution; 7 — in the cyclohexanone solution and 8 — in the benzoic alcohol solution): 1,2 — PVC-poly-α-olefins blend; 3,4 — PVC-polyethylene wax blend; 5,6 — PVC-dodecane blend and 7–9 — PVC-PMMA blend.

PVC-TPU blends

Liquid phase degradation of PVC blends with TPU showed that in this case the PVC dehydrochlorination rate V_{HCl} increases with increase of TPU content in blend in solvents with both high (B = 242 cm^{-1}, cyclohexanone) (Fig. 8, Curve 1), and low basicity (B = 20 cm^{-1}, 1,2,3-trichloropropane) (Fig. 8, Curve 2). However, the degree of dependence of V_{HCl} of PVC on the TPU content in blend is different being considerably greater in cyclohexanone, than in 1,2,3-trichloropropane and is connected to the different nucleophilic properties of the solvents.

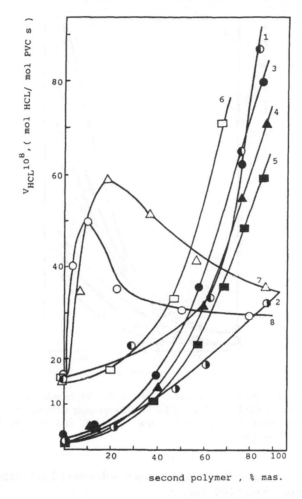

Figure 8   The dependence of the thermal degradation rate of the PVC in solution on the second polymer content in the blend (N_2, 2% polymer solution, $T = 423$ K); (1,6,7,8 – in the cyclohexanone solution, 2–5 – in the 1,2,3-trichloropropane): 1,2 – PVC-TPU blend; 3,6 – PVC-SNR-40; 4,7 – PVC-SNR-26 and 5,8 – PVC-SNR-18.

PVC-nitrile rubbers blends

Degradation of diluted solutions (2 mass. %) of PVC, in 1,2,3-trichloropane ($B = 20$ cm^{-1}) with SNR indicate an increase of degradation rate with increase in rubber grades: SNR-18, SNR-26 and SNR-40 (Fig. 8, Curves 3–5). However, in the degradation of PVC blends with SNR in cyclohexanone ($B = 242$ cm^{-1}), the dehydrochlorination rate increases with second polymer content (Fig. 8, Curve 6). For PVC with SNR-18 and SNR-26, the degradation rate of PVC in solutions is extreme (Fig. 8, Curves 7,8).

At 423 K a 2% polymer solution forms a single phase increasing the concentration up to 25% in 1,2,3-trichloropropane and to 60% in cyclohexanone leads to thermodynamic

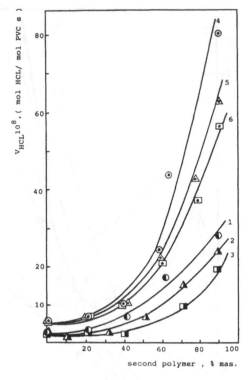

Figure 9   The dependence of the thermal degradation rate of the PVC in solution on the second polymer content in the blend (N_2, T = 423 K); (1–3 in the 1,2,3-trichloropropane solution, 25% polymer solution; 4–6 in the cyclohexanone solution, 60% polymer solution): 1,4 – PVC-SNR-18 blend; 2,5 – PVC-SNR-26 blend and 3,6 – PVC-SNR-40 blend.

incompatibility. In this case both the degradation in solution of 1,2,3-trichloropropane and PVC blends increases with SNR content.

PVC-PMMA blends

In the liquid-phase degradation of PVC blends with PMMA in the cyclohexanone solution ($B = 242$ cm^{-1}), dehydrochlorination inhibition is observed for PMMA contents of 10%, or below. Further increase in PMMA content considerably increases the degradation rate and at 30 mass. % exceeds the dehydrochlorination rate of pure PVC (Fig. 7, Curve 7). For degradation in 1,2,3-trichloropropane solution ($B = 20$ cm^{-1}), the dehydrochlorination rate increases at all proportions (Fig. 7, Curve 8).

   In benzoyl alcohol solution ($B = 208$ cm^{-1}) a more complicated picture is observed. The dependence PVC dehydrochlorination rate is similar to that in cyclohexanone solution. However, the reduction of PVC degradation rate is greater by about 30 times in benzoyl alcohol in comparison to 2–3 times in the case of cyclohexanone. Values of the PVC degradation rate V_{HCl} at higher contents of PMMA are considerably lower than for PVC without solvent (Fig. 7, Curve 9).

CONCLUSION

The HCl elimination rate in the degradation of PVC blends with other polymers is determined by:

(1) the presence of functional groups within macromolecules, (TPU, SNR);
(2) the ability to form hydrochloric complexes in case of using TPU and SNR;
(3) the stability of the second polymer to oxidation by air;
(4) the ability of low-molecular products formed in the degradation to interact with labile carbonylallyl groups or with HCl.

In PVC degradation in solution is dominated by the basicity parameter B which determines the degree of specific interaction of polymer with solvent.

REFERENCES

1. K.S. Minsker and G.T. Fedoseeva, Destrukshija i stabilizashija polivinilkhlorida. — M. Khimija (1979).
2. V.S. Pudov and R.A. Papko, *Vysokomol. Soed.*, T.(B)12.n 3.s.218 (1970).
3. V.S. Pudov, *Plastmassy*, n2, s.18 (1976).
4. Polymernye smesi/pod redakshijei Pola, Nijumena. — M. Mir (1981).
5. J. Piglovski and W. Laskawski, *Angew. Macromol. Chem.*, v.84, p. 163 (1980).
6. S.N. Blagova, L.N. Samoilenko, V.V. Strakhov and N.I. Lytkina, Issledovanije tekhnologii poluchenija iskusstvennykh kog tekhnicheskogo naznachenija. Sbornik trudov ShNIITEI legprom.-M. s.108 (1984).
7. A.A. Berlin, A.G. Kronman, D.M. Janovskij and V.A. Kargin, *Vysokomol. Soed.*, T.(B)2.n 8.s.1188 (1960).
8. Ju.G. Oganesov, A.G. Kuleznev and S.S. Vojushkij, *Vysokomol. soed.*, T.(B)2.n 9.s.691 (1970).
9. A.B. Aivasov, Kh.G. Mindijarov, Ju.V. Zelenev, Ju.G. Oganesov and V.G. Raevskij, *Vysokomol. Soed.*, T.(B)12.n 1.s.10 (1970).
10. S.V. Kolesov, I.V. Neboikova, A.M. Steklova, S.V. Vladychina and K.S. Minsker, *Vysokomol. Soed.*, T.(A)31.n 2.s.430 (1989).
11. A.M. Shargarodskij and G.A. Tishin, *Plastmassy*, n 1.s.29 (1976).
12. A.E. Kulikova, F.A. Ekstrin, G.I. Shilov, E.G. Pomeranskeva and V.I. Kotenkov, *Vysokomol. Soed.*, T.(B)16.n 2.s.140 (19??).
13. N. Grassi, Khimija proskessov destrukshii polimerov. M. Inostrannaja Literatura (1959).
14. S.V. Kolesov, I.V. Neboikova, R.M. Akhmeykhanov, R.I. Ableev and K.S. Minsker, *Plastmassy*, n 11. s.26 (1988).
15. E.L. Akopjan, A.Ju. Karmilov, V.G. Nikolskij, A.M. Khachatrin and N.S. Enikolopjan, *Doklady AN SSSR*, T.291.n 1.s.133 (1986).
16. I.V. Neboikova, S.B. Vladychina, R.M. Akhmetkhanov, S.V. Kolesov and K.S. Minsker, *Plastmassy*, n 1. s.49 (1987).
17. V.K. Bemjakov, A.A. Berlin, N.I. Bukin, V.A. Orlov and O.G. Tarakanov, *Vysokomol. Soed.*, T.(A)10.n 3.s.599 (1968).
18. K.S. Minsker, M.I. Abdullin, R.R. Gizatullin and G.E. Zaikov, *Vysokomol. soed.*, T.(A)27.n 7.s.1428 (1985).
19. V.A. Palm, Osnovy kolichestvennoi teorii organicheskikh reakshij. L. Khimija (1977).
20. K.S. Minsker, M.I. Abdullin, R.R. Gizatullin and A.L. Buchachenko, *Doklady AN SSSR*, T.276.n 5.s.1181 (1984).
21. K.S. Minsker and G.E. Yaikov, Polymer Yearbook. — L. Ed. R.A. Petrick, Harwood Acad. Publ. Chur, p. 203 (1987).
22. M.I. Abdullin, N.P. Zueva, V.S. Martemjanov, B.F. Teplov and K.S. Minsker, *Plastmassy*, n 9. s.33 (1981).
23. K.S. Minsker, M.I. Abdullin, R.R. Gizatullin and N.N. Sigaeva, *Vysokomol. Soed.*, T.(B)25.n 5.s.337 (1983).

CONCLUSION

The HCl elimination rate in the degradation of PVC blends with other polymers is determined by:

(1) the presence of functional groups within other macromolecules, (TP, TPI, TPB);
(2) the ability to form hydrochloric complexes in case of other PVC, PB, PIB;
(3) the solubility of the second polymer to restrain in by sar;
(4) the ability of a macromolecular relation forms in the degradation to interact with labile covalent groups or with HCl.

In PVC degradation in solution is dominated by the basicity parameter if when determining the degree of specific interaction of polymer with solvent.

REFERENCES

[References illegible]

Mathematical Modelling of the Mechanism of Some Aromatic Amine Inhibitors

V.V. KHARITONOV, B.L. PSIKHA and G.E. ZAIKOV*

Institute of Chemical Physics, Chernogolovka, 142432, Moscow reg.
Institute of Chemical Physics, Moscow 117334, Kosygin str.4

1 INTRODUCTION

Selection of highly effective antioxidants

A correct understanding of the processes of antioxidant is important in ensuring the quality of articles produced from hydrocarbon materials. However, a theory for the design of antioxidants with defined activity is still absent. There are no scientifically proven methods for selection of antioxidants for stabilization of organic substances in processing conditions, storing and exploitation.[1–4]

In an oxidizing system tens of different interdependable reactions proceed simultaneously. Efficiency of an antioxidant is determined by the result of a range of simultaneous reactions, in which the *InH* molecule, *In·* radical and other products of transformation participate. Estimation of the antioxidant effect requires quantitative information on both the mechanism of oxidation of material and reactions with *InH* and *In·* participation in oxidative conditions. For an understanding to be achieved the effects of even small changes in oxidation conditions (temperature, concentration of ingredients, special additions and admixtures) can lead to a noticeable change of oxidation mechanism and the absolute and relative rates of reaction.

Values of a couple of parameters are determined from the results (for example, induction period τ). Such methods of estimation allows all the tested antioxidants to be ranked according to their "activity". However, even small changes of conditions (temperature, concentration of antioxidant) lead to unpredictable changes in the "activity".

Analysis performed leads to the following conclusion: new test methods are needed, which should allow efficient determination of all the necessary information on the mechanism of antioxidant action. All the reactions influencing its efficiency should be characterized reliably at a quantitative level.

Mathematical modelling represents an important instrument for investigation of processes of inhibited oxidation. This study is concerned with an investigation of mechanism of acting of four inhibitors with approximately similar aromatic amine class using mathematical models: Naugard 1–4, Naugard R, Naugalube 640, Naugalube 680. The investigation was performed in oxidizing *n*-hexadecane at 140°C.

2   STATEMENT OF THE PROBLEM

The study of inhibited oxidation is usually performed in two stages.

Common scheme of hydrocarbon oxidation is the basis of the first stage of investigation

$$RH + O_2 \xrightarrow{k_{0.0}} 2r^{\cdot} \tag{0.0}$$

$$I \xrightarrow{k_{0.1}} 2e^{\cdot} r_i^{\cdot} \tag{0.1}$$

$$RH + r^{\cdot} \xrightarrow{k_{0.2}} R^{\cdot} + rH \tag{0.2}$$

$$RH + r_i^{\cdot} \xrightarrow{k_{0.3}} R^{\cdot} + r_i H \tag{0.3}$$

$$O_2 + R^{\cdot} \xrightarrow{k_1} RO_2^{\cdot} \tag{1.0}$$

$$RH + RO_2^{\cdot} \xrightarrow{k_{2.1}} R^{\cdot} + ROOH \tag{2.1}$$

$$RH + RO_2^{\cdot} \xrightarrow{k_{2.2}} R^{\cdot} + \text{molecular product} \tag{2.2}$$

$$ROOH \xrightarrow{k_{s1}} \begin{cases} \xrightarrow{k_{3.1}} 2r^{\cdot} & (3.1) \\ \xrightarrow{k_{3.3}} \text{molecular product} & (3.3) \end{cases}$$

$$ROOH + ROOH \xrightarrow{k_{s2}} \begin{cases} \xrightarrow{k_{3.2}} 2r^{\cdot} & (3.2) \\ \xrightarrow{k_{3.4}} \text{molecular product} & (3.4) \end{cases}$$

$$RO_2^{\cdot} + RO_2^{\cdot} \xrightarrow{k_{6.0}} O_2 + \text{molecular product} \tag{6.0}$$

Here RH is an oxidizable medium, I is an initiator, r is a radical of any structure, different from $r_i^{\cdot}$, $R^{\cdot}$, $RO_2^{\cdot}$. The scheme includes:

1. Chain formation; spontaneous (0.0) and initiated (0.1).
2. Chain propagation (2.1)–(2.2). In this case not all the consumed oxygen is transformed to hydroperoxide (2.2).
3. Degenerative branching of hydroperoxides (3.1)–(3.2), and $ROOH$ (3.3)–(3.4).
4. Quadratic chain termination (6.0).

The process achieves a quasistationary concentration of radicals $r_i^{\cdot}$, $r^{\cdot}$, $R^{\cdot}$, $RO_2^{\cdot}$ and is fully characterized by nine kinetic parameters;

1. $k_{0.1}$, and e the constant for initiator decomposition and diffusion
2. $W_0 = 2 \cdot k_{0.0} \cdot [RH]_0$ — the rate of radical formation without $ROOH$ decomposition;
3. $P_2 = \dfrac{k_{2.1}}{k_{2.1} + k_{2.2}}$ — chain propagation, proceeding without $ROOH$ formation;
4. $P_6 = \dfrac{2 \cdot k_6}{(k[RH])^2}$ — quadratic termination of $RO_2^{\cdot}$ radicals;

5. $P_{3.1} = k_{3.1}^2 \cdot P_2^0$ — 1st order degenerative branching by *ROOH*;
6. $P_{3.2}$ — 2nd order degenerative branching by *ROOH*;
7. k_{s1} — 1st order consumption of hydroperoxide;
8. $P_{s2} = k_{s2} \cdot P_2$ — 2nd order consumption of hydroperoxide.

Initiator constants $k_{0.1}$ and e should be determined by independent experiment. The equations for oxidation (0.0)–(6.0) can be presented as follows:

$$v = v_1 - 0,5 \cdot P_6 v_2^2 + 0,5 \cdot w_0 \tag{1}$$

$$v_1 = (1-q) \cdot v_2 + P_6 v_2^2 \tag{2}$$

$$w_s = P_6 v_2^2 \tag{3}$$

$$\dot{y}_p = (1-q) \cdot v_2 - k_{s1} y_p - 2 \cdot P_{s2} y_p^2; \quad y_p(0) = y_p^0 \tag{4}$$

$$\dot{q} = (0,5 \cdot w_0 + v_1) / [RH]_0; \qquad q(0) = 0 \tag{5}$$

Here the following notations has been introduced:

$$v = k_{0.0}[RH][O_2] + k_2[RH][RO_2^{\cdot}] - k_6[RO_2^{\cdot}]^2 \tag{6}$$

the rate of oxygen consumption;

$$v_1 = k_1[O_2][R^{\cdot}] \tag{7}$$

the reaction rate (1.0);

$$v_2 = k_2[RH]_0[RO_2^{\cdot}] \tag{8}$$

the total initial rate of reactions (2.1) and (2.2);

$$y_p = [ROOH] / P_2 \tag{9}$$

the reduced concentration of hydroperoxides;

$$q = ([RH]_0 - [RH]) / [RH]_0 \tag{10}$$

the depth of transformation of oxidizing substance;

$$w_s = w_0 + w_i + w_y \tag{11}$$

the total rate of initiation;

$$w_i = 2 \cdot e \cdot k_{0.1}[I] \tag{12}$$

the initiation rate, conditioned by initiator;

$$w_y = 2 \cdot P_{3.1} y_p + 2 \cdot P_{3.2} y_p^2 \tag{13}$$

the rate of degenerative branching.

These equations assume the concentration of oxygen $[O_2]$ is held constant $[O_2]$ = const. Investigation of oxidation in a particular substance requires that it is established that the supposed scheme is in accordance with the process under investigation.

Methods of identification of the oxidation mechanism and determination of kinetic parameters for oxygen consumption have been described previously.[5] These methods have been used for the investigation of hydrocarbons with different structures and include n-heptadecane[6] and n-pentadecane.[7]

The next stage of investigation is identification of the mechanism of inhibitor action. The following scheme of transformations has been used for investigation of inhibitors of steric hindered phenol and aromatic amines:

$$InH + RO_2^{\cdot} \xrightarrow{k_7} In^{\cdot} + ROOH \tag{7.0}$$

$$ROOH + In^{\cdot} \xrightarrow{k_{-7}} InH + RO_2^{\cdot} \tag{-7.0}$$

$$InH + R^{\cdot} \xrightarrow{k_{7.1}} In^{\cdot} + RH \tag{7.1}$$

$$InH + r_i^{\cdot} \xrightarrow{k_{7.2}} In^{\cdot} + r_i H \tag{7.2}$$

$$In^{\cdot} + RO_2^{\cdot} \xrightarrow{k_{8.0}} InOOR \tag{8.0}$$

$$In^{\cdot} + RO_2^{\cdot} \xrightarrow{k_{8.1}} InH + \text{molecular product} \tag{8.1}$$

$$In^{\cdot} + In^{\cdot} \xrightarrow{k_9} In - In \tag{9.0}$$

$$In^{\cdot} + RH \xrightarrow{k_{10}} InH + R^{\cdot} \tag{10.0}$$

Process (7.0)–(10.0) proceeds in a quasistationary regime and are characterized by the following kinetic parameters:

$$P_{7.0} = \frac{k_7}{k_2 [RH]_0} \qquad P_{7.1} = \frac{k_{7.1}}{k_1 [O_2]} \qquad P_{7.2} = \frac{k_{7.2}}{k_{0.3} [RH]_0}$$

$$P_8 = \frac{k_{8.0}}{k_{8.0} + k_{8.1}} \qquad P_{-7} = k_{-7} \cdot C \qquad P_9 = 2 \cdot k_9 \cdot C^2$$

$$P_{10} = k_{10} \cdot [RH]_0 \cdot C \qquad \text{Notation:} \quad C = \frac{k_2 [RH]_0}{k_{8.0} + k_{8.1}}.$$

The system of equations, describing the whole process of inhibited oxidation (0.0)–(10.0), looks like:

$$v = v_1 - 0,5 \cdot P_6 v_2^2 + 0,5 \cdot w_0 \tag{14}$$

$$v_1 = (1-q) \cdot v_2 + P_6 v_2^2 + P_{7.0} v_2 z - P_{-7} xy + v_2 x \tag{15}$$

$$w_s = P_6 v_2^2 + 2 \cdot v_2 x + P_9 x^2 \tag{16}$$

$$P_{7.0} v_2 z + P_{7.1} v_1 z + P_{7.2} w_i z / (1 - q + P_{7.2} z) = P_{-7} xy + v_2 x + P_9 x^2 + P_{10}(1-q)x \tag{17}$$

$$\dot{y} = P_2(1-q) \cdot v_2 - k_{s1} y - 2 \cdot k_{s2} y^2 + P_{7.0} v_2 x - P_{-7} xy \tag{18}$$

$$\dot{z} = -(P_8 v_2 x + P_9 x^2) \tag{19}$$

$$\dot{q} = (0,5 \cdot w_0 + v_1) / [RH]_0 \tag{20}$$

$$y(0) = y_0; \qquad z(0) = z_0; \qquad q(0) = 0 \tag{21}$$

Along with the notations (6)–(13) the following have been introduced:

$$x = [In^{\cdot}] \cdot (k_{8.0} + k_{8.1}) / k_2 [RH]_0 \tag{22}$$

$$z = [InH] \tag{23}$$

Instead of a reduced concentration of hydroperoxides y_p has been introduced:

$$y = [ROOH] \tag{24}$$

The expression for w_y changes correspondingly:

$$w_y = 2 \cdot k_{3.1} y + 2 \cdot k_{3.2} y^2 \tag{25}$$

In a specific situation it is necessary to determine which of the supposed reactions (7.0)–(10.0) play the main role.

This approach has been applied to the study of oxygen consumption,[8] sterically hindered phenols[9] and aromatic amines.[10]

3   EXPERIMENTAL

n-Hexadecane (HD) was purified by shaking with alkaline potassium permanganate, sulphuric acid, oleum, and washed after each operation with distilled water. HD, was then dried above sodium sulphate, and then calcium hydride and distilled with the aid of an argon gas leak in vacuum. After purification HD contained not more than 10^{-4} mole/l of hydroperoxides and 10^{-5} mole/l of double bonds. Cumyl peroxide recrystallized twice from ethanol was used as initiator.

Kinetic curves for oxygen absorption for HD at 140°C were obtained using a special manometric device.[11] The dependence of the oxidation rate on time $v(t)$ was calculated, and used for identification of the mechanism and kinetic parameters.

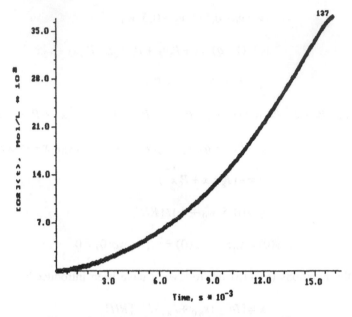

Figure 1   Hexadecane T = 140   Autooxidation $ROOH_o = 0$.

Experiments on the autooxidation (Figs. 1–2) and with different rates of initiation W_i (Figs. 3–4) have been performed. A series of experiments with different content of hydroperoxide were also performed (Figs. 5–6).

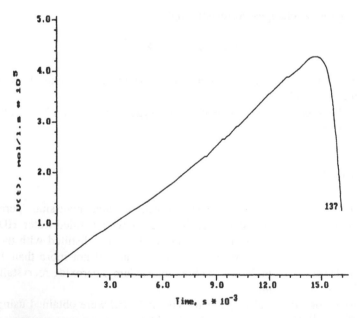

Figure 2   Hexadecane T = 140   Autooxidation $ROOH_o = 0$ $V(t) = d[O_2]/dt$.

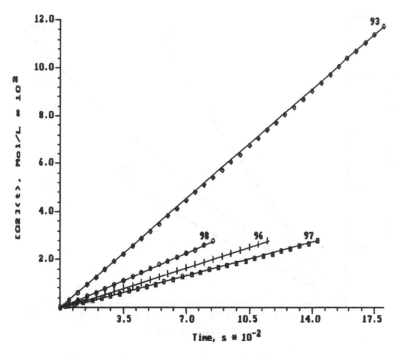

Figure 3   Hexadecane T = 140 $W_i = (1, 2, 4, 16.7)e–6$.

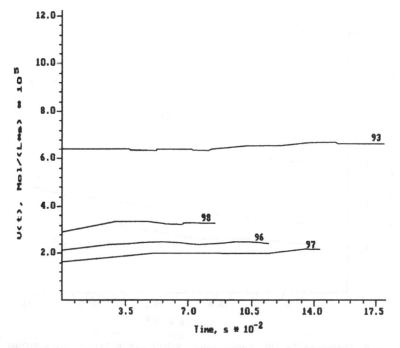

Figure 4   Hexadecane T = 140 $W_i = (1, 2, 4, 16.7)e–6$ $V(t) = d[02]/dt$.

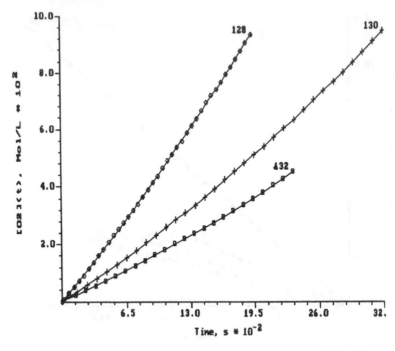

Figure 5   Hexadecane T = 140   Autooxidation $ROOH_o = (4, 8, 21)e{-}2$.

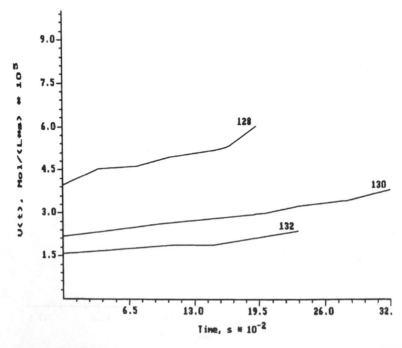

Figure 6   Hexadecane T = 140   Autooxidation $ROOH_o = (4, 8, 21)e{-}2$ $V(t) = d[O2]/dt$.

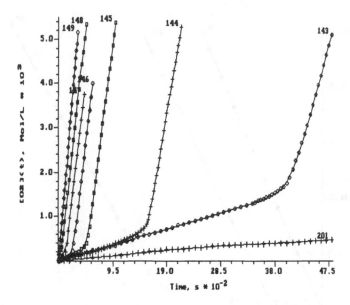

Figure 7   Naugard 1–4   Hexadecane T = 140 W_i = 1e–6 InH = (0, 0.25–50)e–4.

Investigation of the termination mechanism were performed using four inhibitors. Kinetics of oxygen absorption was measured using the following initial conditions:

1. The initial concentration of inhibitor $[InH]_0$ at an initiation rate of $W_i = 1 \cdot 10^{-6}$ mole/l · sec (Figs. 7, 8, 11–16) and for Naugard 1–4 a study with $W_i = 2,5 \cdot 10^{-7}$ mole/l · sec was performed (Figs. 9–10).

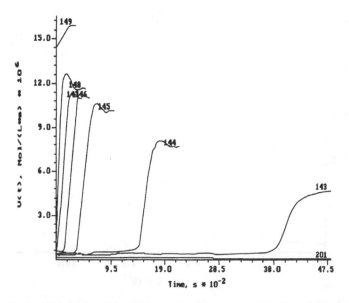

Figure 8   Naugard 1–4   Hexadecane T = 140 W_i = 1e–6 InH = (0, 0.25–50)e–4.

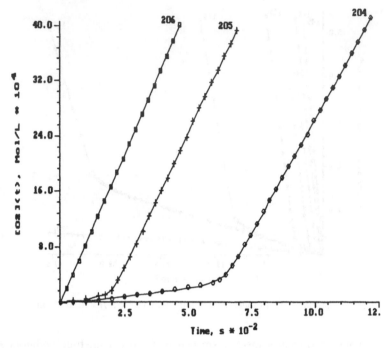

Figure 9   Naugard 1–4   Hexadecane T = 140 $W_i = 0.25e{-}6$ $InH = (0, 2.5, 5)e{-}5$.

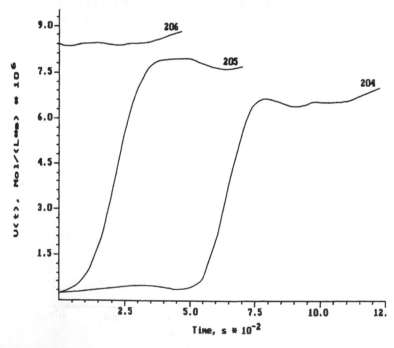

Figure 10   Naugard 1–4   Hexadecane T = 140 $W_i = 0.25e{-}6$ $InH = (0, 2.5, 5)e{-}5$ $V(t) = d[02]/dt$.

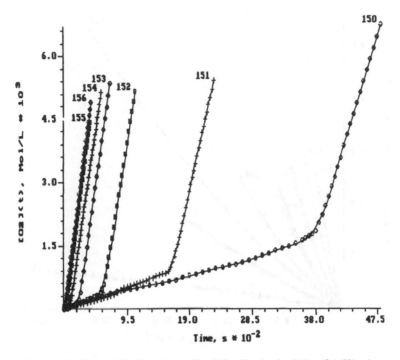

Figure 11   Naugard R   Hexadecane T = 140 W_i = 1e–6 InH = (0.1–50)e–4.

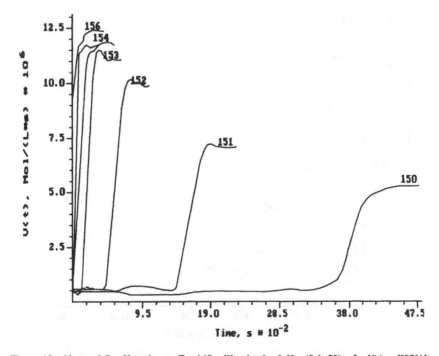

Figure 12   Naugard R   Hexadecane T = 140 W_i = 1e–6 InH = (0.1–50)e–5 $V(t)$ = d[02]/dt.

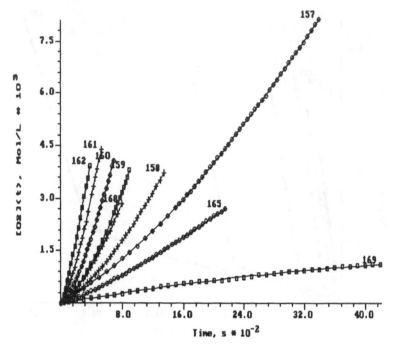

Figure 13   Naugalube 640   Hexadecane T = 140 W_i = 1e–6 InH = (0.2–500)e–4.

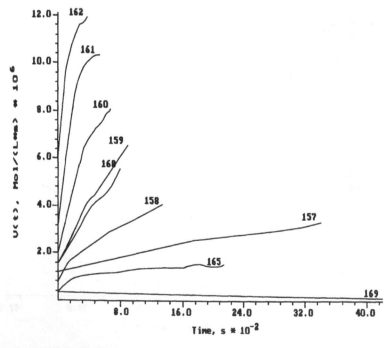

Figure 14   Naugalube 640   Hexadecane T = 140 W_i = 1e–6 InH = (0.2–500)e–4 $V(t)$ = d[02]/dt.

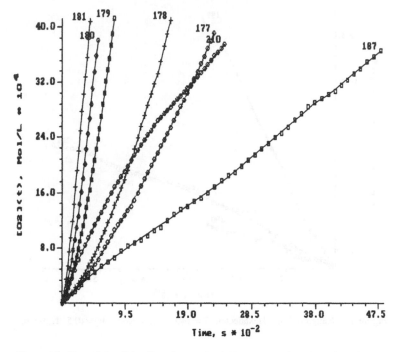

Figure 15   Naugalube 680   Hexadecane T = 140 W_i = 1e–6 InH = (0.4–500)e–4.

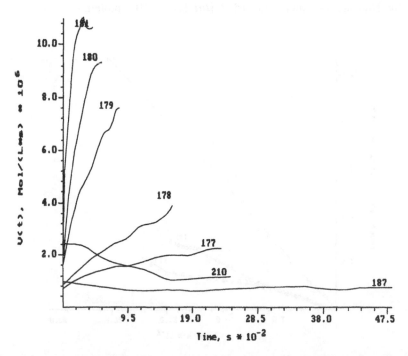

Figure 16   Naugalube 680   Hexadecane T = 140 W_i = 1e–6 InH = (0.4–500)e–5 $V(t)$ = d[02]/dt.

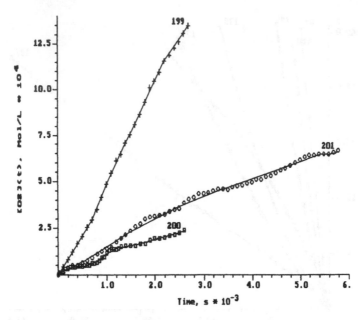

Figure 17   Naugard 1–4   Hexadecane T = 140 InH = 5e–3 W_i = (0.5, 1, 3)e–6.

2. The rate of inhibition W_i at higher concentrations of inhibitor [InH] were studied (Figs. 17–20). For Naugalube 640 and Naugalube 680 [InH]$_o$ = 5 · 10^{-2} mole/l was used, and for Naugard 1–4 and Naugard R [InH]$_o$ = 5 · 10^{-3} mole/l.

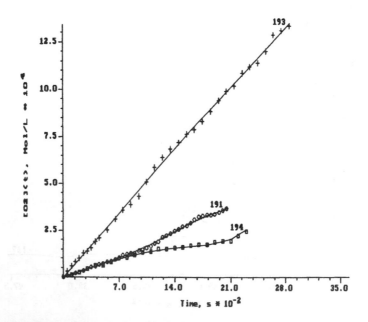

Figure 18   Naugard R   Hexadecane T = 140 InH = 5e–3 W_i = (0.5, 1, 3)e–6.

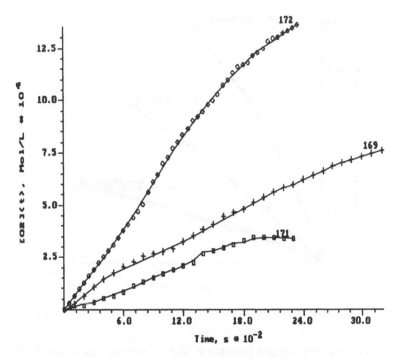

Figure 19   Naugalube 640   Hexadecane T = 140   *InH* = 5e–3 W_i = (0.5, 1, 3)e–6.

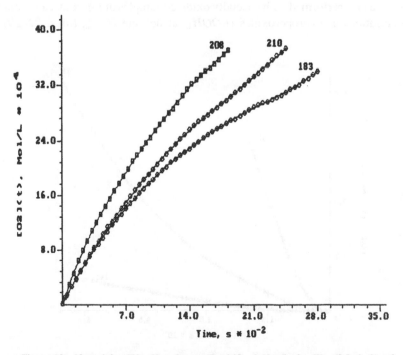

Figure 20   Naugalube 680   Hexadecane T = 140   *InH* = 5e–2 W_i = (0.5, 1, 3)e–6.

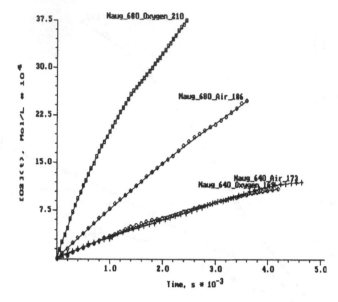

Figure 21   Naugalube 640 & Naugalube 680   Oxygen & Air $W_i = 1e-6$ $InH = 5e-2$.

3. Variation of the partial oxygen pressure at high concentrations of $[InH]_o$, Fig. 21 for Naugalube 640 and Naugalube 680 were performed for comparison.

Studies were also performed with specially oxidized samples of n-hexadecane containing high concentrations of hydroperoxides $[ROOH]_o$ at definite $[InH]_o$ (Figs. 22–29).

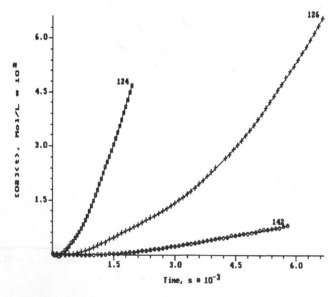

Figure 22   Naugard 1–4   Hexadecane   T = 140 $InH = 8e-4$ $ROOH = (45, 103, 218)e-3$.

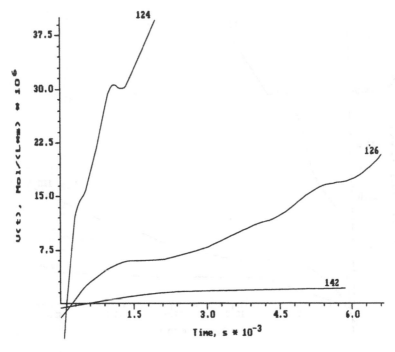

Figure 23   Naugard 1–4   Hexadecane $InH = 8e–4$ $ROOH = (45, 103, 218)e–3$ $V(t) = d[O2]/dt$.

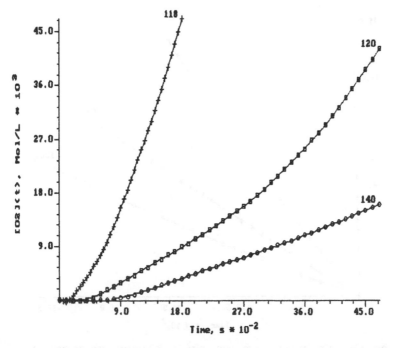

Figure 24   Naugard R   Hexadecane T = 140 $InH = 5e–4$ $ROOH = (60, 100, 203)e–3$.

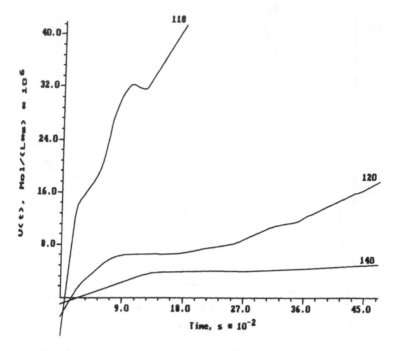

Figure 25   Naugard R   Hexadecane $InH = 5e{-}4$ $ROOH = (60, 100, 203)e{-}3$ $V(t) = d[02]/dt$.

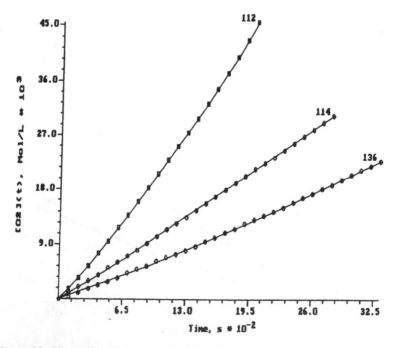

Figure 26   Naugard R   Hexadecane   T = 140 $InH = 5e{-}4$ $ROOH = (50, 96, 198)e{-}3$.

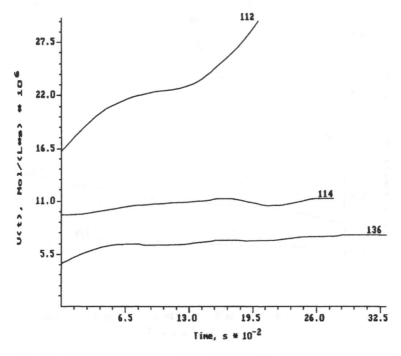

Figure 27   Naugalube 640   Hexadecane InH = 5e–4 $ROOH$ = (50, 96, 198)e–3 $V(t)$ = d[02]dt.

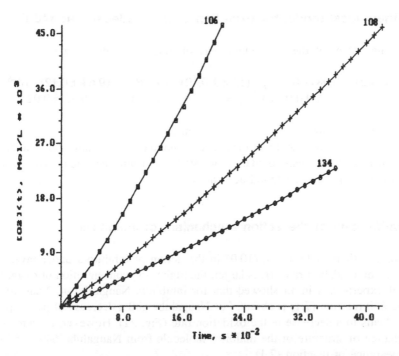

Figure 28   Naugalube 680   Hexadecane T = 140 InH = 5e–4 $ROOH$ = (50, 101, 209)e–3.

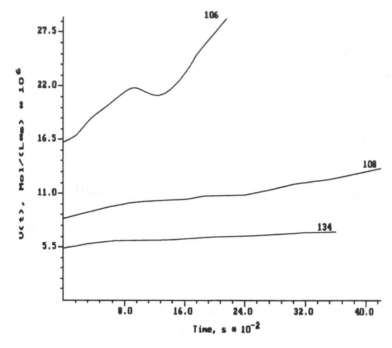

Figure 29   Naugalube 680   Hexadecane $InH = 5e{-}4$ $ROOH = (50, 101, 209)e{-}3$ $V(t) = d[O2]/dt$.

4.1   Mathematical model for oxidation of *n*-hexadecane at 140°C

The following values of the parameters were obtained from these data:

$$P_6 = (4.21 \pm 0.08) \cdot 10^3 \quad W_0 = (1.72 \pm 0.10) \cdot 10^{-8} \quad P_{3.1} = (9.64 \pm 0.12) \cdot 10^{-6}$$
$$P_{3.2} = (1.82 \pm 0.02) \cdot 10^{-5} \quad k_{s1} = (2.56 \pm 0.17) \cdot 10^{-5} \quad P_2 = 0.82 \pm 0.02$$

The reaction (3.4) is not important in the scheme.

Figs. 30–33 show a comparison of the experimental data on oxidation with calculations. The mechanism of hexadecane oxidation at 140°C has been identified correctly and the kinetic parameters have been found accurately.

4.2   Identification of the action mechanism of inhibitors

The role of reactions (7.1), (7.2), (10.0) in the action of inhibitors under investigation is explored. The oxidation rate $v(t)$ is larger, the higher the concentration of oxygen $[O_2]$. A series of experiments in air showed that for inhibitors Naugard 1–4, Naugard R and Naugalube 640 reaction (7.1) is absent. For Naugalube 680 a decrease of partial pressure of oxygen leads to a decrease in the oxidation rate (Fig. 21). However, it seems strange that the change of structure of the inhibitor molecule from Naugalube 640 to 680 leads to the appearance of reaction (7.1).

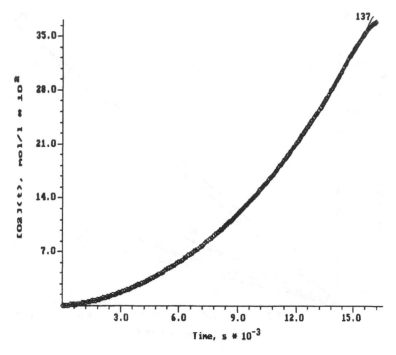

Figure 30   Hexadecane T = 140   Autooxidation $ROOH_o = 0$ Experiment & Calculation.

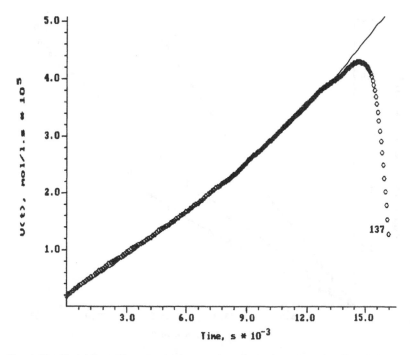

Figure 31   Hexadecane T = 140   Autooxidation $ROOH_o = 0$ Experiment & Calculation.

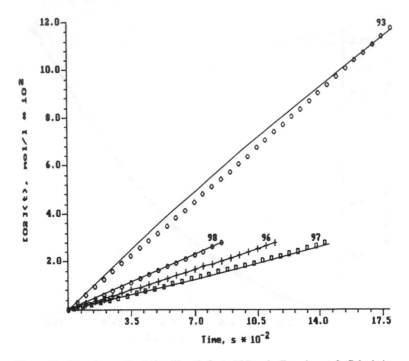

Figure 32   Hexadecane T = 140 W_i = (1, 2, 4, 16.7)e–6   Experiment & Calculation.

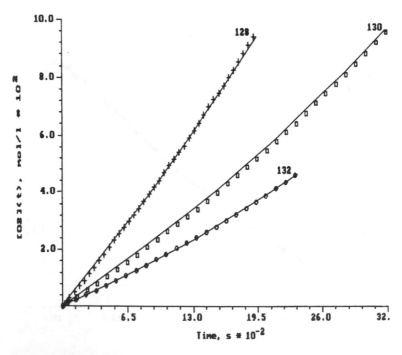

Figure 33   Hexadecane T = 140 $ROOH_o$ = (4, 8, 20)e–2   Experiment & Calculation.

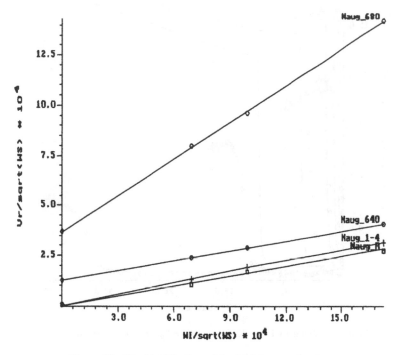

Figure 34   The identification of the inhibitors mechanisms.

Identification of reactions (7.2), (10.0) are performed in a series of experiments with different initiation rates w_i (Figs. 17–20). At large concentrations of inhibitor the dependence of $v(0) / \sqrt{w_s}$ on $w_i / \sqrt{w_s}$ is represented by a straight line with slope $1/(1 + P_{7.2}[InH]_o)$ and intercept $P_{10} / \sqrt{P_9}$, Fig. 34.

Treatment of the data gives the following parameters:

$P_j \backslash InH$	Naugard 1–4	Naugard R	Naugalube 640	Naugalube 680
$P_{7.2}$	$9 \cdot 10^2$	$1.1 \cdot 10^3$	$1 \cdot 10^2$	$0.14 \cdot 10^2$
$P_{10} / \sqrt{P_9}$	$9 \cdot 10^{-6}$	$8 \cdot 10^{-6}$	$1.3 \cdot 10^{-4}$	$3.7 \cdot 10^{-4}$

4.3   Investigation of action mechanism of Naugard 1–4 and Naugard R inhibitors

Qualitative analysis allows the inhibitors to be divided into two groups: (Naugard 1–4 — Naugard R) and (Naugalube 640 – Naugalube 680). Let us consider — Naugard 1–4.

First estimate the parameters $P_{7.0}$, P_8, P_9 from the initial parts of the $v(t)$ dependences with small concentrations of inhibitor $[InH]_o = 2 \cdot 10^{-5} - 2 \cdot 10^{-4}$ mole/l; $P_{7.0} = 2 \cdot 10^4$, $P_8 = 1$, $P_9 \simeq 1 \cdot 10^{-6}$, $P_{10} \simeq 1 \cdot 10^{-8}$.

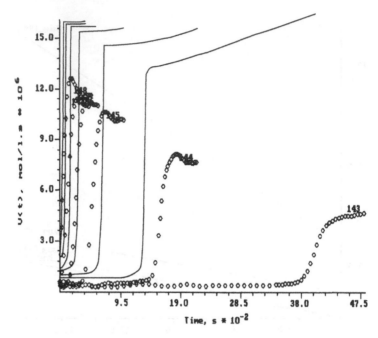

Figure 35   Naugard 1–4 $W_i = 1e{-}6$ $K_w = K_p = K_i = 0$ $V(t) = d[O2]/dt$.

Deviation of calculated from experiment data shows that the number of parameters used are insufficient, Fig. 35.

The most characteristic differences in behavior are:

1. Increasing the inhibitor concentration $[InH]_o$ increases, the induction period.
2. The initial oxidation rates become smaller than w_i for concentrations of $[InH]_o \geqslant 2 \cdot 10^{-4}$ mole/l — unexplainable by the supposed mechanism.
3. Final oxidation rates, are smaller than the calculated ones.
4. A stronger dependence of the final oxidation rate on inhibitor concentration is observed than calculated.

Naugard 1–4 exists as an equilibrium between two states with equilibrium constant, proportional to the quantity of inhibitor $[InH]_o$:

$$I \xleftarrow[\quad k_w^-\quad]{\quad k_w^+[InH]_o\quad} I_2 \qquad (\text{W.1})$$

Initiator I_2 decomposes to radicals with constants k_{i2} and e_2, which is different from $k_{0.1}$ and e. As a result the real rate of initiation depends on time and $[InH]_0$ as follows:

$$w_i(t) = w_i^o \cdot \exp\left(-k_{0.1}\frac{1 + K_p[InH]_o}{1 + K_w[InH]_o} \cdot t\right) \qquad (26)$$

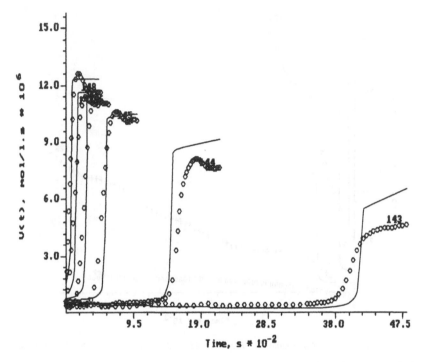

Figure 36   Naugard 1–4 $W_i = 1e{-}6$ $K_w = 1.55e5$ $K_p = 2.2e5$ $K_i = 0.77e5$ $V(t) = d[O2]/dt$.

Notations introduced here are the following:

$$w_i^o = 2 \cdot k_{0.1} e \cdot [I]_o \cdot \frac{1 + K_i [InH]_o}{1 + K_w [InH]_o} \tag{27}$$

$$K_w = k_w^+ / k_w^- \tag{28}$$

$$K_p = K_w \cdot k_{i2} / k_{0.1} \tag{29}$$

$$K_i = K_w \cdot (k_{i2} e_2) / (k_{0.1} e) \tag{30}$$

It follows from (27)–(30) that $[InH]_o = 0$ initiation is defined by the $k_{0.1}$ and e parameters; whereas when $[InH]_o$ increases, the initiation parameters change and in the limit as $[InH]_o \to \infty$ become equal to k_{i2}, e.

The rate of initiation depends on the inhibitor concentration and is quantitatively characterized by three parameters: K_w, K_p, K_i. Description of the experimental data can be improved considerably (Fig. 36), if proper values of these parameters have been selected. The ratios $k_{i2}/k_{0.1}$ and e_2/e are defined simply with $K_w > 5 \cdot 10^4$. The calculation, Fig. 36, was performed with values of parameters: $K_w = 1{,}55 \cdot 10^5$, $K_p = 2{,}1 \cdot 10^5$, and $K_i = 0{,}77 \cdot 10^5$.

Consider now the experimental data for Naugard 1–4 — autooxidation of preliminarily oxidized samples, Figs. 22–23.

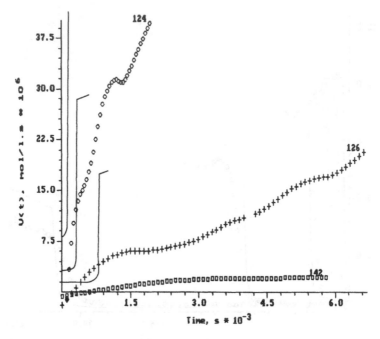

Figure 37   Naugard 1–4   Autooxidation $ROOH$ = (45, 103, 218)e–3 InH = 8e–4   No product.

The experimental data shows that:-

1. Qualitatively the $v(t)$ dependence for oxidized samples during autooxidation is different from that during their initial oxidation.
2. In initial period, oxidation is followed by gas evolution ($v < 0$), the intensity of which is higher the larger the hydroperoxide content.

These features indicate that the present experiments cannot be described by parameters, found from the initial oxidation. Fig. 37 shows a comparison of the $v(t)$ dependence with the calculated values and indicates the possibility of a very fast inhibitor process. In reality most of the inhibitor is used during the initial heating ($\simeq$ 200 sec) before experimental data are collected. All the observed inhibition is achieved by a transformation product formed from interaction with hydroperoxide, according to the reaction:

$$In^{\cdot} + ROOH \xrightarrow{\ k^{(1)}_{-7}\ } InOOH + RO^{\cdot} \qquad (-7.1)$$

Let us consider another feature of the experiment — gas evolution. The dependence of the initial rate on $[ROOH]_o$.

Gas evolution occurs only in the initial period, so its source has to disappear with time. Several hypotheses as to the physical nature of this phenomenon can be considered. Gas evolution is connected with decomposition of hydroperoxides, having specific structures. Decomposition of this hydroperoxide, in the presence of Naugard 1–4 is followed by gas evolution.

TABLE 1

Values of kinetic parameters of investigated inhibitors in oxidizing n-hexadecane at T = 140°C

$P_j \backslash InH$	Naugard 1–4	Naugard R	Naugalube 640	Naugalube 680
$P_{7.0}$	$2 \cdot 10^4$	$1.5 \cdot 10^4$	$5.6 \cdot 10^3$	$5.3 \cdot 10^3$
$P_{7.2}$	347.5	359.2	40.6	10.0
P_8	1.0	1.0	1.0	1.0
P_9	$\approx 1 \cdot 10^{-6}$	$\approx 1 \cdot 10^{-6}$	$\approx 1 \cdot 10^{-6}$	$\approx 1 \cdot 10^{-6}$
P_{10}	$\lesssim 1 \cdot 10^{-8}$	$\lesssim 1 \cdot 10^{-8}$	$1.8 \cdot 10^{-7}$	$1.3 \cdot 10^{-7}$
P_{-7}	$3 \cdot 10^{-4}$	$5 \cdot 10^{-4}$	$1 \cdot 10^{-2}$	$1 \cdot 10^{-2}$
$P_{-7}^{(1)}$	$\approx 1 \cdot 10^{-5}$	$\approx 1 \cdot 10^{-5}$	$\approx 1 \cdot 10^{-5}$	$\approx 1 \cdot 10^{-5}$
$P_7^{(1)}$	$1.0 \cdot 10^3$	$1.0 \cdot 10^3$	$0.6 \cdot 10^3$	$0.6 \cdot 10^3$
$P_8^{(1)}$	0.05	0.05	0.03	0.03
K_w	$1.55 \cdot 10^5$	$1.55 \cdot 10^5$	$2 \cdot 10^2$	$2 \cdot 10^2$
$k_{i2}/k_{0.1}$	1.29	0.89	≈ 1.0	≈ 1.0
e_2/e	0.385	0.493	0.450	0.440
k_g	0.15	0.40	—	—
P_{p1}	—	—	$2.0 \cdot 10^3$	$1.5 \cdot 10^3$
k_{p2}	—	—	$1.8 \cdot 10^{-3}$	$2.5 \cdot 10^{-3}$
$w_z(0)$	—	—	—	$2.05 \cdot 10^{-6}$
K_z	—	—	—	$7.5 \cdot 10^{-4}$

$$ROOH_2 + ROOH_2 \xrightarrow{\;k_g\;} \text{gas} \qquad\qquad (G.1)$$

Thus, the parameters $P_7^{(1)}, P_8^{(1)}, k_g$ are required from experimental data with unknown concentrations of $[InOOH]_o$ and $[ROOH_2]_o$. The parameters attained as a result of data fitting can be used to describe satisfactorily corresponding experimental data. The initial parts of dependences of $[O_2](t)$ for calculation and experiment are compared in Fig. 44.

Final values of P_j in Table 1, and comparison of corresponding calculated and experiment data are made in Figs. 38–43.

4.4 Investigation of the mechanism of action of inhibitors Naugalube 640 and 680

The main qualitative difference between the first and second group lies in the behavior of $v(t)$ with different initial concentrations $[InH]_o$, Figs. 13–16. The second group of inhibitors have no induction period: the rates increase monotonously, from the very beginning. In the scheme (7.0)–(10.0) this behaviour can be explained by a large value of the rate constant k_{-7}.

The parameters $P_{7.0}, P_8, P_9$ and P_{-7} can be estimated from the initial parts of $v(t)$ dependences in the series, and parameters $P_{7.2}$ and P_{10} from the results of preliminary analysis (Item 4.2). Coincidence of calculation with experiment data occurs initially, after which the calculated value rises sharply, i.e. calculated inhibition ends much earlier, than in experiment. The most probable product with such properties in the scheme (7.0)–(10.0) is $InOOR$, formed in the reaction:

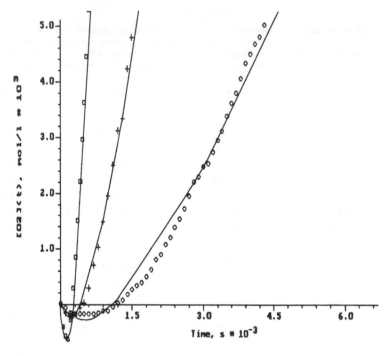

Figure 38   Naugard 1–4   Autooxidation   *InH* = 8e–4   Experiment & Calculation.

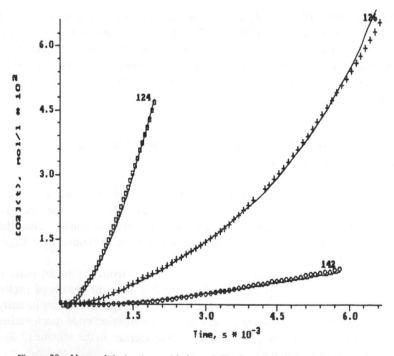

Figure 39   Naugard 1–4   Autooxidation   *InH* = 8e–4   Experiment & Calculation.

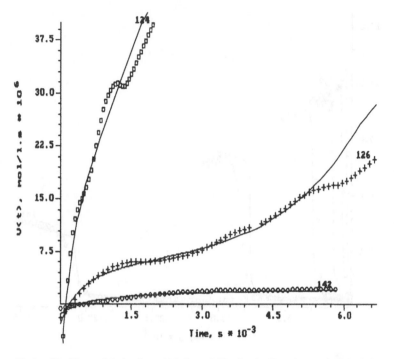

Figure 40 Naugard 1–4 Autooxidation *InH* = 8e–4 Experiment & Calculation.

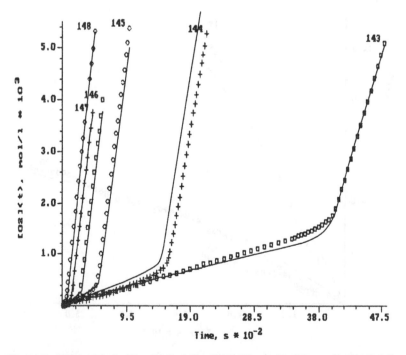

Figure 41 Naugard 1–4 W_i = 1e–6 *InH* = (0.25–50)e–4 Experiment & Calculation.

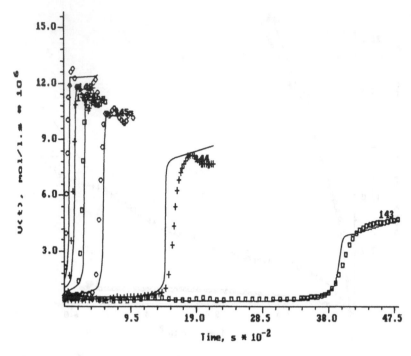

Figure 42   Naugard 1–4 W_i = 1e–6 InH = (0.25–50)e–4   Experiment & Calculation.

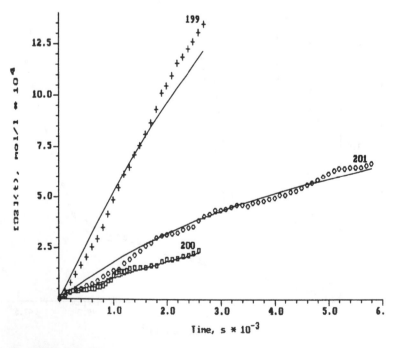

Figure 43   Naugard R InH = 5e–3 W_i = (0.5, 1, 3)e–6   Experiment & Calculation.

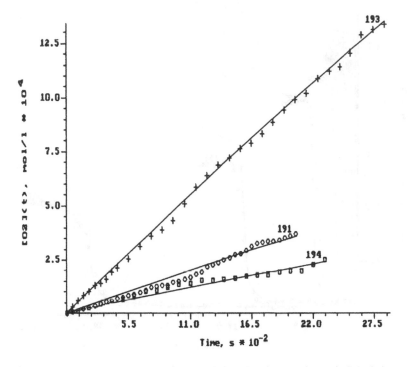

Figure 44   Naugard R InH = 5e–3 W_i = (0.5, 1, 3)e–6   Experiment & Calculation.

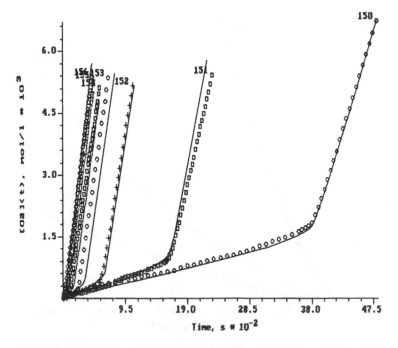

Figure 45   Naugard R W_i = 1e–6 InH = (0.1–50)e–4   Experiment & Calculation.

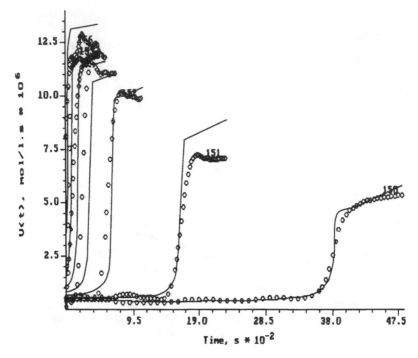

Figure 46   Naugard R W_i = 1e–6   *InH* = (0.1–50)e–4   Experiment & Calculation.

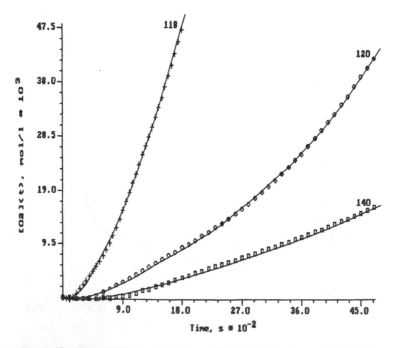

Figure 47   Naugard R   AutoOxidation   *InH* = 5e–4   Experiment & Calculation.

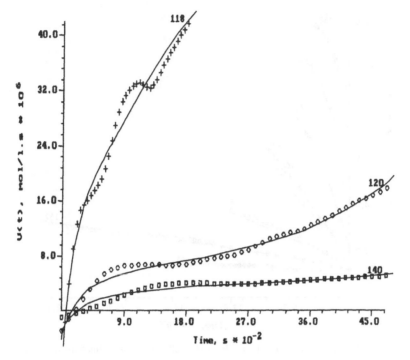

Figure 48   Naugard R   Autooxidation InH = 5e–4   Experiment & Calculation.

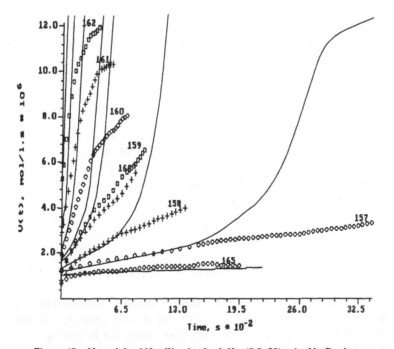

Figure 49   Naugalube 640 W_i = 1e–6 InH = (0.2–20)e–4   No Product.

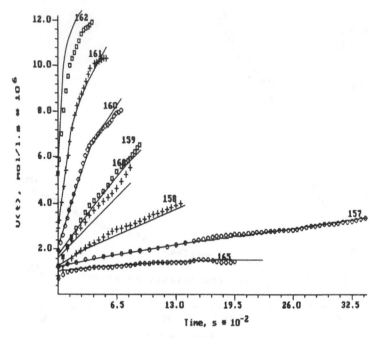

Figure 50   Naugalube 640 $W_i = 1e{-}6$ $InH = (0.2{-}20)e{-}4$ $InOOR$: Inhibitor & Initiator.

$$In^{\cdot} + RO_2^{\cdot} \xrightarrow{\ k_{8.0}\ } InOOR \tag{8.0}$$

Several alternative mechanisms for the retardant action have been considered, the choice was the following:

$$InOOR + R^{\cdot} \xrightarrow{\ k_{p1}\ } In^{\cdot} + \text{molecular product} \tag{P.1}$$

The principal stage is the formation of $In^{\cdot}$ radical in the reaction that provide a nearly unlimited duration of retardance. From the point of view of the experiment there is no difference between which radical interacts with the product: $R^{\cdot}$ or $RO_2^{\cdot}$. Consideration of inhibited autooxidation of the initial sample radical $R^{\cdot}$ appears favourable. Beside inhibiting the product $InOOR$ can initiate:

$$InOOR \xrightarrow{\ k_{p2}\ } 2^{\cdot} r^{\cdot} \tag{P.2}$$

Incorporation of reactions $(P.1)$–$(P.2)$ allows agreement between experiment, and calculation to be improved, Fig. 50.

Inhibitors of the Naugalube group exhibit a weaker inhibitor concentration dependence than the Naugard group, K_w for Naugalube 640 is appreciably smaller, than for the first two inhibitors. Calculations, Figs. 49–50, were made using the following values of the constants: $K_w = 2 \cdot 10^2$, $K_p = 2 \cdot 10^2$ and $K_i = 0{,}9 \cdot 10^2$.

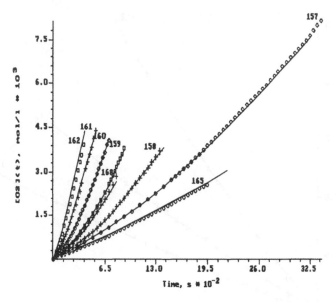

Figure 51   Naugalube 640 W_i = 1e–6 InH = (0.2–20)e–4   Experiment & Calculation.

Comparison of experimental data for Naugard 1–4 and Naugard R indicates that oxidation in the presence of Naugalube 640 occurs without gas evolution. In other qualitative aspects this series for inhibitors are similar. In accordance with this mechanism braking is performed mainly by the product of reaction (–7.1) and its influence is characterized by two parameters $P_7^{(1)}$ and $P_8^{(1)}$; which respectively define the intensity and duration of retardation action.

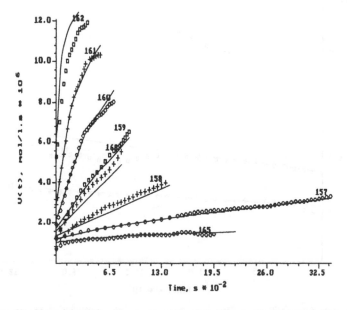

Figure 52   Naugalube 640 W_i = 1e–6 InH = (0.2–20)e–4   Experiment & Calculation.

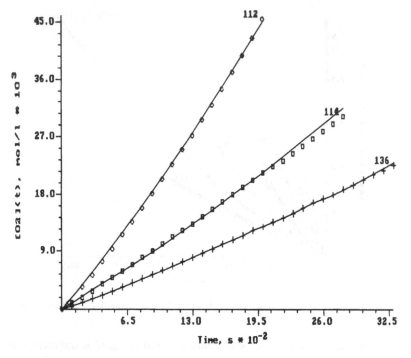

Figure 53   Naugalube 640   Autooxidation   *InH* = 5e–4   Experiment & Calculation.

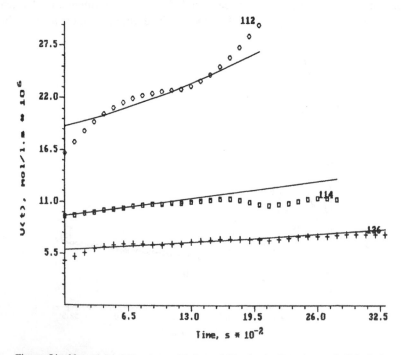

Figure 54   Naugalube 640   Autooxidation   *InH* = 5e–4   Experiment & Calculation.

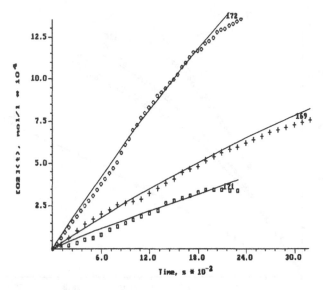

Figure 55   Naugalube 640 InH = 5e–2 W_I = (0.5, 1, 3)e–6   Experiment & Calculation.

Experimental data Naugalube 680 are practically similar to those for Naugalube 640 in two series of experiments: initiated oxidation of different $[InH]_o$ (Figs. 15–16) and autooxidation of oxidized samples (Figs. 28–29). The general difference between them is observed at concentration $[InH]_o$ = 5 · 10^{-2} mole/l: the initial rate of oxidation increasing sharply in experiments with Naugalube 680, Fig. 20. If you determine values of kinetic parameters from the first two series of experiments and calculate dependences $[O_2](t)$ for the third one, the calculated values will lie lower than the experimental ones, Fig. 56.

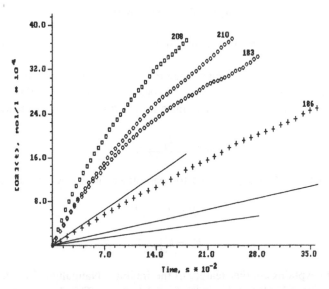

Figure 56   Naugalube 680 InH = 5e–2 W_i = (0.5, 1, 3)e–6   186: (Air W_i = 1e–6).

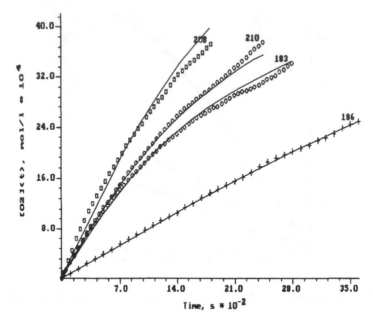

Figure 57   Naugalube 680 $InH = 5e-2$ $W_i = (0.5, 1, 3)e-6$ Admixture-Initiator   186: Air.

An explanation of the difference between Naugalube 680 and 640 is that it contains a mixed initiating character. The amount in the system is proportional to the inhibitor concentration, at small $[InH]_0$ it does not have a practical effect and only if $[InH]_0$ increases from $5 \cdot 10^{-3}$ up to $5 \cdot 10^{-2}$ mole/l does its effect become apparent, Fig. 15.

The simplest mechanism of initiation can be written as:

$$w_z = w_z(0) \cdot \exp(-K_z t) \tag{31}$$

where $w_z(0)$ is the initial rate of initiation and K_z is the constant for the proportion of initiator/inhibitor in each system.

It is possible using values of these parameters to satisfactorily describe the experiment, Fig. 57. For the description of experiment 186, performed in air, the values of parameters $w_z(0)$ and K_z should be decreased by five times for the following mechanism of initiation:

$$I_z + O_2 \xrightarrow{k_z} 2e^{\cdot}r^{\cdot} \tag{Z.1}$$

According to reaction (Z.1) expressions for parameters $w_z(0)$, K_z are as follows:

$$w_z(0) = 2e^{\cdot}k_z^{\cdot}[I_z]_0 \tag{32}$$

$$K_z = k_z^{\cdot}[O_2] \tag{33}$$

Reaction (Z.1) explains another feature of the inhibitor Naugalube 680, decrease of the oxygen partial pressure leads to a decrease of oxidation rate. This decrease appears because

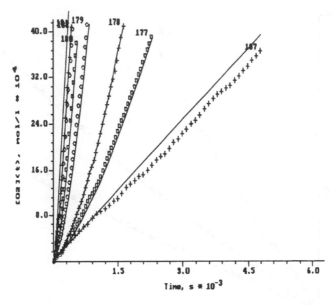

Figure 58 Naugalube 680 W_i = 1e–6 InH = (0.2–50)e–4 Experiment & Calculation.

of reaction (Z.1), and not (7.1), an effect which could hardly be explained by changes in structure in the inhibitors under consideration.

Comparison of experimental and calculated data for Naugalube 680 are presented in Figs. 57–61. The calculation was made after final optimization of parameters. Corresponding values of P_j are presented in Table 1.

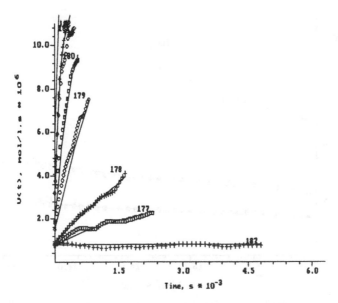

Figure 59 Naugalube 680 W_i = 1e–6 InH = (0.2–50)e–6 Experiment & Calculation.

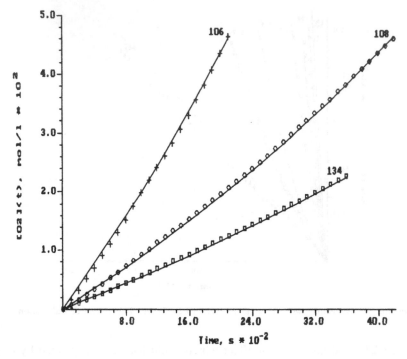

Figure 60   Naugalube 680   Autooxidation   *InH* = 5e–4   Experiment & Calculation.

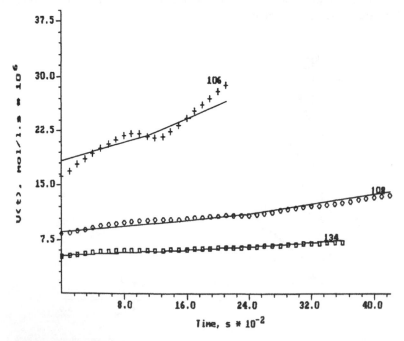

Figure 61   Naugalube 680   Autooxidation   *InH* = 5e–4   Experiment & Calculation.

5 DISCUSSION

1. Reactions characteristic of all the inhibitors investigated

The retardant mechanism of autooxidation of oxidized samples of hexadecane by the product of transformation of initial inhibitor should be attributed to reaction (–7.1), and its effect characterized by two kinetic parameters $P_7^{(1)}$ and $P_8^{(1)}$, which define the intensity and duration of retardance.

The effect on the rate of initiation by cumylperoxide appeared to be a general property of all the inhibitors.

2. Reactions, characteristic of inhibitors of the Naugard group

Autooxidation of oxidized samples of hexadecane in presence of inhibitors of the Naugard group in the initial stage is accompanied by gas evolution according to reaction (G.1).

3. Reactions, characteristics of inhibitors of the Naugalube group

The important role in the initial oxidation is played by the products of transformation of the inhibitor *InOOR*, formed according to the reaction (8.0) and has both inhibiting (*P*.1) and initiating (*P*.2) properties.

4. Reactions, characteristic for some inhibitors

The reaction (Z.1) observed in the presence of large concentrations of Naugalube 680 inhibitor ($[InH]_o = 5 \cdot 10^{-2}$ mole/l) should be attributed to them. Taking into account all these additional reactions the system of equations, describing considered process of inhibited oxidation looks as follows:

$$v = v_1 - 0{,}5 \cdot P_6 v_2^2 + 0{,}5 \cdot w_0 - k_g \cdot y_2^2 \tag{34}$$

$$v_1 = (1-q) \cdot v_2 + P_6 v_2^2 + P_{7.0} v_2 z - P_{-7} xy + v_2 x + 2 \cdot P_7^{(1)} v_2 z_1 \tag{35}$$

$$w_s = P_6 v_2^2 + 2 \cdot v_2 x + P_9 x^2 + 2 \cdot P_7^{(1)} v_2 z_1 \tag{36}$$

$$P_{7.0} v_2 z + P_{7.2} w_i z / (1-q+P_{7.2}z) + P_{p1} v_1 z_2 = (P_{-7} + P_{-7}^{(1)}) xy + v_2 x + P_9^2 x + P_{10}(1-q)x \tag{37}$$

$$\dot{y} = P_2 (1-q) \cdot v_2 - k_{s1} y - 2 \cdot k_{s2} y^2 + P_{7.0} v_2 z + P_7^{(1)} v_2 z_1 - (P_{-7} + P_{-7}^{(1)}) xy \tag{38}$$

$$\dot{z} = -(P_8 v_2 x + P_9 x^2) + P_{p1} v_1 z_2 - P_{-7}^{(1)} xy \tag{39}$$

$$\dot{q} = (0{,}5 \cdot w_0 + v_1 + P_{p1} v_1 z_2) / [RH]_o \tag{40}$$

$$\dot{z}_1 = P_{-7}^{(1)} xy - P_7^{(1)} P_8^{(1)} v_2 z_1 \tag{41}$$

$$\dot{z}_2 = P_8 v_2 x - P_{p1} v_1 z_2 - k_{p2} z_2 \tag{42}$$

$$\dot{y}_2 = -k_g y_2^2 \tag{43}$$

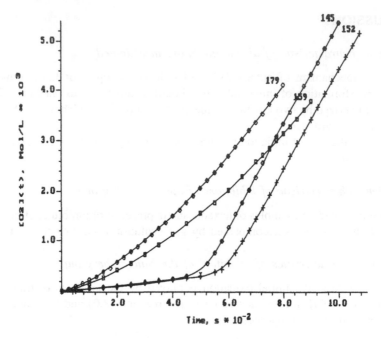

Figure 62 Hexadecane $T = 140$ $W_i = 1e{-}6$ $InH = 2e{-}4$.

$$y(0) = y_0; \qquad z(0) = z_0; \qquad q(0) = 0; \tag{44}$$

$$z_1(0) = z_1^0; \qquad z_2(0) = z_2^0; \qquad y_2(0) = y_2^0; \tag{45}$$

Along with the former notations the following ones are introduced here:

$$z_1 = [InOH]; \qquad z_2 = [InOOR]; \qquad y_2 = [ROOH_2] \tag{46}$$

$$P_{-7}^{(1)} = k_{-7}^{(1)} \frac{k_2[RH]_o}{k_{8.0} + k_{8.1}}; \qquad P_{p1} = \frac{k_{p1}}{k_1[O_2]}; \tag{47}$$

Let us compare kinetic parameters for inhibitors of the Naugard and Naugalube groups.

1. Values of $P_{7.0}$ and $P_{7.2}$, which define the efficiency of retardance, are several times higher for Naugard than for Naugalube group. Moreover, Naugard inhibitors are sufficiently stronger in decreasing the rate of initiation: equilibrium constant K_w is 2–3 degrees higher than for Naugalube. As a result, at similar concentrations of $[InH]_o$ the Naugard group decreases the rate of oxidation stronger and hold it constant for a long time.
2. A large value of the P_{-7} parameter decreases their efficiency strongly in presence of hydroperoxides: the rate of oxidation grows fast with increase of $[ROOH]$ concentration. The stage with constant rate disappears from the kinetic curve, and is "linearized", Fig. 62.

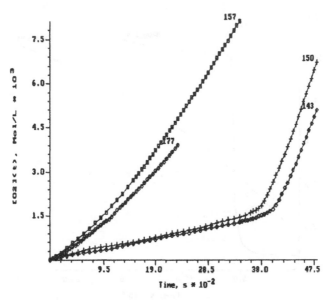

Figure 63   Hexadecane   T = 140 W_i = 1e–6 InH = 1e–3.

3. The presence of an active product *InOOR* for inhibitors of Naugalube group increases the duration of retardance: slow growth of oxidation rate is being observed for a long period of time.

4. Comparatively large P_{10} value for inhibitors of Naugalube group leads to the fact that initial rate of oxidation decreases until a definite level with the growth of $[InH]_o$ grows with the increase of concentrations of Naugalube and Naugard inhibitors, Figs. 62–63.

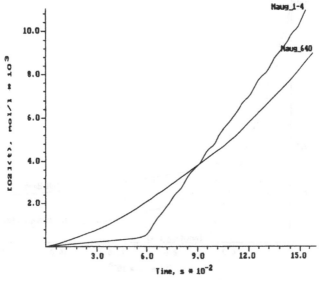

Figure 64   Hexadecane   T = 140 W_i = 1e–6 InH = 2e–4   Calculation.

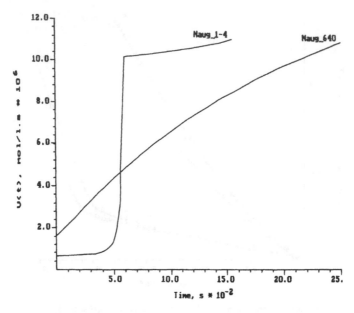

Figure 65   Hexadecane   T = 140 W_i = 1e–6   InH = 2e–4   Calculation.

5.   The process of autooxidation retardance of oxidized samples is defined by *InH* product. $P_7^{(1)}, P_8^{(1)}$ parameters of this product for all the inhibitors are close, consequently, kinetic curves are close. The difference is present only at the initial stage, where gas evolution is observed in the presence of Naugard inhibitors.

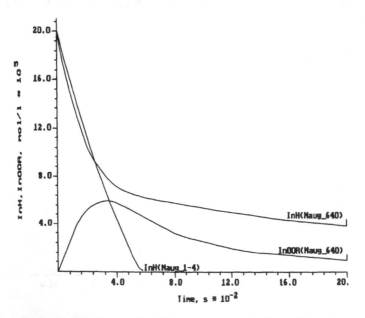

Figure 66   Hexadecane   T = 140 W_i = 1e–6   *InH* = 2e–4   Calculation.

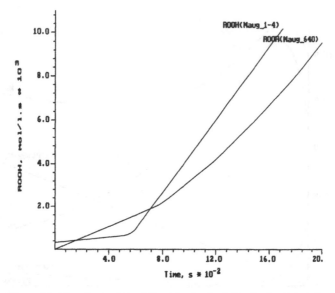

Figure 67 Hexadecane T = 140 W_i = 1e–6 InH = 2e–4 Calculation.

6 CONCLUSIONS

1. Kinetic rules of inhibited oxidation of n-hexadecane have been investigated at 140°C in presence of the following inhibitors: Naugard 1–4, Naugard R, Naugalube 640, Naugalube 680. The investigation was performed in conditions of autooxidation and initiated oxidation at different rates of initiation in a wide interval of concentration changes of inhibitors and hydroperoxides, at different oxygen partial pressures.

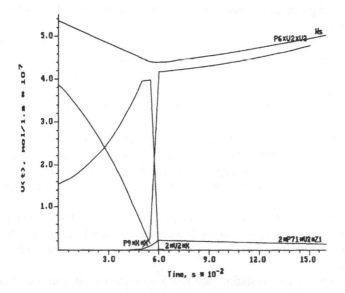

Figure 68 Naugard 1–4 Hexadecane T = 140 W_i = 1e–6 InH = 2e–4 Calculation.

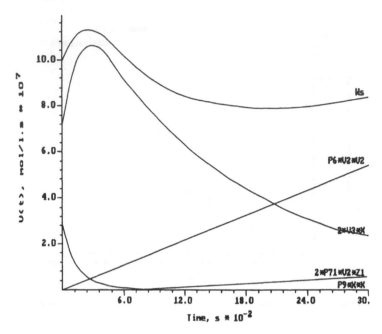

Figure 69   Naugalube 640   Hexadecane   T = 140 W_i = 1e–6   *InH* = 2e–4   Calculation.

2. Mathematical model of the investigated process has been worked out that characterizes quantitatively its main rules, describes all the existing experimental data.

3. Results of the investigation divide the inhibitors into two groups: (Naugard 1–4–Naugard R) and (Naugalube 640 — Naugalube 680). Inhibitors in the first group have large efficiencies of retardance; large values of parameters $P_{7.0}$ and $P_{7.2}$. The efficiency of the second group of inhibitors is lower, because of lower $P_{7.0}$ and $P_{7.2}$ and larger P_{-7} and P_{10}.

4. The mechanism of retardance of autooxidation of preliminary oxidized samples differs qualitatively from that in conditions of initiated oxidation. In oxidized medium inhibiting is caused mainly by the product of transformation of initial inhibitor (supposedly *InOH*). Properties of this product are weakly dependent on the structure of the initial inhibitor: product of all the inhibitors has low intensity and very long duration of retardance.

5. In the presence of Naugard group inhibitors autooxidation of oxidized samples is followed by gas evolution at a certain initial stage. The supposition has been put forward that gas evolution appears during decomposition of hydroperoxide of a definite structure under the influence of inhibitor molecules able to structurize oxidizing medium.

6. The influence of inhibitors on initiating properties of cumylperoxide has been found. This property is the most potent for Naugard group and lower — for Naugalube and is connected with structurizing ability of nitrogen containing molecules of these inhibitors.

7. Sharp increase of the oxidation rate has been observed at a maximal concentration of Naugalube 680 inhibitor $[InH]_o = 5 \cdot 10^{-2}$ mole/l.

REFERENCES

1. N.M. Emanuel, G.E. Zaikov and Z.K. Maizus, Oxidation of Organic Compounds. Effect of Media. Oxford, Pergamon Press, p. 426 (1984).
2. N.M. Emanuel, G.E. Zaikov and V.A. Kritsman, Chemical Kinetics and Chain Reactions. Historical Aspects. N.Y., Nova Science Press, p. 468 (1993).
3. V.L. Rubailo, S.A. Maslov and G.E. Zaikov, Liquid-Phase Oxidation of Unsaturated Compounds, N.Y., Nova Science Press, p. 227 (1993).
4. K.S. Minsker, S.V. Kolesov and G.E. Zaikov, Degradation and Stabilization of Polymers on the Base of Vinylchloride, Oxford, Pergamon Press, p. 576 (1988).
5. V.V. Kharitonov and B.L. Psikha, *Khimicheskaya fizika*, Vol. 6, No 2, p. 218 (1987).
6. V.V. Kharitonov and B.L. Psikha, *Khimicheskaya fizika*, Vol. 8, No 1, p. 85 (1989).
7. N.F. Trofimova and B.L. Psikha, *Khimicheskaya fizika*, Vol. 11, No 11, p. (1992).
8. V.V. Kahritonov and B.L. Psikha, *Doklady Akad. nauk SSSR*, Vol. 269, No 4, p. 892 (1983).
9. V.V. Kharitonov and B.L. Psikha, Proceeding of VIII All Union Symp. on Combustion and Explosion, Chernogolovka, Moscow region, p. 111 (1986).
10. B.L. Psikha, Ph.D. Thesis, Dept. of Inst. of Chem. Phys., Chernogolovka, Moscow Region (1980).
11. V.V. Kharitonov, B.N. Zhitenev and A.I. Stainslavskii, Soviet sertificat No 582481, Bulletin of invention No 44 (1977).

Significance and Application of the Dynamic Rheological Method in Kinetorheology
Part II. High-Conversion Radical Polymerization of Some Water-Soluble Ionogen Monomers and the Gel Effect

D.A. TOPCHIEV and YU.G. YANOVSKY*

Institute of Petrochemical Synthesis of Russian Academy of Sciences, Leninskii Prospekt 29, 117912, Moscow, GSP-1, Russia
Institute of Applied Mechanics of Russian Academy of Sciences, Leninskii Prospekt 32A, 117334, Moscow, Russia

INTRODUCTION

The application of water soluble polymers and composites is continuously expanding. Investigations are encouraged into the synthesis of water soluble polymers by novel and traditional methods aimed at producing polymers of a predetermined chemical composition, structure, molecular weight and propagation of chemical chains in macromolecules.

Traditionally water soluble polymers are synthesized by the radical polymerization of a broad range of water soluble monomers. The quantitative investigation of such reactions have shown that the control of such processes very often do not fit the framework of "ordinary" radical reactions. Some of the factors that influence the polymerization and copolymerization of water soluble monomers: the functional (including ionogen) groups and propagating chains (ion pairs, associates, free ions); solvent: the pH and ionic strength of the solution; the nature and concentration of low-molecular weight salts and neutralizing agents.

Over a period of 15–20 years, many research centers have been engaged in the systematic investigation of the kinetics and mechanisms of radical polymerization of water soluble monomers methacrylates.[1-6] In contrast, reactions of water-soluble vinyl monomers alkylacrylates (metacrylates) have been comprehensively studied in recent years.[7,8]

Polymerization of water soluble diallyl quaternary salts monomers have been investigated and the initial kinetic of radical polymerization of a series of dialkyldiallyl ammonium in aqueous and organic media have indicated the following general features:

The presence of ionogen groups in the diallyl monomer leads to a degradation transfer of the chain onto monomer (in all cases, an initiation rate order of 0.5 was detected), typical of allyl monomers. These reactions proceed fairly rapidly forming high-molecular product. The corresponding constants for effective chain transfer onto monomer were determined.[4] A mechanism to explain the absence of degradation chain transfer onto monomer was suggested.[3,4]

In all cases (irrespective of the radical initiator, nature of solvent, temperature, nature of the N-alkyl substituents and counterion in the monomer quaternary salt), the sharp and non-linear increment in the initial polymerization rate with increasing initial monomer concentration is due to a decrease in the rate constant for bimolecular chain rupture. This is due to the increase in viscosity of the initial solution, diffusion control of chain reaction rupture due to the viscosity of the monomer is evident in the lowest, practically, "zero" degree of conversions. The reaction is characterized by an order of unit over a broad range of monomer concentrations (0,1–5,5 mol/l) and temperature (18–80°C) in H_2O and CH_3OH.

POLYMERIZATION OF N,N-DIMETHYL-N,N-DIALLYL AMMONIUM CHLORIDE (DMDAAC) MONOMER IN WATER SOLUTIONS.

Kinetics

Kinetic studies of the polymerization of DMDAAC in water were carried out calorimetrically. The Tian-Calvet microcalorimeter used under isothermal conditions, measures the heat of chemical reaction with a sensitivity of 0,00072 Wt/mm and measurement error of 3,5%. The heat of polymerization of DMDAAC in water was 65,1 kJ/mol. The degree of monomer-to-polymer conversion was determined by PMR data (Varian-60 NMR spectrometer). The calorimetric method employed in the above produces reproducible kinetic measurements at temperatures $T \geqslant 60°C$ and initial monomer concentration $[M] \geqslant 4,5$ mol/l.

The calorimetric and PMR investigations produced three series of integrated kinetic curves in the conversion-time coordinates in which the monomer concentration (at $T°C =$ const. and $[I] =$ const.) and the temperature (at $|I| =$ const. and $|M| =$ const.), the initiator concentration (at $T°C =$ const. and $[M] =$ const.), were varied respectively, (Fig. 1).

The kinetic curves are S-like in shape which is typical of gel-effect reactions. The length of the linear segment of the kinetic curves and the corresponding initial polymerization rate values, estimated from the calorimetric data are in satisfactory agreement with the parameters obtained from the dilatometric measurements.[3]

Unlike the polymerization reactions of alkylmethacrylates in bulk, the process is characterized by linear kinetic curves preceding the onset of auto-acceleration. A possible reason for this effect is that the aqueous solutions of quaternary salts of the monomer are characterized by extremely high values of the relative viscosity in comparison with those of other water soluble monomers at the same concentrations ($[M] = 4,5–5,5$ mol/l). This fact leads to diffusion control of the bimolecular depropagation of the initiation reaction. The maximal reaction rate is attained at comparatively low monomer-to-polymer conversion values, i.e. at comparatively low concentration of polymer.

In such systems, the onset of acceleration (gel-effect) are created at comparatively low degree of polymerization. The resulting kinetic data can be processed and illustrated in the form of the dependence of the reduced polymerization rate on conversion, Fig. 2.

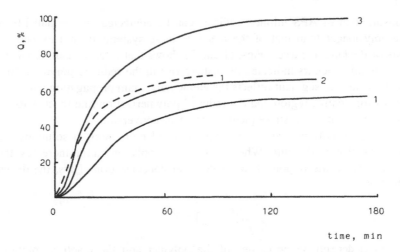

Figure 1   Monomer-to-polymer Q-conversion vs. DMDAAC polymerization time at 60° and 80°C, [APS] = $5 \cdot 10^{-3}$ mol/l. Curves 1–3 represent monomer concentrations $[M]$ = 4,5; 5,0; 5,5; respectively.

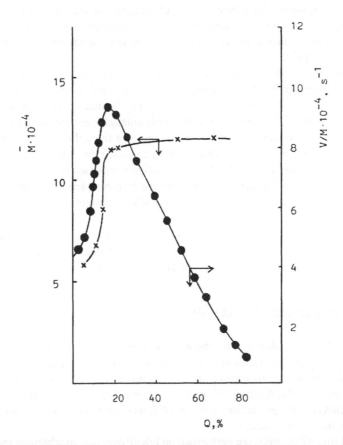

Figure 2   Reduced polymerization rate (1) and corresponding molecular weight values of the forming polymer (2) vs. DAC polymerization conversion time in water solutions at 60°C. [APS] = $5 \cdot 10^{-3}$, $[M]$ = 5,5 mol/l.

The consumption of APS initiator radicals can be neglected, as the half-life of APS is considerably longer than that of the reaction, these systems do not exceed 5 hours.

Analysis of the kinetic curves, Figs. (1 and 2) shows that after a weak initial segment, the anti-acceleration occurs indicated by a sharp rise in the reduced polymerization rate. The third characteristic segment reflects the elementary chain propagation reaction passes into the diffusion control region and the reduced polymerization rate reaches its minimal value under conditions of a subsequent increase in conversion.

The gel-effect in polymerization reactions of vinyl monomers in solutions has been investigated less than in the bulk. When the degree of polymerization increases, the auto-acceleration shifts to the region of lower polymer concentration. Thus, the dependence is described by:

$$\overline{P}_n^\alpha \cdot [C_p] = K,$$

where K and α depend on properties of the polymer and the reaction solutions. The parameters K and α do not depend on the monomer and initiator concentrations and molecular weight of the polymer formed.

During polymerization, the concentration of polymer is increased until the reaction volume is filled with swollen macromolecules coils, subsequently a decrease in the mean size down to an unperturbed θ-size occurs after which partial mutual penetration and entanglement takes place.

The auto-acceleration process occurs when the conditions for the appearance of a solid steric fluctuation entanglement polymer network in the reaction solution are achieved. An increase in the degree of polymerization allows the formation of a network at lower polymer concentrations. The formation of the steric entanglement network should facilitate a sharp fall in the translation diffusion after the onset of the gel-effect and the magnitude of the constant for the bimolecular chain depropagation rate. Under conditions of high monomer-to-polymer conversion, the degree of polymer polymerization should appreciable influence the polymerization kinetics. The change in chain depropagation rate constant and polymerization rate should correspond to the $\overline{MM}$ dependence of the polymer formed on the degree of conversion.[4]

The above example indicates that the calorimeter data is limited due to the fact that there is a "threshold" initial monomer concentration $[M]$ below which the sensitivity of the method is insufficient at $T = 60°C$ and is $\geqslant 4,5$ mol/l.

KINETORHEOLOGICAL ESTIMATES

The majority of studies devoted to rheological estimates of the kinetics of chemical processes, including polymerization have been on viscous systems.[7,8]

Dynamic rheological investigations estimate simultaneously viscous and elastic properties of the system under study. Analysis of the real and imaginary components of the complex dynamic modulus G^*-storage and loss moduli, G' and G'', respectively provided information about the structure of the sample.

Rheological investigations were performed on DKhP dynamic mechanical spectrometer.[9] The experiments were carried out in an inert gas atmosphere at a constant circular deformation frequency (lg $\omega = 0,2$ s^{-1}).

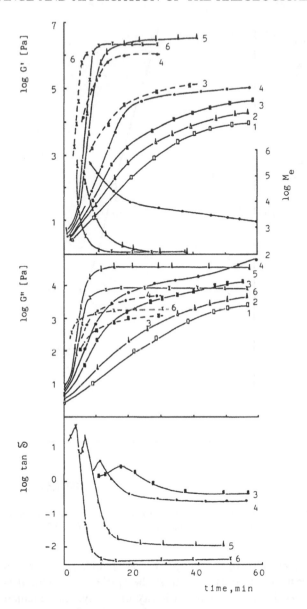

Figure 3 G', G'' module and tan δ vs. time in the process of DMDAAC polymerization at 60°C and 80°C (– – –), lg $\omega = 0{,}2$ s^{-1}, initiator concentration [APS] = 5 · 10^{-3} mol/l. Curves 1–6 represent monomer concentrations: 3,0; 3,5; 4,0; 4,5; 5,0; 5,5 mol/l, respectively.

The results of kinetorheological studies, Fig. 3 are for DMDAAC — polymerization in water at 60°C and 80°C, respectively. The curves represent different monomer concentrations and describe all the stages from the onset to the completion of the polymerization process. Comparison of those data with the above kinetic model shows that they are qualitatively similar, having a S-like shape, but distinguish from each other by the intensity of the gel-effect depending on the value of M. Comparison of Figs. 1

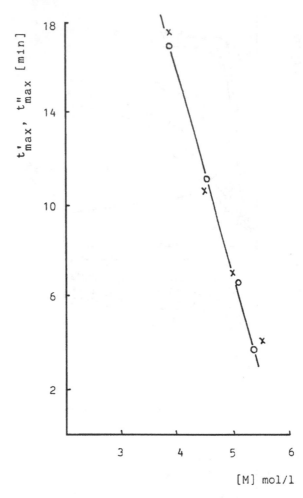

Figure 4   Time for achieving a maximal polymerization rate, $t'_{max}(x)$ and time for achieving a maximal tangent for losses $t''_{max}(o)$ vs. initial monomer concentration $[M]$, at 60°C $[APS] = 5 \cdot 10^{-3}$ mol/l.

and 3 shows that the kinetorheological dependences are obtained over a broader monomer concentration range than that from the kinetic studies; $|M| = 2,5–5,5$ mol/l as compared with 4,5–5,5 mol/l. This is due to the high sensitivity of the dynamic method in comparison with the calorimetric method. The experimental viscosity, G''/ω value, at $t = 0$ compares well with the values from the Newtonian macroscopic viscosity of the initial reaction systems, determined by Ubellode viscometer. The initial monomer solutions are typical Newtonian fluids, as there is no elastic component present at $t = 0$ (Fig. 3).

Analysis of the tan $\delta(t)$ dependences, Fig. (3) indicated by the position of the maxima on the time scale, coincides with changes in the time-dependent polymerization reaction rate, Fig. (4).

The G'-storage modulus of the system characterizes the density of entanglement network nodes formed,[10] and supports the influence of the gel-effect or the slow-final polymerization stages.

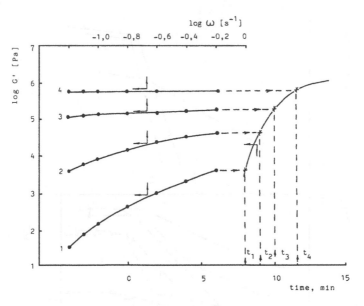

Figure 5 G' values vs. polymerization time t, for DMDAAC $[M]$ = 5,0 mol/l, [APS] = 5 · 10^{-3} mol/l at 60°C, lg ω = 0,2 s^{-1} and G'' values vs. ω (Curves 1–4) correspond to reaction mixture compositions of PDMDAAC and DMDAAC simulating the reaction system at different high conversions at times, $t_1 - t_4$, at 60°C.

Experimental evidence for the existence of a physically entangled network is a high elasticity plateau in the $G'(\omega)$ curves.[11] Elasticity theory allows determination of the density of the topological entanglements.[10–12] Experiments were carried out to analyze the frequency dependences of the solutions, Fig. 5. The frequency varied over the range lg ω = (−1, 2 to 0, 2 s^{-1}). Curves 3 and 4 are typical of a polymer with an elasticity plateau, corresponding to the presence of an entanglement network and were used to estimate the values of the dynamic M_e between entanglement nodes, Fig. 6. According to the data

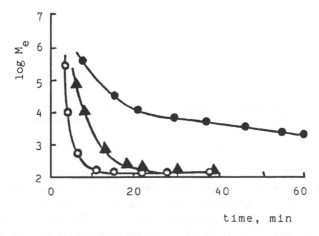

Figure 6   Molecular weight of dynamic segment, M_e vs. DMDAAC polymerization time, [APS] = 5 · 10^{-3} mol/l at 60°C. Curves 1–3 represent monomer concentrations: 4,5; 5,0; 5,5 mol/l, respectively.

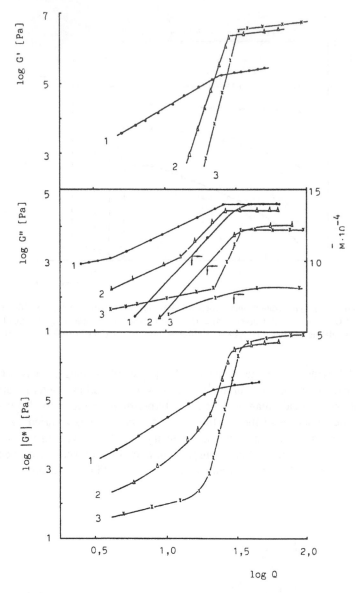

Figure 7 G', G'', $|G^*|$ module vs. Q-conversion and molecular weight of the forming polymer $\overline{M}$ during DMDAAC polymerization 60°C, [APS] = 5 · 10^{-3} mol/l). Curves 1–3 represent monomer concentrations: 4,5; 5,0; 5,5.

obtained from $G'(t)$, Fig. 3 and $M_e = f(t)$ curves, Fig. 6, the horizontal segments corresponds to the slow-final polymerization stage when the formation of the network is practically completed. The rheological data mirror the mechanism involved in the structuralization process. The dependences of the rheological data on G', G'' moduli and absolute values for the complex modulus $|G^*| = \sqrt{(G')^2+(G'')^2}$ on the monomer-to-polymer conversion are presented in Fig. 7.

It is important to emphasize that those dependences are characterized by the linear segments of different polymerization phases (initial steady-state stage, gel-effect region, slow-finish polymerization). Thus, at low conversions, Q, one can visualize a linear dependence of the G'' (loss modulus) parameter characterizing the system viscosity on Q. With a rise in conversion, one can observe a definite linear dependence on Q but of another parameter G' (storage modulus) which determines the elastic properties of the system. These straight segments have different slopes corresponding to different monomer concentrations. The slow-final polymerization stages are responsible for the horizontal behaviour.

At high conversion, the major contribution to $|G^*|$ values is due to the G' value (storage modulus) which is several orders of magnitude higher than G'' (loss modulus) values. It is also clear that the complex dynamic modulus $|G^*|$ dependence on Q is less informative in terms of separate polymerization phases and reflects only the total effect of the changing viscoelastic properties of the system with conversion.

Taking into account that the $|G^*|/\omega$ parameter is the complex (macroscopic) viscosity of the system correlates with the effective values measured during continuous deformation in steady-state flow, we may conclude that the individual estimates of the viscous and elastic parameters of the system are more informative than the complex value in the terms of the kinetic features of the radical polymerization process.

It is believed that in the auto-acceleration stage, besides entanglement network packing there must occur an increase in the degree of polymerization of the polymer formed with an increase in conversion. Indeed, the formation of the entanglement network should condition the decrease in the translation diffusion rate of the macromolecules.

Analysis of the change in the rheological characteristics in the process of DMDAAC polymerization has led us to a suggested physical model of the structuralization process involving the stage of formation of the fluctuation entanglement network and increase in the $\overline{MM}$ of the polymers. The data obtained allow us to establish appropriate analytical expressions related to the rheological and kinetic parameters of the reaction system at each stage of polymerization on:

$$G' = K_1 \cdot Q^n 1, \qquad G'' = K_2 \cdot Q^n 2$$

where K and n are constants.

It is possible to determine the Q-values for the polymerization reaction. In the region of higher conversion a maximum polymerization rate can be determined from the position of the peak or the time dependence segment of the loss tanget, tan δ.

CONCLUSION

In the case of polymerization reactions of water soluble and in particular ionogen monomers, the influence of a number of factors on the reaction rate constants, can be revealed only by a comprehensive kinetic and rheophysical study of polymerization.

REFERENCES

1. D.A. Topchiev, G.T. Nazhmetdinova, A.M. Krapivin, V.A. Shreider and V.A. Kabanov, *Vysokomolek. Soed.*, v. 25A, N. 3, p. 636–641 (1983).
2. V.A. Kabanov, D.A. Topchiev and G.T. Nazhmetdinova, *Vysokomolek. Soed.*, v. 24B, p. 51–53 (1984).
3. D.A. Topchiev, Yu.A. Malkanduyev, Yu.V. Korshak, A.K. Mikitaev and V.A. Kabanov, *Acta Polymerica*, **36**, N. 7, p. 373–374 (1985).
4. V.A. Kabanov and D.A. Topchiev, *Vysokomolek. Soed.*, v. 30A, N. 4, p. 678 (1988).
5. G.B. Butler and R.J. Angelo, *J. Am. Chem. Soc.*, **79**, 3128 (1957).
6. R.M. Ottenbritte and D.D. Shilady, Polymeric Amines and Ammonium Salts (IUPAC) Ed: by Goethals E.J. Oxford, N.Y., p. 143 (1980).
7. D.N. Emelyanov, I.E. Kononova, A.V. Ryabova and N.F. Simonova, *Vysokomolek. Soed.*, v. 17B, N. 2, p. 163 (1975).
8. B.A. Korolev, M.B. Lochinov, V.E. Dreval', V.P. Zybov, G.V. Vinogradov and V.A. Kabanov, *Vysokomolek. Soed.*, v. 25A, N. 11, p. 2430 (1983).
9. L.P. Ulyanov, Yu.G. Yanovsky, V.M. Neimark and S.I. Sergeenkov, *Zavodskaya Labor.*, v. 39, N. 11, p. 1402 (1973).
10. Yu.G. Yanovsky, A.B. Zachernyuk, *et al.*, *Vysokomolek. Soed.*, v. 29A, N. 4, p. 829 (1986).
11. Yu.G. Yanovsky and G.V. Vinogradov, *Vysokomolek. Soed.*, v. 21A, N. 11, p. 85 (1980).
12. W. Kuhn, *Helv. Chim. Acta*, t. 30, 839 (1947).

Compression Moulding of Thermoplastic Elastomer Composites

GABRIEL O. SHONAIKE

Dept. of Polymer Eng., Kyoto Institute of Technology, Matsugasaki, Sakyo-ku, Kyoto 606, Japan

1 INTRODUCTION

Fibre reinforced thermoplastic composites are extensively used for various applications in aerospace and automobile, construction where high performance materials are essential. These materials possess several advantages over thermoset composites, including processability, short moulding cycles, increased shelf life, damage tolerance, good physical properties, excellent chemical and corrosion resistance, material quality consistency and high economic return on recycling.[1-7] Another important advantage of thermoplastics is that they do not require time for cure to occur.[8] The major disadvantage of thermoplastics is that they have high viscosities which makes impregnation into reinforcing fibre more difficult. However, within the last couple of years, various impregnation techniques such as commingled yarn, coated yarn, co-woven fabric and film stacking have been developed[9-12] to eliminate the impregnation problem. With all these techniques, impregnation of thermoplastic resin into the reinforcing media is improved. As a result of the development of good impregnation and interfacial characteristics, thermoplastic composites offer a combination of high toughness and good environmental resistance.[13]

Most research has carried out in the area of thermoplastic elastomer composites using rigid semi-crystalline matrix resin whilst flexible thermoplastic elastomers have been ignored. Non-impregnated fibre reinforced natural or synthetic rubbers are however commercially available for pneumatic tires, flexible couplings, air springs and due to their lightweight and flexibility are suitable for making air-tight components.[14] These non-impregnated elastomer composites are difficult to process and often lead to layer separation and by impregnation, the above disadvantages can be eliminated.

In this article, a new mould which was especially designed to prevent loss of matrix resin during compression moulding will be mentioned. Secondly, the results of impregnated fibre-reinforced polyester-polyether and ethylene-vinyl-acetate elastomers will be discussed.

2 FABRICATION TECHNIQUE

Under normal fabrication conditions, loss of matrix resin is inevitable due to the high viscosity of the thermoplastic matrix resin. This loss can be greatly reduced by using a combination of a specially constructed mould and blend-setting of matrix resin and

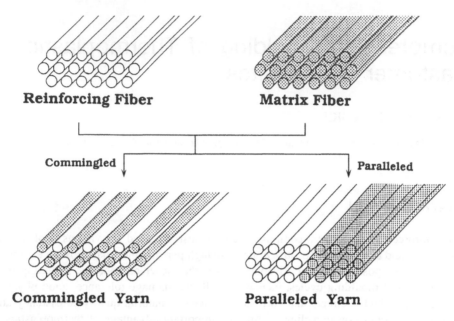

Figure 1   Schematic arrangement of matrix-fibre yarn.

reinforcing fibre. Fig. 1 shows the two methods of blend-setting, i.e. parallel yarn and commingled yarn arrangements. In the parallel arrangement, matrix and reinforcing fibres are uni-directionally wound layer upon layer on a metallic frame, whilst the commingle process involves mixing the matrix with reinforcing fibres before winding on the frame. Adequate tension of the pre-mix is required as it prevents wobbling of the matrix-fibre mixture. Mixing of reinforcing fibre and matrix in commingled process is more effective and requires less impregnation time and pressure than the parallel arrangement. In the latter situation it takes a longer time and requires a greater pressure for the matrix resin to flow into the reinforcing fibre. After blend-setting (commingled or parallel) on the metal frame, it is placed in the mould shown in Fig. 2.

The major advantage of this mould is that it prevents loss of matrix resin during compression moulding. The mould width is 50 mm, length is 330 mm and the compression moulding region is 250 mm, and each side has 40 mm serrations. The matrix-fibre which had been blend-set is placed in between the top and bottom parts of mould which had been heated to the required temperature and the desired pressure is applied. The serrations at each end of the mould provide a temperature difference between the moulding section and free ends, and produces spontaneous solidification of the matrix resin. With this type of mould, the matrix resin is frozen-in during fabrication.

3   LABORATORY TRIAL

The as-received pellets of ethylene-vinyl-acetate copolymer were supplied by Toyobo Co Ltd. The pellets were spun into fibre using a laboratory type extruder with fine spinneret. The glass fibre was obtained from Asahi Glass Co. Ltd. The matrix elastomer and glass

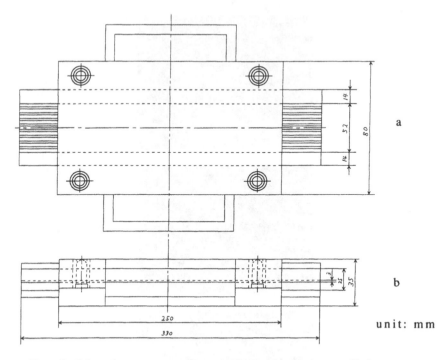

Figure 2   Schematic arrangement of compression mould: (a); top part (b); bottom part.

fibre were parallel pre-mix to give a volume fraction of 60:40. Samples were compression moulded using a pressure of 6 MPa, a temperature of 120°C and impregnation times of 5, 10, 15, 20 and 25 minutes. Optical microscopy was performed on small portions of composite cut and polished using various grades of emery paper and finally polishing on soft cloth using 0.3 and 0.03 μm diamond paste. The void contents were measured using an optional ignition method. The effect of impregnation condition on tensile properties were investigated.

3.1 Degree of impregnation

Three methods can be used to check the degree of impregnation of matrix resin into the reinforcing agent. These are:

(1)   observation of micrograph cross-sections.
(2)   void contents determination.
(3)   transverse tensile strength measurement.

3.1.1 Micrographs of cross-section

The micrographs of cross-section of glass fibre reinforced ethylene-vinyl-acetate elastomer composites, Fig. 3. for various impregnation times was varied between 5 and 25 minutes. The micrographs contain voids characterized by thick dark bubbles which can be seen

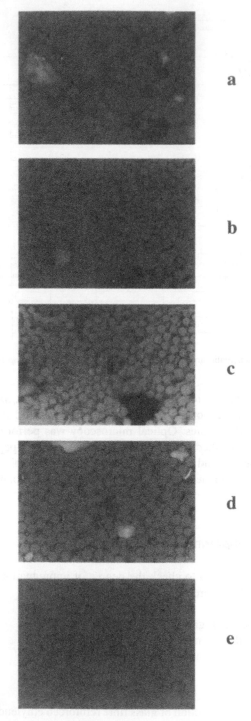

Figure 3   Optical micrographs of cross-section of glass fibre reinforced ethylene-vinyl-acetate elastomer composite: (a) 5 min, (b) 10 min, (c) 15 min, (d) 20 min and (e) 25 min.

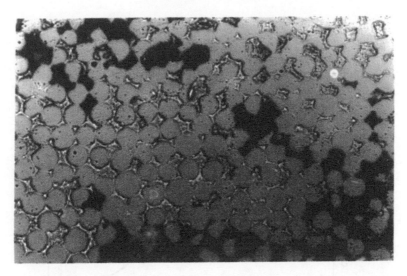

Figure 4   Optical micrograph of cross-section of glass fibre reinforced polyester-polyether elastomer composite at 5 min impregnation time.

clearly in 5, 10 and 15 minutes impregnation time. The shape of voids are large and localised. This is in contrast to our previous investigation[15,16] on glass fibre reinforced polyester-polyether composite, where the shape is small and distributed on the surface of the sample. The shape of void (or diameter) depends on the type of matrix resin. One of the main reasons why the size of the voids is small in reinforced polyester-polyether elastomer composite may be due to the presence of crystallites in the hard polyester segment which suppress formation of large voids. The type of voids obtained after 5 minutes impregnation time in glass fibre reinforced polyester-polyether composite are shown in Fig. 4.

3.1.2   Void content/impregnation time

The influence of impregnation time on the quantity of voids in reinforced ethylene-vinyl-acetate was explored in Fig. 5. The void contents were highest using 5 minutes impregnation time, at about 8%. There is no significant difference between the void contents using 5 and 15 minutes but above 15 minutes a dramatic reduction occurred. The void contents were reduced to about 3% after 20 minutes impregnation time and a further reduction to around 2.5% after 25 minutes. The requirement to use 25 minutes raises the question of whether degradation of matrix resin may be occurring.

3.1.3   Transverse tensile strength/impregnation time

The transverse tensile strength is an important measure of the degree of impregnation, the transverse strength being influenced by the matrix resin whilst the strength in the longitudinal direction is affected by the reinforcing glass fibre. The tensile strength increases with increasing impregnation time up to around 15 minutes before leveling off, Fig. 6.

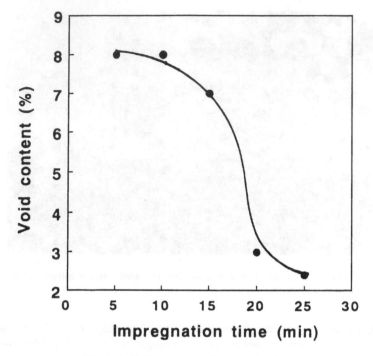

Figure 5   Void contents vs. impregnation time.

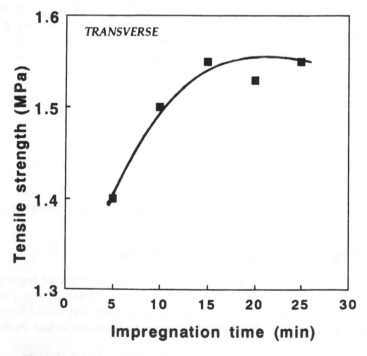

Figure 6   Transverse tensile strength as a function of impregnation time.

4   INDUSTRIAL VIABILITY

These laboratory trials indicate that this new tooling system may be of commercial value. Due to the flexibility of impregnated elastomer composites, impregnated time may be produced and can be applied to various textile structures such as weaving, knitting and braiding.

REFERENCES

1. I.Y. Chang and J.K. Lees, *J. Thermoplas. Compos. Mater.*, **1**, 277 (1988).
2. H. Wittich and K. Friedrich, *J. Thermoplas. Compos. Mater.*, **1**, 221 (1988).
3. T. Matsuo, *J. Jap. Soc. Polym. Proc.*, **5**, 225 (1993).
4. W.J. Cantwell, P. Davis and H.H. Kausch, *Compos. Struct.*, **14**, 151 (1990).
5. T. Matsuo, *Adv. Compos. Lett.*, **1**, 171 (1992).
6. G.O. Shonaike, M. Matsuda, H. Hamada, Z. Maekawa and T. Matsuo, *Compos. Interf.*, **2**, 157 (1994).
7. D.C. Leach, Advanced Composites (ed. I. Patridge), Elsevier Pub. Essex, Chap. 2, pp. 43–109 (1989).
8. I.Y. Yang and J.K. Lees, *J. Thermoplas. Compos. Mater.*, **1**, 277 (1988).
9. H. Hamada, G.O. Shonaike, M. Matsuda, Z. Maekawa and T. Matsuo, *Proc. 25th Inter. SAMPE Tech. Conf.*, Philadelphia, pp. 1008–1017 (1993).
10. L. Ye, V. Klinkmuller and K. Freidrich, *J. Thermoplas. Compos. Mater.*, **5**, 32 (1992).
11. T. Lynch, *SAMPE J.*, **25**, 17 (1989).
12. A.P. Majidiu, M.J. Rotermund and L.E. Taske, *SAMPE J.*, **24**, 12 (1989).
13. G.P. Desio and L. Rebenford, *Proc. ANTEC '89* p. 1416 (1989).
14. T. Akasaka, Textile Structural Composites (eds. T.W. Chou and F.K. Ko), Vol. 3, Elsevier Pub. Amsterdam, pp. 279–330 (1989).
15. G.O. Shonaike and T. Matsuo, *Compos. Struct.* (accepted).
16. G.O. Shonaike and T. Matsuo, *J. Reinf. Plast. Compos.* (in press).

New Metal Chelates as Antioxidant Stabilizers for Polymers and Low Molecular-Weight Substances

V.G. VINOGRADOVA and A.B. MAZALETSKII

Institute of Chemical Physics of the Russian Academy of Sciences, Russia, Moscow 117334, St. Kosygina 4

INTRODUCTION

The majority of polymers and plastics undergo thermal and thermooxidative degradation due to action at elevated temperatures of dioxygen in the course of processing, use and storage. Degradation is also stimulated by light and mechanical strain and this occurs in the processing of polymers in melt and solutions.

Thermooxidative degradation of polymers in melts and solutions is established to be a radical-chain process. Oxidation of polyolefins occurs through reactions of the $—CH_2—$ groups.[1,2] The basic reactions, determining this type of degradation are similar to those taking place in the case of liquid phase oxidation of hydrocarbons by dioxygen. Significant success has been achieved in the mechanisms of liquid phase oxidation of hydrocarbons of various types thanks to the efforts of Russian scientists, N.M. Emanuel and his coworkers and the works of G. Russel, F. Mayo, K.U. Ingold.[3-6]

The initial stages of oxidation of hydrocarbons (*RH*) and/or polymers (*PH*) are as follows:

$$I \xrightarrow{\ RH\ } R^{\cdot}\ (W_i) \quad \text{chain initiation}$$

$$\left.\begin{array}{l} R^{\cdot} + O_2 \longrightarrow RO_2^{\cdot} \\ RO_2^{\cdot} + RH \longrightarrow ROOH + R^{\cdot} \end{array}\right\} \text{chain propagation}$$

$$ROOH \longrightarrow RO^{\cdot} + OH^{\cdot} \xrightarrow{\ RH\ } R^{\cdot} \quad \text{degenerate chain branching}$$

$$\left.\begin{array}{l} R^{\cdot} + R^{\cdot} \longrightarrow R-R \\ R^{\cdot} + RO_2^{\cdot} \longrightarrow ROOR \\ RO_2^{\cdot} + RO_2^{\cdot} \longrightarrow ROH + R-CO-R+O_2 \end{array}\right\} \text{chain termination}$$

At high concentrations of dissolved oxygen, chain termination is determined by recombination of peroxy radicals $RO_2^{\cdot}$. In this case the dependence of the oxidation rate on the substrate concentration and the initiation rate is given by:

$$W_{ox} = W_i^{0.5}[RH].$$

This equation can not be applied to oxidation of the polymer solutions under definite conditions. Depending on the oxidation conditions and the properties of macromolecules, the oxidation rate may be expressed by a general equation:

$$W_{ox} = W_i^n[PH]^m,$$

where $1 \geqslant n \geqslant 0.5$; $1 \geqslant m \geqslant 0$, if chain termination is due to bimolecular interaction of two free radicals [i.e. quadratic termination] n is 0.5 and m is 1. In the case of linear termination n is 1 and m is 1.

Emanuel and Buchachenko[7] emphasized that kinetic control of polymer oxidation in solutions "depends on the thermodynamic properties of the solvent".

It follows even from the above simplified scheme that the reactions of alkyl and peroxy radicals play a very important role in the oxidation and hence in degradation and stabilization of polymers. In the case of polymer oxidation the reactions of alkyl radicals are substantial, for instance stepwise transfer of carbon-centered free valency along the polymer chain:[8]

$$P^{\cdot} + PH \longrightarrow PH + P^{\cdot}$$

Noticeable degradation of polymers (C—C bond scission), occurs at early stages of the process which is not characteristic for hydrocarbon oxidation. In polyethylene solutions this results from both: monomolecular decomposition of $PO_2^{\cdot}$ radicals and bimolecular interaction of two $PO_2^{\cdot}$ radicals.[9]

The control of thermooxidative degradation of polymers, making it different from oxidation of hydrocarbons, predetermine the conditions for polymer stabilisation and specific properties of antioxidants and stabilizers for polymer protection. When the polymers are oxidized the stabilizers have to deactivate peroxy radical $PO_2^{\cdot}$ as well as alkyl radical $P^{\cdot}$. Most universal stabilizers, such as aromatic amines and phenols, selectively react with the former.

It is very important for polymer stabilization that the radicals produced from inhibitors (phenoxy and aminyl radicals) can initiate and/or participate in polymer degradation:[10]

$$In^{\cdot} + PH \longrightarrow InH + P^{\cdot}$$

$$In^{\cdot} + POOH \longrightarrow InH + PO_2^{\cdot}$$

When the polymers are in the melt or solution state, their effective protection can be achieved, if the stabilizer satisfies the following conditions. Firstly, it is capable of terminating both $PO_2^{\cdot}$ and $P^{\cdot}$ radicals. Secondly, the intermediate particles and/or stabilizer transformation products should not participate in polymer degradation:

antioxidant stabilizer $\mathcal{A}$ $\quad \dfrac{\quad PO_2^{\cdot} \quad}{\quad P^{\cdot} \quad}$ molecular products without peroxide groups or inactive radicals

METAL CHELATES AS INHIBITORS AND STABILIZERS

Chelate compounds of metals of variable valency and those of zinc are used as stabilizers for polymer protection. Owing to a variety of ligands practically any known property of stabilizer can be achieved: (i) a high compatability with polymer, (ii) low volatility, (iii) required redox characteristics, (iv) suitable stabilizer transformation products, etc.

There is considerable experience in the successful application of transition metal complexes as stabilizers. Transition metal complexes with sulphur containing ligands (dithiocarbamates and dithiophosphates) widely used as antioxidants in rubbers and plastics. It is concluded that a primary function of this class of complexes in both thermal oxidation and photo-oxidation of poly-olefins is to destroy hydroperoxides in an ionic process. The sulphur-containing products of ligand reaction such as sulphur acids play the main role.[11,12]

Apart from this function, complexes of transition metals and those of some metals of constant valency show the ability to scavenge effectively alkyl and peroxy radicals of hydrocarbons and polymers. The possibility of retarding liquid phase oxidation of hydrocarbons by metal complexes, terminating the chains carrying peroxy radicals, was shown many years ago.[13,14] The transition metal dithiocarbamates dithiophosphates and xanthates were among the earliest of antioxidants to be investigated. Their antioxidant action in cumene, styrene, etc. has been examined in detail.[11,15-20]

The mechanisms of peroxy radicals trapping by Co(III) dimethylglyoxime, when oxidizing cumene has been investigated.[21] The Cu(II), Ni(II) and Zn(II) complexes of 3-hydroxy-2-iminomethylbenzo[b]thiophene were found to be antioxidants in the case of n-decane oxidation.[22]

Interesting effect of repeated chain termination by the salts and complexes of Cu(II), Co(II) and Co(III) was studied recently.[23,24] This phenomenon results from the capability of peroxy radicals, formed in the course of oxidation of primary and secondary alcohols, amines and some other organic substrates, to oxidize and reduce metal ions in alternating reactions. Such redox reactions provide continuous inhibitor (metal ion in a proper oxidation number) regeneration.

ANTIRADICAL ACTIVITY OF METAL COMPLEXES IN HYDROCARBON OXIDATION

Antiradical activity of transition metal complexes containing S, Se, N and O atoms in the chelate centre have been studied, Table 1.

The experiment consisted in studying oxidation of ethylbenzene, unsaturated hydrocarbons (nonene-1, styrene, cyclohexene) and n-decane at 50-70°C initiated by azo-bis-isobutyronitrile (AIBN). In some cases oxidation of ethylbenzene and nonene-1 was performed at 120°C. Hydrocarbons have been thoroughly purified from hydroperoxides. Concentration of chelates varied from $5 \cdot 10^{-4}$ to $5 \cdot 10^{-3}$ mol/l. Only the initial stage of oxidation was considered. The rate of hydrocarbon oxidation was shown to be independent on dioxygen concentration in the oxidizing gas mixture containing from 20 to 100% of O_2 by volume. Suppression of hydrocarbon oxidation by chelate complexes under given experimental conditions indicates that they terminate the oxidation chains by deactivating peroxy radicals.

TABLE 1

Type of the complex	Structure
N,N-dialkyldithiocarbamate $M(dtc)_n$	R_2N—C with S, S to M/n

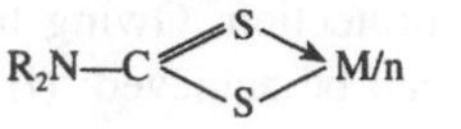

M = Zn, Pb, Ni, Cu, Sb
R = i-C_3H_7, C_2H_5, n-C_4H_9, 2-furyl, C_6H_{12}, PhCH(CH_3), 2-thienyl

N-phenymercaptobenzimidazole $M(mbiz)_n$

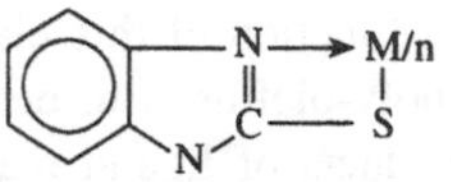

M = Cu(I), Cu(II), Ni, Zn

α-thiopicoline anilide $M(tpa)_n$

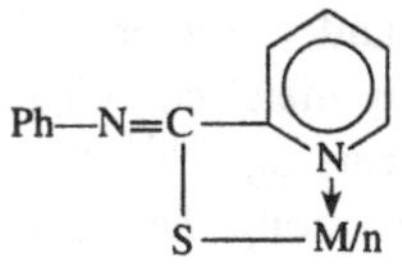

Ph—N=C ... N ... S——M/n

M = Cu(II), Co(II), Ni, Vo, Zn

8-mercaptoquinolinate (thiooxynate) $M(tox)_n$

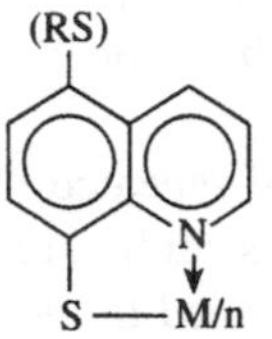

(RS)

S——M/n

M = Cu(II), Co(III), Ni, Ag, Zn
R = H, C_nH_{2n+1} (n = 3–5)

8-hydroxyquinolinate (oxynate) $M(ox)_n$

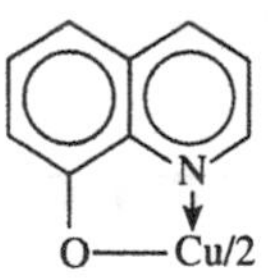

O——Cu/2

Complexes of 5-thiols of N-phenyl- and N-methylpyrazole
$M(I,S)_nR$
$M(II,S)_nR$

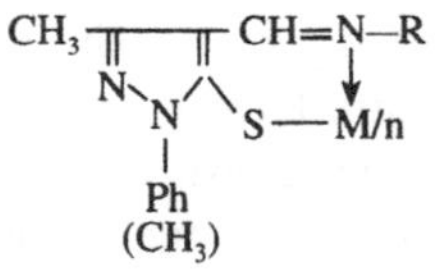

CH_3 ... CH=N—R
N—N ... S—M/n
Ph
(CH_3)

M = Cu(II), Co(II), R = CH_3
M = Zn, R = CH_3, C_3H_7, C_4H_9
M = Cu, R = CH_3, C_6H_5

Thiophene $M(III,S)_nR$
Furane $M(IV,S)_nR$

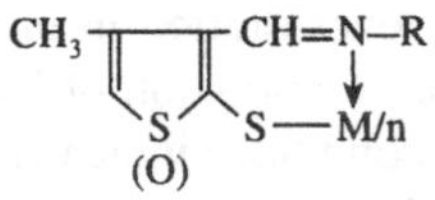

CH_3 ... CH=N—R
S ... S—M/n
(O)

M = Zn, R = CH_3

TABLE 1   *continued*

Type of the complex	Structure
Complexes of N-phenyl-5-selenolpyrazole $M(I,Se)_n$	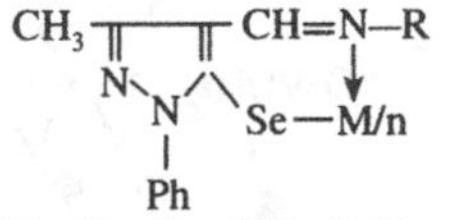 CH_3 — CH=N—R N—N Se—M/n Ph M = Zn, R = CH_3, C_3H_7
Complexes of o-vanillale-o-anisidine $M(van)_n$	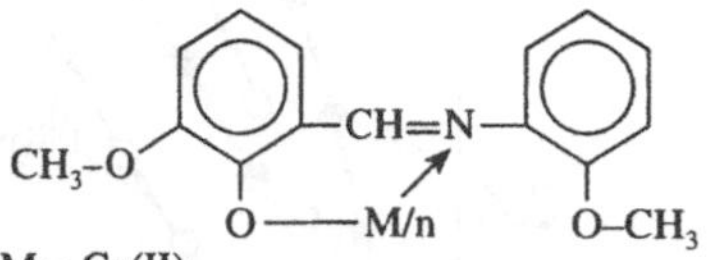 CH_3–O — CH=N — O — M/n O–CH_3 M = Co(II)
Acetylacetonate $M(acac)_2$	M = Co(II), Cu(II)

The inhibiting activity of metal complexes was characterized: (i) by stoichiometric inhibition coefficients estimated from the induction period τ by using an equation $f_\tau = (W_i \cdot \tau)/C_0$, where W_i is the chain initiation rate, and C_0 is the initial concentration of metal complex (under certain conditions f_τ means the number of radicals deactivated by one molecule of the inhibitor and its transformation products); (ii) by inhibition constants (k_{inh}) evaluated from the initial rate of inhibited oxidation and corresponding rate constants of the chain termination reaction:

$$RO_2^\bullet + ML \xrightarrow{\ k_{inh}\ } products.$$

Under experimental conditions only complexes, containing strong electron-donating S and/or Se atoms or S and N atoms in the chelate active centre, exhibiting an inhibiting action were used. The compounds with chelate centres of $M(O,N)_n$ and $M(O,O)_n$ types either catalysed the process [for instance, $Co(van)_2$ and $Co(acac)_2$] (Fig. 1, curve 2) or do not influence the oxidation rate [$Cu(oxin)_2$ and $Cu(acac)_2$].

Fig. 1 shows typical kinetic curves of dioxygen absorption upon ethylbenzene oxidation in the presence of complexes of different types. Their inhibiting activity varied widely. Complexes of Zn (and Ag) decreased the rate of dioxygen absorption as compared with oxidation in the absence of the complex. On the other hand, sharp induction periods to the onset of dioxygen absorption were observed in the presence of the most effective inhibitors such as Cu(II) complexes [$Cu(mbiz)_2$, $Cu(dtc)_2$, $Cu(tox)_2$ and $Cu(I,S)_2CH_3$] and $Ni(dtc)_2$ (Fig. 1, curves 5–7). Most of complexes under investigation exhibited an intermediate inhibiting activity, Fig. 1 (curve 4).

Inhibition coefficients f_τ for some of complexes are summarized in Table 2. The values of f_τ for dithiocarbamates are equal to 3–5 and exceed the number of ligands in the chelate molecule. Those for other complexes are not more than the number of ligands in the initial complex and make up 1 to 2. It is worth noting that f_τ for ethylbenzene and nonene-1 oxidation are close to one another.

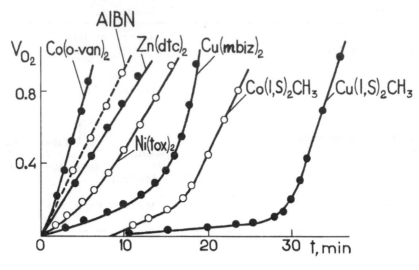

Figure 1   Kinetic curves of dioxygen consumption in ethylbenzene oxidation in the presence of various metal chelates ($C_0 = 1 \cdot 10^{-3}$ mol/l, $W_i = 1 \cdot 10^{-6}$ mol/l · s, 75°).

It should be underlined that inhibiting activity is inherent only in complexes with ligands coordinated to heavy metal ions. In fact under experimental conditions free ligands (LH) and their Na-salts did not display such an activity.

TABLE 2

Inhibition coefficients for various metal complexes in the course of ethylbenzene and nonene-1 oxidation
(75°, $W_i = 10^{-6}$ mol/l·s)

Type of Complex	$f_\tau = (W_i \cdot \tau / C_0)$	
	ethylbenzene	nonene-1
$Cu(dtc)_2$	4.7	4.5 (2,8 styrene)
$Ni(dtc)_2$	3.1	2.8
$Sb(dtc)_2$	4.5	—
$Zn(dtc)_2$	—	—
$Bi(dtc)_2$	0.9	
$Pb(dtc)_2$	2.6	1.0
$Cu(tpa)_2$	1.2	
$Cu(tox)_2$	2 (2 styrene)	
$Ni(5-S-tox)_2$	0.8	
$Cu(I,S)_2R$	1.9	2.0
$Co(I,S)_2R$	2.0	
$Zn(I,Se)_2R$	2.0	
$Zn(V,S)_2R$	1.8	
$Ni(I,S)_2R$	1.0	
$Cu(I)mbiz$	1.0	
$Cu(II)(mbiz)_2$	1.2	

MECHANISM OF INHIBITING ACTION

Peroxy radicals $RO_2^{\cdot}$, which are chain-carrying particles in the oxidation of hydrocarbons, can play a role of oxidizing agents and their interaction with the metal complexes may be due to the following reactions:

$$RO_2^{\cdot} + M^n L_n \begin{cases} \longrightarrow ROOH + L'ML_{n-1}^{\cdot} & \text{abstraction of H atom} & (1) \\ \longrightarrow (RO_2)M^{n+1}L_{n-1} & \text{oxidative addition} & (2) \\ \longrightarrow (RO_2)M^n L_{n-1} + L^{\cdot} & \text{radical substitution} & (3) \\ \longrightarrow RO_2^- + M^n L_n^+ & \text{electron transfer} & (4) \end{cases}$$

The majority of complexes used do not carry labile hydrogen atoms in the ligands and therefore reaction (1) is hardly probable in contrast to Co(III)dimethylglioxime which has such atoms and capable of reacting with $RO_2^{\cdot}$ via (1).[21]

Oxidative addition (2) may proceed in the case of complexes where metals take a higher oxidation state in comparison with the initial one. Complex $Co(I,S)_2R$ was found to interact with $RO_2^{\cdot}$ radical according to reaction (2). The reaction of electron transfer (4), takes place in non-polar hydrocarbon medium, probably as an intermediate stage of oxidative addition of ligand (2) or ligand substitution (3):

$$Cu(dtc)_2 + RO_2^{\cdot} \rightleftharpoons [Cu(dtc)_2 \cdots RO_2^{\cdot}] \rightleftharpoons \{[Cu(dtc)_2] \cdots RO_2^- \}.$$

The $Ni(dtc)_2$ complex may react in the same way. There is information in literature that formation of cations $Cu(dtc)_2^+$ and $Ni(dtc)_2^+$ is possible.[31,32]

The data obtained for the majority of complexes investigated are likely to indicate that metal complexes interact with $RO_2^{\cdot}$ radicals in a reaction of ligand substitution (3). This mechanism of inhibition, according to which each ligand deactivates one peroxy radical:

$$ML_n + RO_2^{\cdot} \longrightarrow ML_{n-1}(RO_2); \quad ML_{n-1}(RO_2) + RO_2^{\cdot} \longrightarrow ML_{n-1}(RO_2)_2$$

agrees with the fact that f_τ and "n" coincide in many cases.

The values of f_τ equal to 1 can also be explained within the framework of the mechanism of ligand substitution if one assumes that the primary product of chelate transformation is poorly active in the reaction with peroxy radicals as compared with the initial complex.

In the above reaction the metal oxidation number in the complex is assumed to be constant during the process. X-ray electron spectroscopy has shown for the products of $Ni(I,S)_2R$ transformation in the reaction with $RO_2^{\cdot}$ radicals that the oxidation number for Ni remains constant and the environment of Ni(II) ion becomes more electronegative (presumably because of insertion of oxygen atoms).

This mechanism of inhibition was qualitatively confirmed by a significant increase in the inhibiting activity of $Cu(dtc)_2$, $Cu(tpa)_2$, $Cu(I,S)_2R$, $Cu(tox)$, $Ni(dtc)_2$ and some other complexes when a free ligand (LH or NaL) was introduced into the system. Fig. 2a shows the corresponding data for $Co(tox)_2$. Such an effect may be explained by regeneration of the initial complex from the products of its transformation

$$ML_{n-1}(RO_2) + NaL(HL) \longrightarrow ML_n$$

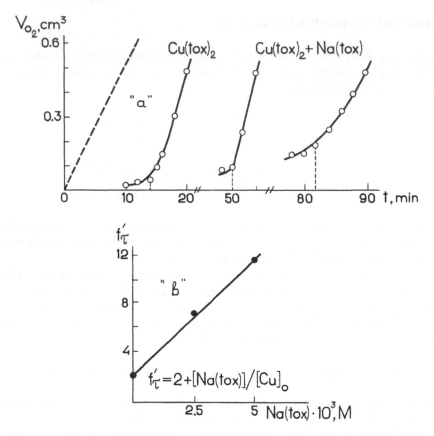

Figure 2   (a) Kinetic curves of dioxygen consumption in ethylbenzene oxidation in the presence of Cu(tox)$_2$ and Cu(tox)$_2$ + Na(tox) (75°, $W_i = 1 \cdot 10^{-6}$ mol/l · s, C_0 of Cu(tox)$_2$ $5 \cdot 10^{-4}$ mol/l, C_0 of Na(tox) $2.5 \cdot 10^{-3}$ mol/l and $5 \cdot 10^{-3}$ mol/l, respectively. (b) Dependence of effective value of stoichiometric inhibition coefficient for Cu(tox)$_2$ on the Na(tox) additives.

If the introduced ligand is completely consumed for the regeneration of the initial chelate, the following linear relationship should be valid

$$f_\tau' = 2 + [LH] / C_o$$

where f_τ' is the stoichiometric inhibition coefficient as calculated for the initial amount of the metal complex. Fig. 2b shows that such a dependence is observed in the experiment.

In accord with the discussed mechanism, the products of dimerization of the radical form of ligand:

$$\overset{\cdot}{L} + \overset{\cdot}{L} \longrightarrow L{-}L$$

(where $L{-}L$ is disulphide) have been detected. Disulphides were found in the reactions of Cu(dtc)$_2$, Ni(dtc)$_2$, Sb(dtc)$_3$, Cu(tox)$_2$ and Zn(tox)$_2$ with $R\overset{\cdot}{O}_2$ radicals.[26,33–35] In the case of thiooxinates and Sb(dtc)$_3$ disulphides were accumulated in stoichiometric amounts as complexes were consumed.

For Cu(dtc)$_2$ (copper diethyldithiocarbamate) an intermediate product was detected by the ESR method which assumed to be Cu(dtc)RO_2, where RO_2 was substituted for dtc-ligand.

Howard et al.[17-19] have detected using the ESR method, a sequence of intermediate products appearing in the course of reaction of Cu(dtc)$_2$ with $RO_2^{\cdot}$ or $ROOH$. One of them showed an ESR spectrum which corresponded to the spectrum of an intermediate found in our reaction system. The authors allowed radical substitution of ligand as one of the possible mechanisms for the reaction of Cu, Ni, Zn dithiocarbamates with $RO_2^{\cdot}$ radicals. Moreover, it was shown that styrene polyperoxide, produced after prolonged autooxidation of styrene in the presence of zinc-di-iso-propyldithiophosphate did contain Zn(i-PrO)$_2$PS and (i-PrO)$_2$PS$_2$ end groups.

It was also proposed[36] that the reaction of Cu(dtc)$_2$ with strong electron acceptor NO$_2$ proceeds as a radical exchange process resulting in formation of the mixed-ligand complex Cu(dtc)(NO$_3$) and thiuramdisulphide (tds).

High inhibiting activity of Cu(dtc)$_2$, Ni(dtc)$_2$ and Sb(dtc)$_3$ with the coefficients of inhibition exceeding the number of ligands needs an additional explanation. The $f_\tau > n$ ratios were shown[28] for three above inhibitors to be due to thiuramdisulphide-induced regeneration of initial complexes from oxygen-containing reaction products:

$$CuL_2' + dtc \longrightarrow Cu(dtc)_2 + 2L'$$

Thus the bulk of data testifies to the interaction of most of investigated complexes with peroxy radicals by a ligand substitution reaction. The question is how monodentate ligand (RO_2) displaces a bidentate one (dtc). Most probably a multistep process takes place. At the first stage, preliminary coordination of $RO_2^{\cdot}$ with the metal ion of the chelate active centre occurs. This stage is favoured by square-planar structure of chelate centres of Cu(dtc)$_2$, Cu(dtc)$_2$, Cu(tox)$_2$ and Ni(dtc)$_2$ where coordination sites 5 and 6 remain unoccupied, as well as by distorted octahedric structure of Sb(dtc)$_3$. At the same time, completely substituted symmetric octahedric complexes Co(dtc)$_3$ and Cr(dtc)$_3$ are known to be extremely inert in ligand exchange reactions. Therefore, incorporation of $RO_2^{\cdot}$ into the coordination sphere of these complexes does not occur and, as a consequence they do not inhibit oxidation of hydrocarbons.

It may be proposed that the second stage of interaction of $RO_2^{\cdot}$ radicals with the complexes is electron transfer from the chelate centre to $RO_2^{\cdot}$. The arguments supporting this point of view are as follows, hydrocarbon peroxy radicals $RO_2^{\cdot}$ can play a role of effective oxidizing agents; second, Cu$^{(2+)}$ and Ni$^{(2+)}$ may be oxidized to Cu$^{(3+)}$ and Ni$^{(3+)}$, stabilized due to the presence of electron donating sulphur atoms in the chelate centre.[31,37] Two mechanisms may be put forward. According to the first one metal ion donates an electron, which is compensated due to electron-rich sulphur atoms. In this case the metal valence remains constant. Another mechanism suggests that the chelate centre as a whole loses an electron, $RO_2^{\cdot}$ radical transforms into anion RO_2^{-}, and oxidized ligand splits off in the form of sulphur-centered radical, dtc$^{\cdot}$. The radicals dimerize to disulphides as reaction products.

$$Cu(dtc)_2 + RO_2^{\cdot} \longrightarrow [Cu(dtc)_2 \ldots RO_2^{\cdot}]$$

$$[Cu(dtc)_2 \ldots RO_2^{\cdot}] \longrightarrow \{[Cu(dtc)_2]^+ \ldots RO_2^-\}$$

$$\{[Cu(dtc)_2]^+ \ldots RO_2^-\} \longrightarrow Cu(dtc)RO_2 + dtc^{\cdot}$$

$$Cu(dtc)RO_2 + RO_2^{\cdot} \longrightarrow Cu(RO_2)_2(dtc^{\cdot}) \longrightarrow dimer$$
$$\searrow CuL_2'$$

$$dtc^{\cdot} + dtc^{\cdot} \longrightarrow thiuramdisulphide\ (tds)$$

Scheme 1

If electron transfer from the chelate centre to $RO_2^{\cdot}$ takes place there must be a correlation between inhibiting activity of complexes, on the one hand, and their redox characteristics, on the other.

ELECTROCHEMICAL METHODS FOR INVESTIGATION OF METAL COMPLEX REACTIONS

In principle, electrochemical methods are suitable for simulating the redox steps in organic reactions. They allow: (i) quantitative characterisation of the redox properties of chemical compounds, (ii) estimation of the electronic effects of substituents, (iii) generation of radicals and radical-ions. Half-wave potentials for voltammeric oxidation and reduction ($E_{1/2}^{ox}, E_{1/2}^{red}$) on the inert electrodes can be considered characteristics of redox properties of complexes. Though $E_{1/2}$ are factually not the standard redox potentials, they may be used to characterize the redox properties of molecules in solutions. It was demonstrated, that the $E_{1/2}^{ox}$ of the first one-electron step of metal dithiocarbamate oxidation are in good correlation with the first ionization potentials of the complexes in the gas phase.[37] Sometimes $E_{1/2}$ are the only available characteristics redox properties of metal compounds. Relationship between chemical compounds reactivity and their electrochemical potentials is demonstrated in more detail if both $E_{1/2}^{ox}$ and $E_{1/2}^{red}$ are known.[38]

To confirm the proposed mechanism of interaction of sulphur-containing metal chelates with peroxy radicals, half-wave potentials corresponding to reversible one-electron processes have been measured voltammetrically on a rotating platinum electrode. It has been found that undissociated chelate centre of complexes takes part in the electrochemical processes.

Some correlation has been revealed between the values of k_{inh} and $E_{1/2}^{ox}$ of chelates. For dithiocarbamates of bivalent metals, k_{inh} rose with $E_{1/2}^{ox}$ decrease, i.e. with an increase in electron-donor properties of chelate complexes. For the complexes with one and the same metal ion (for instance Cu(II)) in the chelate centre it can be seen that a rise of k_{inh} with decreasing $E_{1/2}^{ox}$ is retained when passing from oxygen-containing to more electron-donating sulphur-containing chelate centers.

Fig. 3 summarizes the data for chelate complexes of various classes. General control for more than 20 metal chelate complexes can be seen according to which k_{inh} values

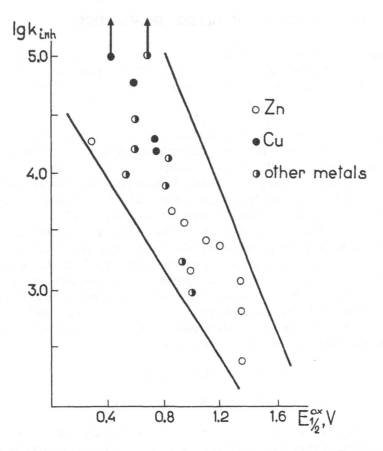

Figure 3   Dependence of inhibition constant k_{inh} for various metal chelates on the $E_{1/2}^{ox}$ of complexes (75°, ethylbenzene).

increase when $E_{1/2}^{ox}$ decrease. The activation energy of inhibition falls down almost linearly with $E_{1/2}^{ox}$ lowering thereby ensuring k_{inh} growth.

Two points are worth noting. First, $E_{1/2}^{ox}$ values should not be too low. The complexes with too low $E_{1/2}^{ox}$ (such as Mn(dtc)$_3$, Fe(dtc)$_3$) are readily oxidized with dioxygen and hence inhibit hydrocarbon oxidation inefficiently. Second, k_{inh} does not depend on $E_{1/2}^{ox}$ only, it is also determined by the geometry of the chelate centre and ligand structure as a whole. As was mentioned, Co(dtc)$_3$ and Cr(dtc)$_3$, in contrast to Pb(dtc)$_2$ and Bi(dtc)$_3$ do not exhibit high inhibiting activity in hydrocarbon oxidation though $E_{1/2}^{ox}$ values for all these complexes are very close. Probably, low inhibiting activity in first two cases is due to some steric hindrance of central metal ions in chelate centers of complexes.

Correlation between $E_{1/2}^{ox}$ and k_{inh} indicates that the limiting stage of interaction of sulphur-containing metal chelates with peroxy radicals consists in the electron transfer from electron-donating chelate centers to peroxy radical. The latter becomes deactivated via formation of an peroxy-anion. In the case of complexes, containing ligands with functional group which may be easily oxidized, the inhibition mechanism via the oxidation of this group in a first stage is not inconceivable.

OXIDATION OF POLYETHYLENE GLYCOL IN SOLUTION

Polyethylene glycols $HO(CH_2CH_2O)_nH$ of various degrees of polymerization as well as their derivatives are used as complex-forming agents, stationary phases in gas chromatography and plastisizers of polymer materials. Solutions of polyethylene glycol (PEG) in water and organic solvents as well as liquid oligomers of ethylene glycol can be used for various technical purposes, as heat carriers.

When exposed to atmospheric oxygen, the heterochain molecules of PEG with activated C—H bonds (in reference to C—H bonds in carbon chain polymers, e.g. polyolefins) suffer a radical-chain oxidation in the melt and solution state.[39,40] The kinetic control and mechanism of liquid-phase oxidation of PEG and related polyethers have not been studied as extensively as hydrocarbons. Only traditional antioxidants, i.e. hindered phenols and bis-phenols, were studied as antioxidant stabilizers for polyethers.[41]

The process of PEG chain oxidation involves free peroxy radicals $RO_2^{\cdot}$ of two types characterized by different chemical properties: (i) terminal α-hydroperoxy radicals ("alcohol-type") and (ii) non-terminal α-alcoxyperoxy radicals ("ether-type"). The former are known to exhibit dual reactivity, having both one-electron oxidizing and one-electron reducing properties.[23] As a result, the phenomenon of repeated chain termination can be observed, e.g. in the processes of liquid-phase oxidation of primary and secondary aliphatic alcohols in the presence of variable valence metal compounds. The mechanism of such phenomenon is based on the following sequence of redox reactions:

$$M^{n+} + RO_2^{\cdot} \longrightarrow M^{(n+1)+} + RO_2^{-}$$
$$M^{(n+1)+} + RO_2^{\cdot} \longrightarrow M^{n+} + \text{molecular products},$$

where M is the metal of variable valence. This type of chain termination leads to prolonged periods of effective inhibition of alcohols oxidation; in the absence of irreversible inhibitor transformation the above redox mechanism leads to value of stoichiometric inhibition coefficient $f_\tau = \infty$ (so called "negative catalysis of chain termination"). The "ether-type" peroxy radicals of low-molecular organic substances do not possess the one-electron reducing properties, and therefore such a mechanism does not take place upon ether oxidation. Evidently, the role of "alcohol-type" peroxy radicals in the gross-process of PEG oxidation depends on the degree of PEG polymerization. The problem of efficient PEG stabilization by variable valence metal compounds needs extensive experimental work to be solved.

We have investigated the ability of oxygen-, nitrogen- and sulphur-containing chelates of Cu, Co, Zn and Bi to stabilize PEGs of various molecular masses in the case of their oxidation in concentrated solutions. When choosing the new metal complexes as potential antioxidant stabilizers, attention was focused on the nitrogen-containing chelates, since oxygen-containing analogues are known usually to catalyze oxidation of organic substances. The sulphur containing complexes, known as efficient inhibitors of hydrocarbons oxidation, participate often in the oxidative side-reactions, leading to oxidative transformation of sulphur ($S^{II} \rightarrow S^{IV} \rightarrow S^{VI}$) accompanied by the loss of antiradical activity of metal compounds.

New bi- and tricyclic chelate complexes I–XXIII with active chelate centers $M(N_4)$, $M(N,N)_2$, $M(O,N)_2$, $M(N,NH)_2$ and $M(S,N)_2$ were objects of our work (Table 3). Complexes

TABLE 3

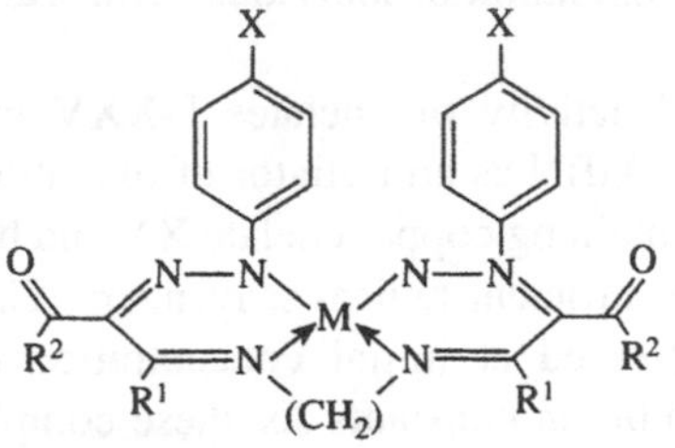

$M = Cu$ $n = 2$, $X = H$, $R^1 = CH_3$, $R^2 = Ph(I)$, $CF_3(II)$,

$n = 4$, $X = H$, $R^1 = CH_3$, $R^2 = CF_3(III)$,

$n = 4$, $R^1 = CH_3$, $R^2 = Ph$, $X = H(IV)$, $OCH_3(V)$, $Br(VI)$, $NO_2(VII)$,

$M = Co$ $n = 2$, $X = H$, $R^1 = CH_3$, $R^2 = Ph(VIII)$; $n = 4$, $X = H$, $R^2 = Ph$, $R^1 = H(IX)$, $CH_3(X)$; $R^2 = CF_3$,

$R^1 = CH_3(XI)$, $Ph(XII)$

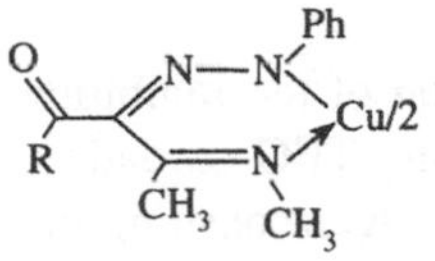

$R = Ph(XIII)$, $CF_3(XIV)$

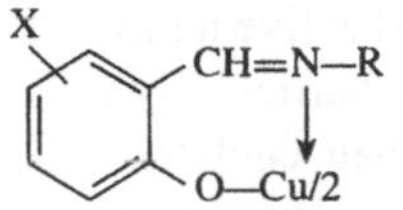

$X = H$, $R = n\text{-}C_4H_9(XV)$, $sec\text{-}C_4H_9(XVI)$,

$tert\text{-}C_4H_9(XVII)$, (XVIII),

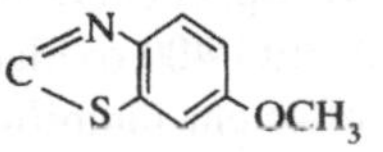

$X = 3\text{-}OCH_3$, $R = Ph(XIX)$

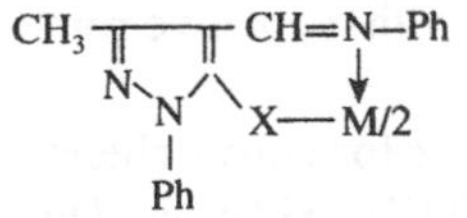

$M = Cu$, $X = S(XX)$, $O(XXI)$, $NH(XXII)$,

$M = Co$, $X = NH(XXIII)$

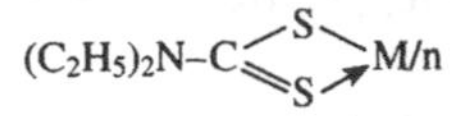

$M = Zn$, $n = 2(XXIV)$,

$M = Bi$, $n = 3(XXV)$.

were synthesized in alcoholic medium from the metal acetates and corresponding organic ligands — Shiff bases or hydrazonimines in H-forms. The inhibiting action of well-known sulphur-containing antioxidants Zn and Bi diethyldithiocarbamates XXIV and XXV was studied for comparison. These complexes show up the evident antioxidant properties upon liquid-phase oxidation of hydrocarbons and aliphatic alcohols.

PRELIMINARY TESTING OF COMPLEXES

In a course of extensive investigation of radical-chain oxidation of organic substances a standard scheme of testing chemical compounds for their ability to play a role of antioxidant stabilizers for polymer materials was applied.[42] At the beginning an analysis was carried out of the antiradical activity of candidate-compounds in "model" reaction of initiated low-temperature liquid-phase oxidation of individual hydrocarbon by dioxygen at atmospheric pressure.

We studied the antiradical activity of chelates I–XXV in the "model" reaction of ethylbenzene oxidation (75°, AIBN as an initiator of oxidation). Under these conditions only nitrogen- and sulphur-containing copper chelate XX and bicyclic nitrogen-containing copper chelate XIII operate as efficient radical scavengers. The chain oxidation of ethylbenzene is completely suppressed at initial concentration of these complexes $C_0 \simeq 10^{-4}$ mol/l. The values of inhibition constants for these complexes were estimated to be $k_{inh} \geq 10^5$ l/mol·s. Zn and Bi dithiocarbamates exhibit low and moderate activity respectively, whereas other complexes either do not change the rate of hydrocarbon oxidation or act as a catalyst.

The inhibiting properties of chelates I–XXV were studied in the reactions of initiated low-temperature oxidation of triethylene glycole (TEG) and oligomeric PEG-400 under the conditions, similar to those of ethylbenzene oxidation. For these two oxidizing substances, all the chelates under study exhibited antiradical properties.

In the presence of most of Cu and Co chelates no sharp completion of the inhibition period τ, followed by the non-inhibited reaction, was observed, during TEG oxidation. After the period of completely suppressed chain oxidation the rate of dioxygen consumption increased gradually for a long time.

Cobalt and copper chelates IX, XX and XXIII were found to be the most active inhibitors of TEG and PEG-400 oxidation. The TEG oxidation is characterized by particularly long periods of efficient inhibition in the presence of some copper and cobalt chelates and, by high values of stoichiometric inhibition coefficient f_τ for these complexes. The complexes XXII and XXIII are characterized by values $f_\tau \simeq 5$, while for chelate IX f_τ reaches, $f \approx 7.5$. This indicates the redox mechanism of "repeated chain termination" with a positive route of irreversible metal complex transformation.

In the case of PEG-400 oxidation the f_τ values decrease from 4 to 2 for cobalt chelate VIII and from 5 to 3.5 for copper chelate XXII as compared with TEG oxidation. This fact demonstrates the decreasing role of "alcohol-type" peroxy radicals in the gross-process of PEG oxidation with increasing degree of polymerization.

At the same time the f_τ values for zinc chelate XXIV (Zn is not prone to one-electron redox-reactions) makes up approximately 2 for both TEG and PEG-400 oxidation. This value corresponds well to the mechanism of ligand substitution, proposed earlier for such complexes in the case of hydrocarbons oxidation:

$$ZnL_2 + RO_2^\cdot \longrightarrow ZnLRO_2 + 1/2\ L_2$$

$$ZnLRO_2 + RO_2^\cdot \longrightarrow \text{products}$$

Thermooxidative stability of the metal complexes is one of their most important properties, which determine the possibility of using these compounds as stabilizers for high-temperature oxidative processes. We studied the action of some complexes in the process of high-temperature (175°) autooxidation of viscous low-molecular oxygen-containing substance — ester of pentaerithrite and aliphatic carbon acids C_5–C_9. Under these conditions nitrogen-containing tricyclic cobalt chelates VIII–XII as well as copper sulphur- and nitrogen-containing chelate XX are not stable and rapidly destroyed. Oxygen- and nitrogen-containing copper chelates XV–XIX operate as catalysts, while nitrogen-containing copper and cobalt chelates XIII, XXII and XXIII play a role of efficient antioxidants. Taking into account the results of preliminary testing one may expect, that some of our complexes will be effective in the process of PEG oxidation.

THE KINETIC CONTROL OF PEG OXIDATION

Liquid-phase oxidation of PEG-1500 (average degree of polymerization $n \simeq 33$), high-molecular PEG-14000 ($n \simeq 320$) and dimethyl ether of polyethylene glycol DMEP-5000 ($n \simeq 110$).

Initiated low-temperature oxidation was performed in chlorobenzene solution at 75°. The concentrations of polymeric substances were as follows: PEG–1500 — 10 mol/l, DMEP-5000 — 6.7 mol/l and PEG-14000 — 5 mol/l, where concentrations were calculated per unit of polymeric chain. The process was carried out at the atmospheric pressure of oxidizing gas mixture, containing 20–100% of O_2 by volume. The viscous solutions were vigorously stirred under the conditions providing a kinetic regime of dioxygen consumption. The process was initiated by AIBN thermolyse, the rate of dioxygen consumption was measured with a volumetric device equipped with the automatic pressure control.

The initiation rate W_i was measured by the inhibitor technique with the use of copper chelate XX as an inhibitor. It was shown previously that the stoichiometric inhibition coefficient f_τ for chelate XX was equal to 2 in the processes of low-temperature oxidation of hydrocarbons and secondary aliphatic alcohols.

The initiation constant k_i for AIBN may decrease considerably with an increase in the solution viscosity due to the "cage-effect", which consists in a lowered probability of escaping the solvent "cage" by the free radicals formed upon AIBN decomposition. In fact, the k_i value in PEG-1500 solution makes up $7 \cdot 10^{-5}$ s^{-1}, accounting for approximately 70% of the k_i value in the case of pure chlorobenzene. However, in DMEP-5000 and PEG-14000 solutions, which are characterized by even higher macroviscosity as compared with PEG-1500, an unexpected increase in k_i value (up to $9 \cdot 10^{-5}$ s^{-1} and $1 \cdot 10^{-4}$ s^{-1}, respectively) was revealed. Probably, this results from structurization of the high-molecular PEG solutions and "ousting" a low-molecular initiator into the phase of low-molecular solvent.

The relationship $W_{ox} \sim [O_2]^0 \cdot W_i^{1/2}$ is accomplished for the rates of PEG-1500, DMEP-5000 and PEG-14000 oxidation in solution,[40] where lower concentrations of PEG and higher temperatures were used. This relationship is typical for the processes of radical-chain liquid-phase oxidation of organic substances, characterized by quadratic chain termination on peroxy radicals $RO_2^\cdot$.

The value of oxidation parameter $k_p/(2k_t)^{1/2}$ (where k_p is the rate constant of chain propagation by peroxy radicals and $2k_t$ is the rate constant of quadratic chain termination by peroxy radicals), calculated per unit of polymeric chain, reaches $1.5 \cdot 10^{-3}$ (l/mol · s)$^{1/2}$ for PEG-1500, $1.9 \cdot 10^{-3}$ (l/mol · s)$^{1/2}$ for DMEP-5000 and $2.5 \cdot 10^{-3}$ (l/mol · s)$^{1/2}$ for PEG-14000. The oxidation proceeds at the kinetic conditions of "long chains": for example, the kinetic chain-length of the PEG-1500 oxidation at $W_i = 7 \cdot 10^{-7}$ l/mol · s makes up ca. 20.

The values of termination constants $2k_t$, an important kinetic characteristics of oxidation process, were estimated by use of a "standard" inhibitor — 2,4,6-tritertbutyl phenol. This phenolic antioxidant (peroxy radical scavenger) is known to have an inhibition constant k_{inh} practically independent on the nature of oxidized organic substance (hydrocarbons, alcohols, ketones etc.). The k_{inh} value at given temperature can be calculated from Arrhenius plot: $\lg k_{inh}$ (in $l/mol \cdot s$) $\simeq 7.4 - \dfrac{20 \text{ kJ}}{\theta}$. We have estimated the $2k_t$ values to be equal to: $\simeq 1 \cdot 10^8 \, l/mol \cdot s$ for PEG-1500, $\simeq 7 \cdot 10^7 \, l/mol \cdot s$ for DMEP-5000 and $\simeq 4 \cdot 10^7 \, l/mol \cdot s$ for PEG-14000 on the basis of experimentally observed values of inhibition parameter $k_{inh}/(2k_t)^{1/2}$. These values are large enough to conclude, that the high rates of PEG oxidation are accounted for by the high reactivity of PEG peroxy radicals in the reaction of hydrogen abstraction from C—H bond in PEG molecule (the k_p value).

An effort was made to estimate quantitatively the role of "alcohol-type" peroxy radicals in PEG-1500 gross-oxidation. It has been supposed that para-benzoquinone (BQ) is capable of carrying out the role of "selective trap" for the α-hydroxyperoxy radicals inherent in strong reducing properties.[43] At initial concentrations C_0 of about 10^{-2} mol/l, BQ reduces sufficiently the rate of initiated low-temperature PEG-1500 oxidation, but it lowers the rate of DMEP-5000 oxidation to approximately the same extent. When initial concentrations of BQ are high, only linear chain termination on BQ takes place and the relationship $W_{ox} \sim [O_2]$. W_i is fulfilled in both cases.

This fact indicates that upon oxidation of viscous solutions of PEG-1500 and DMEP-5000 the main inhibition reaction of BQ is its interaction not with α-hydroxyperoxy radicals $RO_2^{\cdot}$, but with alkyl radicals $R^{\cdot}$, taking part in gross-oxidation:

$$BQ + R^{\cdot} \xrightarrow{\ k_{inh}\ } \text{chain termination}$$

BQ is known to be an alkyl radical scavenger of moderate activity in the polymerization processes, which are realized with participation of growing alkyl radicals $\sim R^{\cdot}$.[44]

In the case of DMEP-5000 oxidation the dependence of W_{ox} on C_0 (BQ) at $[O_2] = const$ and the dependence W_{ox} on $[O_2]$ at C_0(BQ) $= const$ allows one to estimate the value of inhibition parameter k_{inh}/k_1 $[O_2]$, where k_1 is the rate constant for the reaction $R^{\cdot} + O_2 \rightarrow RO_2^{\cdot}$. The high value of this parameter ($\simeq 30$) signifies that BQ is an efficient alkyl radical scavenger for DMEP-5000 oxidation.

Autooxidation of PEG-1500 was studied in o-dichlorobenzene solution at 95° and atmospheric pressure of O_2, the PEG-1500 concentration was 10 mol/l.

The kinetic curves of dioxygen consumption at small degrees of PEG-1500 transformation corresponded to the equation: $[O_2]_{cons.} = at + \varphi t^2$, typical for the processes of organic autooxidation with degenerate chain branching. The initial rate of PEG-1500 auto-oxidation (a) made up $1 \cdot 10^{-6}$ mol/l $\cdot$ s and the autoacceleration factor φ was equal to about 10^{-9} mol/l $\cdot s^2$. The amount of hydroperoxides $ROOH$ formed coincides approximately with the amount of O_2 consumpted. This fact correlates with the commonly accepted mechanism of radical-chain autooxidation of organic substances in liquid phase.

INITIATED OXIDATION OF PEG-1500 IN THE PRESENCE OF METAL CHELATES

Upon oxidation of PEG-1500 which occupied a position between the oligomers and high-

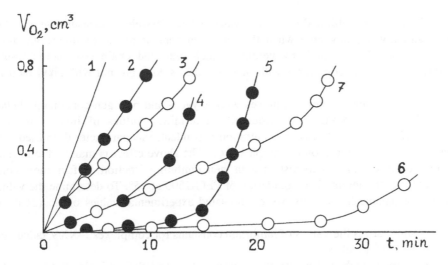

Figure 4   Kinetic curves of dioxygen consumption in the initiated PEG-1500 oxidation (75°, $W_i = 7.4 \cdot 10^{-7}$ mol/l·s): (1) without chelate additives, (2–7) in the presence of chelates: (2) I $(6 \cdot 10^{-4}$ mol/l), (3) IV $(3 \cdot 10^{-4}$ mol/l), (4) VIII $(1 \cdot 10^{-3}$ mol/l), (5) XIII $(3 \cdot 10^{-4}$ mol/l), (6) XXIII $(6 \cdot 10^{-4}$ mol/l), (7) IX $(1.9 \cdot 10^{-4}$ mol/l).

molecular PEGs respectively to their molecular masses chelate compounds of all classes exhibit various inhibiting activity. The characteristic kinetic curves of dioxygen consumption in the course of PEG-1500 oxidation in the presence of chelates, Fig. 4.

Sulphur- and nitrogen-containing copper chelate XX, nitrogen-containing bicyclic copper chelate XIII, nitrogen-containing bicyclic cobalt chelate XXIII and bismuth dithiocarbamate XXV appeared to be the most active inhibitors. In the presence of this chelates at an initial concentration C_0 of about $5 \cdot 10^{-4}$ mol/l the chain oxidation of PEG-1500 was completely suppressed. The stoichiometric inhibition coefficient f_τ for copper complex XIII was equal to 2–3 and increased gradually when C_0 rised. The f_τ values were ca. 2 for cobalt chelate XXIII and ca. 3 for bismuth complex XXV.

Tricyclic nitrogen-containing copper and cobalt chelates, bicyclic nitrogen-containing copper chelates XIV and XXII as well as bicyclic copper chelates with oxygen- and nitrogen-containing chelate centers XV–XIX, XXI were less effective inhibitors. In their presence at initial concentrations up to 10^{-3} mol/l chain oxidation is only partially suppressed. The values f_τ for copper and cobalt chelates are equal to 1–3 with the only exception of cobalt chelate IX, for which the f_τ value is higher: $f_\tau \approx 5$ at $C_0 = 2 \cdot 10^{-4}$ mol/l. The rate of inhibited oxidation in the presence of IX is practically constant during the entire period of inhibition (curve 7 in Fig. 4). This fact agrees with the redox-mechanism of repeated chain termination.

Using the initial rates of PEG-1500 oxidation, inhibited by chelates (W_{inh}), we determined or estimated the values of inhibition constants k_{inh}, which characterized quantitatively the ability of complexes to terminate the chains by interacting with free radicals:

$$W_0/W_{inh} - W_{inh}/W_0 = k_{inh}/(2k_t W_i)^{1/2} \cdot C_0,$$

where W_0 is the initial rate of non-inhibitor oxidation.

The initial rates of inhibited oxidation, when it was possible to measure, were found to be practically unchangeable when the dioxygen content in the oxidizing gas mixture varied from 20 to 100% by volume, which indicated chain termination results from interaction of the complexes with peroxy radicals $RO_2^{\cdot}$ but not with alkyl radicals $R^{\cdot}$.

In the case of some copper complexes with oxygen- and nitrogen-containing chelate centers (XIV–XVII, XXII), experimentally determined values of inhibition parameter $k_{inh}/(2k_t)^{1/2}$ appeared to be concentration-dependent: this parameter decreased with increasing initial concentration of complexes C_0. The above effect is mainly due to catalytic function of complexes coexisting with their inhibiting function. In fact, these complexes catalyse non-initiated autooxidation of PEG-1500 at 75°. To determine the values of inhibition parameter correctly, we extrapolated experimental values of $k_{inh}/(2k_t)^{1/2}$ to $C_0 = 0$.

Table 3 summarizes the values of inhibition constants for complexes I–XXV, calculated using $2k_t = 1 \cdot 10^8$ l/mol $\cdot$ s.

To date a great number of practically important "synergistic" antioxidant mixtures for various systems, suffering oxidation, are known. A peculiarity of these binary or multi-component mixtures is that their inhibiting action is much higher, then arithmetic sum of individual components. As a rule, the synergistic system include inhibitors of different types, as follows:

(i)   free radical scavenger plus hydroperoxide decomposer,
(ii)   photostabilizer plus hydroperoxide decomposer,
(iii) acceptors reacting with free radicals of different nature (peroxy- and alkyl radicals, for example)
(iv)   radical scavenger plus regenerating component etc.[11,45-47] Some of these systems include only peroxy radical scavengers, e.g. hindered phenol plus non-hindered secondary aromatic amine or hindered plus non-hindered phenols.[48,49]

We have discovered an interesting synergistic antioxidative mixture including equimolar quantities of copper chelates XX and XXII. It is seen from Fig. 5, that addition of relatively ineffective inhibitor XXII to the strong antioxidant XX results in significant prolongation of the period of retarded oxidation. The induction period for the above synergistic mixture corresponds to the stoichiometric inhibition coefficient $f_\tau \approx 8$, calculated by using C_0 for the chelate XX, while the f_τ for XX alone is equal to 2. Such an effect takes place in mixtures, containing chelates XX and XXI, but to a lower degree.

It is worth noting, that in the course of oxidation, inhibited by the mixture of chelates XX and XXII (Fig. 5), the characteristic colour of chelate XX disappears gradually and after 20 min oxidation the solution gets coloured like the chelate XXII alone. Then the decolouration of the solution continues and complete bleaching is attained at the end of the induction period. Thus, there are reasons to suppose, that some inhibition, taking place after a 20 min interval, is accounted for by the presence of minor amounts of the highly efficient inhibitor — chelate XX. The latter is regenerated partially by a reaction between the chelate XXII and conversion products of chelate XX.

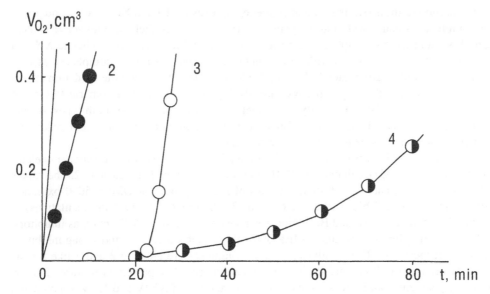

Figure 5   Kinetic curves of dioxygen consumption in the initiated PEG-1500 oxidation (75°, $W_i = 7.4 \cdot 10^{-7}$ mol/l · s): (1) without chelate additives, (2–4) in the presence of chelates: (2) XXII ($5 \cdot 10^{-4}$ mol/l), (3) XX ($5 \cdot 10^{-4}$ mol/l), (4) mixture of XXII and XX ($5 \cdot 10^{-4}$ mol/l of each chelate).

INITIATED OXIDATION OF PEG-14000 AND DMEP-5000 IN THE PRESENCE OF METAL COMPLEXES

Upon oxidation of high-molecular PEG-14000 and DMEP-5000 the reactions of "alcohol-type" peroxy radicals are expected to be either insignificant or not proceed at all. In this case chain carrying "ether-type" peroxy radicals do not exhibit one-electron reducing properties. The chelates of variable valence metals, tested in the reactions of PEG-14000 and DMEP-5000 oxidation, may be classified as follows:

(i)   compounds not inhibiting oxidation (though they provided an inhibiting activity upon oxidation of PEG-1500, TEG and PEG-400),
(ii)   compounds inhibiting oxidation of all PEGs investigated.

The first group includes all tricyclic cobalt complexes with chelate center $Co(N_4)$ (compounds VIII–XII) and most of the bicyclic copper complexes with chelate center $Cu(O,N)_2$. The only exceptions are compounds XVIII and XIX, containing strong electron-donor substituents in the ligands. It can be supposed, that the complexes of the first group are not able to terminate the chains by reacting with "ether-type" peroxy radicals. The inhibiting properties of these complexes in oligometric PEGs and PEG-1500 oxidation may be attributed to the interaction with "alcohol-type" peroxy radicals. In this case experimentally determined inhibition constants are complex values, that may be described by the expression: $k_{inh}^{exp} = \alpha \cdot k_{inh}$, where $0 < \alpha < 1$ is the coefficient, representing the share of "alcohol-type" peroxy radicals in the gross-process of PEG oxidation.

In addition to the above mentioned copper complexes XVIII and XIX, the second group of complexes is composed of all bicyclic and tricyclic copper chelates with active centers $Cu(N,N)_2$ and $Cu(N_4)$, sulphur- and nitrogen-containing bicyclic copper chelate XX and bicyclic cobalt chelate XXIII with active center $Co(N,NH)_2$. All these complexes, as well as Zn and Bi dithiocarbamates XXIV and XXV, are able to terminate the oxidation chains in reactions with "ether-type" peroxy radicals. Upon oxidation of oligomeric PEGs and PEG-1500 these chelates are likely to interact with peroxy radicals of both types. At any rate, copper chelates I and XX inhibit efficiently the initiated cyclohexanol oxidation, which is mediated by α-hydroxyperoxy radicals.

The most active inhibitors of PEG-14000 and DMEP-5000 oxidation are copper chelates XIII, XVIII, XX and cobalt chelate XXIII, the f_τ values of which are close to 2. The values of inhibition constants k_{inh} of some chelates of second group for DMEP-5000 oxidation are summarized in Table 3 (the value $2k_t$ of $7 \cdot 10^7$ l/mol·s is used in calculations).

Sulphur-containing Zn and Bi dithiocarbamates XXIV and XXV function as inhibitors, when oxidizing PEG-14000 and DMEP-5000; they exhibit moderate and strong inhibiting activity, respectively. The f_τ values for these complexes are close to 2 and 3 like for the case of oligomeric PEGs and PEG-1500 oxidation. The data presented in Table 3 show definitely a significant increase in the antiradical activity of XXIV and XXV as compared to their effects on the hydrocarbon oxidation. This fact testifies to the high activity of PEG and DMEP peroxy radicals in the termination reactions with metal chelates in contrast to hydrocarbon peroxy radicals. The enhanced reactivity of PEG peroxy radicals leads to elevated rate constants of other reactions, mediated by PEG peroxy radicals ($2k_t$, k_p).

CORRELATION BETWEEN ONE-ELECTRON REDOX PROPERTIES OF METAL CHELATES AND THEIR ANTIRADICAL ACTIVITY IN PEG OXIDATION

With the object of estimating quantitatively the redox-properties of metal chelates we have measured the values of half-wave potentials of electrochemical one-electron oxidation ($E_{1/2}^{ox}$) and reduction ($E_{1/2}^{red}$) of chelates in solution on the rotating platinum electrode. The $E_{1/2}$ values vs. saturated calomel reference electrode are presented in Table 4.

Comparison of the data given in Tables 3 and 4 shows, that in some cases a distinct correlation between the antiradical activity of metal chelates and their redox-properties is obvious. For the related tricyclic nitrogen-containing compounds of cobalt(2+), which is known to be one-electron reducing agent, one can see that the k_{inh} values increase when half-wave potentials of oxidation of these complexes decrease (chelates IX–XII, Fig. 6). This fact is clear evidence of the inhibition mechanism, including redox-interaction between the complexes and peroxy radicals:

$$Co^{2+} + RO_2^{\cdot} \longrightarrow Co^{3+} + RO_2^{-}$$

In the case of related tricyclic copper chelates IV–VII, polar substituents X in organic ligands influence both $E_{1/2}^{ox}$ and k_{inh}. Electron-donor substituent OCH_3 gives rise to a considerable decrease in $E_{1/2}^{ox}$ and simultaneously to an increase in k_{inh}, while electron-acceptor substituents Br and NO_2 stimulate an increase in $E_{1/2}^{ox}$ and decrease in k_{inh}

TABLE 4
Values of inhibition constants k_{inh} for chelates I–XXV in PEG-1500 oxidation (75°)

Chelate	$k_{inh} \cdot 10^{-4}$ l/mol·s	Chelate	$k_{inh} \cdot 10^{-4}$ l/mol·s
I	1.5 (1.6)*	XV	4
II	0.3 (0.5)*	XVI	5.3
III	1.1 (1.4)*	XVII	6
IV	4.9	XVIII	≥10
V	6.3	XIX	≥10
VI	2.5		
VII	2.1	XX	≥15
		XXI	8
VIII	5.0	XXII	9
IX	30	XXIII	≥15
X	13		
XI	8.5	XXIV	8 (0.2)**
XII	5	XXV	≥15 (0.7)**
XIII	≥15 (≥15)*		
XIV	7.5		

* DMEP-5000, 75°
** Ethylbenzene, 75°

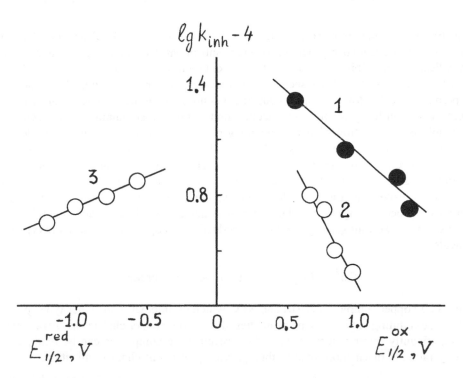

Figure 6   Relationship between the inhibiting activity and redox characteristics of chelates: (1) Co(2+) chelates IX–III, (2) Cu(2+) chelates IV–VII, (3) Cu(2+) chelates XV–XVII, XXI.

TABLE 5

Half-wave potentials (in V) of electrochemical oxidation $E_{1/2}^{ox}$ and reduction $E_{1/2}^{red}$ of chelates I–XXV on the rotating platinum electrode vs. saturated calomel reference electrode (solvent: acetonitrile-chlorobenzene = 1:1; 0.1 mol/l $(C_2H_5)_4NClO_4$ as supporting electrolyte).

Chelate	$E_{1/2}^{ox}$	$E_{1/2}^{red}$	Chelate	$E_{1/2}^{ox}$	$E_{1/2}^{red}$
I	0.72	−0.92	XV	1.09	−1.20
II	0.97	−0.72	XVI	1.08	−1.00
III	1.07	−0.40	XVII	1.07	−0.90
IV	0.76	−0.62	XVIII	0.95	−0.91
V	0.67	−0.64	XIX	0.85	−0.73
VI	0.84	−0.55			
VII	0.96	−0.40	XX	0.74	−0.29
			XXI	1.30	−0.55
VIII	1.00		XXII	0.72	−1.08
IX	0.56		XXIII	0.62	
X	0.91				
XI	1.28		XXIV	1.01	
XII	1.38		XXV	0.83	
XIII	0.74	−0.55			
XIV	1.10	−0.30			

(Fig. 6). When comparing the properties of chelate pairs I and II, III and IV, XIII and XIV one can see that strong electron-acceptor substituent CF_3 similarly affects $E_{1/2}^{ox}$ and antiradical activity of bi- and tricyclic nitrogen-containing copper complexes. The inhibiting activity of all these complexes seems to be predestined by the reaction of their oxidation by peroxy radicals. However, since copper(2+) is not prone to be a one-electron reducing agent, oxidation likely involves the electron-donor nitrogen-containing chelate center as the whole with formation of metal-containing radical cation. This reaction might lead to final oxidation of organic ligands with the oxidation state of metal being unchanged.

As to bicyclic copper chelates with oxygen- and nitrogen-containing chelate center (complexes XV–XVII, XXI), which are active in the reaction with "alcohol-type" peroxy radicals exclusively, as the k_{inh} values increase the better the one-electron reduction of complexes proceeds (Fig. 6). This fact completely agrees with the mechanism of inhibition of these complexes, suggesting one-electron reduction of copper(2+) by α-hydroxyperoxy radicals:

$$Cu^{2+} + RO_2^{\bullet} \longrightarrow Cu^+ + \text{molecular products}$$

Related copper complexes XVIII and XIX, which do interact with "ether-type" peroxy radicals, exhibiting only oxidizing properties, differ from copper chelates as characterized by significantly lower values of $E_{1/2}^{ox}$. As a result these complexes can be oxidized by peroxy radicals much more readily thus providing efficient chain termination.

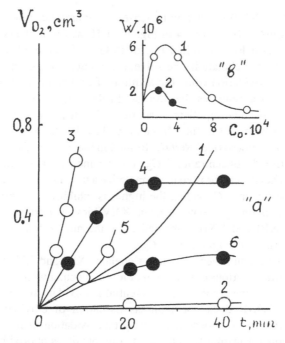

Figure 7   (a) Kinetic curves of dioxygen consumption in the PEG-1500 autooxidation (95°, pO_2 = 760 Torr): (1) without chelate additives, (2–6) in the presence of chelates: (2) XX ($1.2 \cdot 10^{-4}$ mol/l), (3) XXI ($1.3 \cdot 10^{-4}$ mol/l), (4) XXII ($2.7 \cdot 10^{-3}$ mol/l), (5) I ($1.2 \cdot 10^{-4}$ mol/l), (6) IV ($7.4 \cdot 10^{-4}$ mol/l). (b) Dependence of initial rate of PEG-1500 autooxidation W (in mol/l · s) on the initial concentration C_0 (in mol/l) of added chelates: (1) IV, (2) XIII.

AUTOOXIDATION OF PEG-1500 IN THE PRESENCE OF METAL COMPLEXES

Nitrogen-containing tricyclic cobalt chelates VIII–XII appeared to be unstable at the temperature of experiments (95°). They were rapidly destroyed when heating. The products of these complexes decomposition were catalysts of PEG-1500 oxidation.

Fig. 7(a) shows the kinetic curves of dioxygen consumption under PEG-1500 autooxidation in the presence of stable metal complexes. It follows from this figure that the addition of chelates at the initial concentrations of 10^{-4}–10^{-3} mol/l results in substantial changes in the kinetics of autooxidation. All the compounds studied can be classified into three groups, according to their effects:

(i)   compounds that completely inhibit autooxidation — sulphur-containing chelates XX and XXV,

(ii)   compounds, catalyzing oxidation at all C_0 used — copper chelates with oxygen- and nitrogen-containing chelate center (except complexes XVIII and XIX),

(iii) all the other complexes, which catalyze oxidation at low C_0 while inhibit oxidation at high C_0. The latter case is represented by the chelates IV and XIII: characteristic dependence of initial oxidation rates on their initial concentrations are shown in Fig. 7(b).

The initial rate of PEG-1500 oxidation increases in the presence of copper chelate XXII with chelate center $Cu(N,NH)_2$ (curve 4 in Fig. 7(a)). However, in the course of oxidation its rate drops to a level below the initial rate of PEG-1500 autooxidation in the absence of chelates. Preheating the chelate in the solvent, saturated by O_2, leads to lowering the initial rate of catalyzed oxidation while the addition of chelate in the preoxidized solution of PEG-1500 markedly accelerates the process. In the latter case, the time of catalyst-to-inhibitor conversion become significantly shorter. All these phenomena may be accounted for by the reaction of highly efficient inhibitors, yielded by the reaction of chelate XXII with O_2 and/or with hydroperoxides $ROOH$, formed in the course of PEG-1500 oxidation.

The value of "critical concentration" C_{cr}, corresponding to the catalyst-to-inhibitor conversion of metal chelate, can be used to characterize a true inhibiting action of complexes in autooxidation: the lower is C_{cr} value, the higher is inhibiting activity of the complex. Along with sulphur-containing chelates XX and XXV, bicyclic nitrogen-containing copper and cobalt chelates XIII and XXIII appeared to be the most efficient inhibitor of PEG-1500 autooxidation (in the magnitude of C_{cr}).

It was shown that new nitrogen-containing chelates XIII and XXIII exhibited a high inhibition activity in the reactions of oxidation of low-molecular weight oxygen-containing substances (when testing the complexes as potential antioxidant stabilizers). At the same time cobalt chelate XXIII does not suppress oxidation of ethylbenzene, a "model" hydrocarbon. This example demonstrates clearly that oxidation of hydrocarbons is unfit as a "model" reaction for testing the chemical compounds as antioxidant stabilizers for oxygen-containing polymer materials.

The optical antioxidant stabilizers for protection of polymer materials against the thermooxidative degradation should satisfy the specific demands formulated in the Introduction. In particular, the stabilizers have to deactivate both peroxy and alkyl radicals, playing an important role in the destruction of polymers.

Under the experimental conditions described above, metal chelates interact mainly as a peroxy radical scavenger. It should be noted, however, that some of the complexes are capable of interaction with alkyl radicals as well. In the presence of initiator AIBN, sulphur-containing copper chelates $Cu(dtc)_2$ and $Cu(tox)_2$ are consumed in the absence of dioxygen due to reaction with $R^•$radicals. Furthermore, sulphur-containing chelates $Cu(dtc)_2$, $Cu(tox)_2$, $Cu(tpa)_2$, $Cu(I,S)_2R$ inhibit effectively styrene thermopolymerization by scavenging chain-propagating alkyl radicals $\sim R^•$. The characteristic features of these complexes is that they are easily reducible. Due to this circumstance alkyl radicals $R^•$, carrying reducing properties, interact effectively with the above-mentioned metal compounds.

As for chemical properties of chelate transformation products, in most cases they are not active catalysts of oxidation process, i.e. do not provoke the sharp increase of oxidation rate immediately after the completion of induction period, as compared with the rate of non-inhibited oxidation. In addition, the products of $Cu(dtc)_2$ transformation regenerate the initial active form of inhibitor, $Cu(dtc)_2$, in the course of oxidation.

Thus, it may be suggested that metal chelates would be proper antioxidant stabilizers for polymers of various nature.

REFERENCES

1. N. Grassie and G. Scott, Polymer Degradation and Stabilization, Cambridge, University Press, pp. 224 (1985).

2. Yu.A. Shlyapnikov, S.G. Kiryushkin and A.P. Mar'in, Antioxidative Stabilization of Polymers, Chichester, Ellis Horwood, pp. 246 (1993).
3. N.M. Emanuel, E.T. Denisov and Z.K. Maizus, Liquid Phase Oxidation of Hydrocarbons, Plenum, New York, pp. 370 (1967).
4. F.R. Mayo, *J. Amer. Chem. Soc.*, **80**, 2497 (1958).
5. G. Russel, *Chem. Ind.*, **49**, 1483 (1956).
6. J.A. Howard and K.U. Ingold, *Canad. J. Chem.*, **44**, 1119 (1966).
7. N.M. Emanuel and A.L. Buchachenko, Khimicheskaya fizika stareniya i stabilizatsii polimerov, Moscow, Nauka, pp. 360 (1982).
8. N.V. Zolotova and E.T. Denisov, *J. Pol. Sci. A-1*, **9**, 331 (1971).
9. P.A. Ivanchenko, V.V. Kharitonov and E.T. Denisov, *Kinetika i kataliz*, **13**, 218 (1972).
10. E.T. Denisov, Okislenie i destruktsiya karbotsepnich polimerov, Leningrad, Khimiya, pp. 287 (1990).
11. S. Al-Malaika, K.B. Chakraborty and G. Scott, in Developments in Polymer Stabilization-6, London, *Appl. Science Publ.*, 73–120 (1983).
12. D.M. Shopov and S.K. Ivanov, Reaction Mechanism of Inhibitors — Peroxide Decomposers, Sofia, Publishing House of Bulgarian Academy of Science, pp. 187 (1988).
13. A.E. Sementchenko, V.M. Solyanikov and E.T. Denisov, *Zhurnal fizicheskoi khimii*, **47**, 1148 (1973).
14. V.M. Gol'dberg, L.K. Obukhova and N.M. Emanuel, *Neftekhimiya*, **2**, 229 (1962).
15. I.V. Shkhiyantz, V.V. Sher, N.A. Netchitayilo and P.I. Sanin, *Neftekhimiya*, **9**, 616 (1969).
16. J.A. Howard, J. Ohkatsu, J.H. Chenier and K.U. Ingold, *Can. J. Chem.*, **51**, 1543 (1973).
17. J.A. Howard and J.H. Chenier, *Can. J. Chem.*, **54**, 382 (1976).
18. J.H. Chenier, J.A. Howard and J.C. Tait, *Can. J. Chem.*, **55**, 1644 (1977).
19. J.H. Chenier, J.A. Howard and J.C. Tait, *Can. J. Chem.*, **56**, 157 (1978).
20. J.A. Howard and S.B. Tong, *Can. J. Chem.*, **58**, 92 (1980).
21. N.V. Zubareva, E.T. Denisov and A.V. Ablov, *Izvest. Akad. Nauk SSSR, ser. khim.*, N 6, 1342 (1971).
22. L.A. Smurova, T.V. Sirota, A.B. Gagarina, V.P. Litvinov, E.G. Ostapenko, Ya.L. Gol'dfarb and N.M. Emanuel, *Doklady Acad. Nauk SSSR*, **198**, 1378 (1971).
23. A.L. Alexandrov, Thesis (1987).
24. G.A. Kovtun and I.I. Moiseev, *Koordinatsionnaya khimiya*, **9**, 1155 (1983).
25. A.B. Mazaletskii, V.G. Vinogradova and Z.K. Maizus, *Doklady Acad. Nauk SSSR*, **253**, 153 (1980).
26. A.B. Mazaletskii, Thesis (1983).
27. V.G. Vinogradova and Z.K. Maizus, *Neftekhimiya*, **10**, 717 (1970).
28. V.G. Vinogradova, A.B. Mazaletskii and A.N. Zverev, *Neftekhimiya*, **27**, 796 (1987).
29. A.N. Zverev and V.G. Vinogradova, *Neftekhimiya*, **22**, 483 (1982).
30. A.N. Zverev, V.G. Vinogradova, S.V. Larionov and S.V. Semscova, *Zhurnal obschei khimii*, **62**, 482 (1992).
31. H.C. Brinkhoff, *Rec. Trav. Chim.*, **90**, 377 (1977).
32. G.M. Larin and G.A. Zverev, *Doklady Acad. Nauk SSSR*, **290**, 249 (1986).
33. A.N. Zverev, V.G. Vinogradova and Z.K. Maizus, *Izv. Acad. Nauk SSSR, ser. khim.*, N 10, 224 (1975).
34. A.N. Zverev and V.G. Vinogradova, *ibid*, N 1, 23 (1979).
35. A.N. Zverev and V.G. Vinogradova, *ibid*, N 9, 1995 (1979).
36. N.D. Yordanov, V. Terziev and B.J. Zhelyazkova, *Inorg. Chim. Acta*, **58**, 213 (1982).
37. N.A. Ulakhovitch, G.K. Budnikov and L.G. Fomina, *Zhurnal obshchei khimii*, **50**, 1620 (1980).
38. T.V. Megdesieva, K.P. Butin and O.A. Reutov, *Uspekhi khimii*, **57**, 1510 (1988).
39. Yu.A. Mikheev and L.N. Guseva, *Khimicheskaya fizika*, **6**, 1259 (1987).
40. Ph. Grosborne, J. Seree Le Roch and L. Sajus, *Bull. Soc. Chim. France*, N 5, 2020 (1968).
41. U. Hahner, W.D. Habicher and K. Schwetlick, *Polymer Degradation and Stability*, **34**, 119 (1991).
42. N.M. Emanuel, G.P. Gladyshev, E.T. Denisov, V.F. Tsepalov, V.V. Kharitonov and K.B. Piotrovskii, The Testing of Chemical Compounds as Polymer Stabilizers, Preprint of International Symposium on the Methods for Estimation and Practical Application of Stabilizers and of Synergistic Mixtures, Moscow, pp. 70 (1973).
43. E.T. Denisov, *Uspekhi khimii*, **54**, 1466 (1985).
44. E.T. Denisov, Konstanty skorosty gomoliticheskih zhidkofaznyh reaktsii, Moscow, Nauka, pp. 712 (1971).
45. P.I. Levin and V.V. Mikhailov, *Uspekhi khimii*, **39**, 1697 (1970).
46. L.I. Mazaletskaya, Thesis (1983).
47. A.N. Zverev, V.G. Vinogradova and Z.K. Maizus, *Izvest. Acad Nauk SSSR, ser. khim.*, N 11, 2479 (1980).
48. G.V. Karpoukhina, Z.K. Maizus and N.M. Emanuel, *Doklady Acad. Nauk SSSR*, **152**, 110 (1963).
49. L.R. Mahoney and M.A. La Rooge, *J. Amer. Chem. Soc.*, **89**, 5619 (1967).

Spin Probes and Labels. A Quarter of a Century of Application to Polymer Studies

A.L. KOVARSKI

Russian Academy of Sciences, Semenov Institute of Chemical Physics, Moscow 117977, Russia

Spin labels and probes came into the arsenal of physical methods for the investigation of polymer systems. The simplicity of the method and absence of special requirements for the samples has permitted the use of this method for the analysis of trace amounts of polymeric substances. Reviews[1-14] have been written over the last 15–20 years and monographs[15,16] provide a detailed analysis of the theory but are unfortunately only accessible to Russian-speaking readers. The well-known book edited by Berliner[17] and related books[18,19] describe application of this method to biological systems.

1 HISTORY OF THE METHOD

Spin label and probe technique had its origin in the 1960s with the advent of a new class of organic free radicals — nitroxides, and with advances in electron paramagnetic resonance (ESR) spectra analysis. Long-lived paramagnetic molecules-stable radicals — were known long before the discovery of the ESR phenomenon (Zavoyski, Kazan, 1944). Nitroxide disulfonate (Fremy salt) was known in the last century. In the first half of the century, tens of stable organic radicals were synthesized. But they had little to offer as sensors of dynamics information primarily because their lifetimes were limited and ESR spectra were uninformative. Ultrastable radicals based on nitroxide compounds proved to be the most suited to these systems.[20,21] Application of these radicals as sensors of structural and dynamic information was initiated by MacConnel *et al.*[22] and directed to the investigation of biological macromolecules.

Prof. M.B. Neiman in the 1960s–1970s synthesized a large number of nitroxide radicals with varying structures.[23-25] In 1966 the first investigations on the implementation of nitroxide radicals for the investigation of the molecular mobility of synthetic polymers were performed.[26,27]

In the 1970s teams from Japan (Kusumoto *et al.*),[30] Great Britain (Bullock *et al.*),[31] Finland (Tormala *et al.*),[32] USA (Kumler and Boyer),[33] joined in the development of the new method and its application to polymers.

Advancement of the theory of ESR spectra of nitroxide radicals required calculation of the frequencies of their molecular motion.[34,35] Frid and Fraenkel[35] analysed the spectra of the nitroxide probes with rotational frequencies limited by the correlation times of $5 \cdot 10^{-11} < \tau < 10^{-9}$ s. Subsequently, the calculational techniques for τ in the low-frequency range (up to 10^{-7} s) have been developed.[36-38] Further advancement of the theory has provided a way of analysis of the frequencies of rotation of probes around different molecular axis (anisotrophy of rotation)[39-42] and the models of molecular motion (continuous reorientation, jumps on different angles),[43,44] to determine the constant of intermolecular interactions of probes providing data on distances between spin probes (labels) and frequencies of their translational motion.[15,16,45,48]

Late in the early 70s and in the 80s, instrumental techniques have enlarged the frequency range of the method (up to 10^{-4} s). Among these are saturation transfer ESR-spectroscopy, electron spin echo, electron-electron double resonance (ELDOR).[49-56]

2   DYNAMICS OF SMALL MOLECULES IN POLYMER MATRIX

One of the most fruitful and ingenious applications is the investigation of the rotational and translational dynamics of spin probes in polymeric systems. Structural variation of nitroxides enables trends and characteristics of small molecules motion to be found. Among these are the relation between the frequencies and activation parameters of motion and the structure and properties of a polymer, their dependence on shape, size and weight of the molecules, the ratio of rotational and translational frequencies of the molecules, anisotropy of rotation, amplitude of an elementary act of reorientation and so on.

The volumes of the spin probe molecules cover the interval from 150 to 600 Å^3 and cover the majority of low molecular weight molecules in a polymer: catalysis, initiators, plasticizers and so on.[58,60,64]

2.1   Frequencies and activation parameters of rotational motion of particles

Correlation times and rotational frequencies of probes are the parameters most easily determined from ESR-spectra. In the range of $5 \cdot 10^{-11} < \tau < 10^{-9}$ s correlation time can be determined by the following equation:[35,44]

$$\tau = 6.65 \cdot 10^{-10} \, \Delta H_{+1}[(I_{+1} / I_{-1})^{1/2} - 1] \; (\text{s}) \qquad (1)$$

where ΔH_{+1} is the width of a low field spectrum component. $I_{\pm 1}$ is the intensity of low and high field components.

To determine τ in the range of $10^{-9} < \tau < 10^{-7}$ s, the parameters have to be calculated from the spectra. Such parameters can be exemplified as follows:

$$\chi = \frac{H(\tau) - H(\tau \Rightarrow 0)}{H(\tau \Rightarrow \infty) - H(\tau \Rightarrow 0)} \cdot 100 \qquad (2)$$

Here: $H(\tau)$, $H(\tau \Rightarrow 0)$, and $H(\tau \Rightarrow \infty)$ are the positions of the maximum of the low field derivative line for the correlation time τ, for free rotation and for the rigid state.[36]

TABLE 1
Activation parameters for probe motion in polymers

N:	Type of motion	Temp. range	E_P^* (kJ/mol)	V^* (cm³/mol)	V^*/V_W	E_V^*/E_P^*	p_E/p_V
1	Small amplitude rotation	$T > T_g$	20–50	30–70	0.3–0.5	0.3–0.4	>1
		$T < T_g$	10–30	10–30	0.1–0.2	0.7–0.8	<1
2	Large amplitude rotation	$T < T_g$	150–300	150–300	1–2	0.8–0.9	<1
3	Translation	$T > T_g$	50–150	50–150	0.5–2	0.3–0	>1 to <1

The equation (2) is adequate only for the jump-like motion of a particle in the range of $8 \cdot 10^{-10} < \tau < 3 \cdot 10^{-8}$ s. At $\tau > 7 \cdot 10^{-9}$ s the value of the correlation time can be calculated by the following equation:[37]

$$\tau = a(1 - S)^b \tag{3}$$

The coefficients "a" and "b" are dictated by the model of motion, $S = 2A'_{zz}/2A_{zz}$ is the relationship among distances between the external extremes of the spectrum at given temperature and $T \Rightarrow 0$. The models of motion and ways of estimation will be discussed in Sec. 2.3.

The interval of correlation times can be enlarged up to $5 \cdot 10^{-6}$ s using the parameter W:[38]

$$W_i = \Delta_i / \Delta_i^r \tag{4}$$

Where: Δ_i and Δ_i^r are the widths of the lines at half-height at the correlation times τ and $\tau \Rightarrow 0$, correspondingly.

The frequencies of rotational motion of low molecular weight molecules in polymers are dependent on the structure, size, state parameters, structural and physical peculiarities of the matrix. All the characteristics of rotational motion of molecules in polymers are summarized in Table 1.

Values of the correlation time of molecules in polymers in the interval $T_g < T < T_g + 100°C$ range from $\sim 5 \cdot 10^{-9}$ s to 10^{-10} s. In the temperature interval of $T_g > T > T_g - 100°C$, τ changes from $\sim 5 \cdot 10^{-9}$ s to $(2–4) \, 10^{-8}$ s.

Temperature dependence of the rotational correlation times of molecules in different polymeric systems in the coordinates log $\tau - 1/T$ constitute two linear branches with different slopes. The change of the slope occurs in the interval 10–20°C of the glass temperature.

The apparent activation energy above T_g is typically 20–50 kJ/mole and 4–10 kJ/mole at $T < T_g$. The preexponential factor is intimately related to the activation energy and has been given the name of "compensational effect". It has been shown that the effective activation energy is related with the true value by the following equation:[57]

$$E_{ef} = E_{true} + TC \tag{5}$$

where C is the coefficient determined by the media characteristics.

The frequency of molecules rotation decreases as their size increases. In this case the rotation frequency of particles with rigid structure is exponentially dependent on the molecular volume in accordance with the following equation:[16,58-60]

$$\tau = A \exp{(BV)} \tag{6}$$

The coefficients A and B depend only slightly on the polymer and amount to: log A = -13.6 ± 1, $B = (1.2 \pm 0.3) \cdot 10^{-2}$ Å^3.

The rotation frequencies are practically unchanged where the molecular mass varies without the volume changing as a consequence of the introduction of heavy atoms to the molecule. This demonstrates that the volume of a molecule appears to be the key parameter controlling their rotation frequencies. For particles with intramolecular rotation the dependence of τ on V proves to be weak. The reason is that the rotation frequencies of such molecules result from the rotation not of the whole molecule but of the nitroxide fragment.

An increase of molecule size causes an increase in τ, however, significant changes in the values of the effective activation energy have been observed. This can be attributed to the fact that the parameter TC in the Eq. 5 amounts to 20–25 kJ/mole and in many cases is considerably above the value of E_{true} (1–25 kJ/mole) and hence, E_{ef} can be completely defined by the polymer characteristics.

Besides the temperature coefficient of molecular motion frequencies the important parameter is the baric coefficient or the activation volume V^*:

$$V^* = -RT(\partial \ln \tau / \partial P) \tag{7}$$

The activation model as well as the model of free volume identifies the activation volume as the minimal fluctuation of the free volume required for the realization of an elementary act of reorientation of a particle, Table 1. The activation volume of rotation accounts for $(30 \pm 10)\%$ of the Van der Waals volume V_W.[16,58,61-64] The value of V^*/V_W is independent of the molecule size and hence it can be applied to define the volume of the relaxor, using the baric coefficient $\partial \ln \tau / \partial P$. In Sec. 3.4 this parameter is applied for the estimation of the size of macromolecular segments.

The implementation of bulk compression makes possible the determination of the activation energy at constant volume E_V and to calculate on the basis of the hybrid model of Macedo-Litovits[65] the probabilities of the free volume formation p_V and energy fluctuation p_E (Table 1).

2.2   Anisotropy of rotation

The experimental technique for the investigation of the spin probe rotation anisotropy have been published.[15,37,39,41,42,54] The correct technique for this purpose is the high resolution ESR.[56] The data resulting from the investigation of liquids and polymers show that the ratio of the correlation times for rotation of a molecule about two orthogonal axis depends predominantly on the shape of a molecule depends only slightly on the matrix. The values of the activation energies of the molecular rotation around different axis are close, the temperature dependence of the anisotropy is missing.

TABLE 2
Activation parameters for α- and β-relaxation processes in polymers

N:	Process	E_P^* (kJ/mol)	V^* (cm^3/mol)	E_V^* / E_P^*	p_E/p_V
1	α	200–400	150–350	0.7	<1
2	β	20–100	20–40	0.8–0.9	<1
3	$\alpha\beta$	20–30	20–60	0.5–0.7	~1 to <1

2.3   Model of rotation

Molecules having the same rotation frequencies can change their orientation for varying angles. Reorientation of molecules with limiting small angles is called *continuous* or *Brownian rotational diffusion*. The *jumps model* implies that an orientation changes by varying angles and that between jumps is at rest.[69] The ESR spectra parameters sensitive to the model of rotation are:

$$R = \Delta H_+/\Delta H_-$$ (8)

where: ΔH_+ and ΔH_- are the shifts of the spectrum components in low and high fields relatively to their position at $T \Rightarrow 0$ K. The dependence of R on ΔH for Brownian and jump-like diffusion.[15,43]

The analysis techniques for liquids[15] showed that the model of rotation is determined by the size and shape of a particle and by the viscosity of the medium. Large asymmetric particles in viscous media are considered to rotate by a continuous diffusion mechanism.

The experiments on amorphous polymers[11,70] have shown that the nature of the rotational motion in such media is dedicated not only by the molecule but by the density of macromolecular packing. In polymers characterised by a large value of the total volume of pores (PS, PE, PP) probes rotate by jumps while in PMMA and PVC characterized by more dense packing the rotational model is close to the Brownian one.

The method of electron spin echo and electron-electron double resonance (ELDOR) make it possible to define the most probable rotation angles of paramagnetic molecules. Specifically, the ELDOR has shown that the mean value of the reorientation angle in organic glasses falls in the range of 0.1–0.4 rad.[71] The mean value of the angle has shown to increase with temperature and for low temperatures the motion is better described by the Brownian diffusion model.

2.4   Low-frequency rotation of particles

Soon after the first results on ESR-spectroscopy of spin probes in polymers were received, it has been found that the distinctions between the values of the frequency for small molecules rotation determined by this method and by the dielectric relaxation method. Johari and Goldstein[72,73] studies of the dielectric relaxation of liquids have not called proper attention to the importance in elucidation the causes of these differences. These studies have shown that two regions of relaxation corresponding to two frequencies of the rotation motion of particles are observed in glass-forming organic liquids with rigid molecules. Similar effects have been detected for the dielectric relaxation of molecules of anthrone in PS.[74] Research on thermostimulated depolarization of rigid dye molecules in a polymer has validated the existence of two frequencies for rotational diffusion.[64,75]

A correlation between these data and dielectric relaxation have shown that these types of rotational motion are characterized by their temperature and baric dependencies were identical with α- and β-relaxational processes of the macromolecular segmental dynamics in solid polymers in Sec. 3.4.

At this point it is appropriate to note that two types of rotational motion are probably distinguished by the molecular rotation amplitude. It should be noted that the particle rotation frequencies are observed to exhibit the greatest differences in the low temperature region ($T < T_g$), while the differences between the frequencies disappear at $T \gg T_g$. The parameters of the low-frequency rotational dynamics of particles in polymers are given in Table 1.

2.5   Ratio of frequencies of rotational and translational motion

The qualitative relationship between the frequencies of rotational and translational motion is of great importance for the analysis of the kinetics and mechanism of chemical reactions in condensed media.[16,58,60,63,76,77] The ratio of frequencies for both types of motion is determined by the difference in their activation parameters:

$$\ln (v_r/v_t) = A + (\Delta E + P\Delta V^*)/RT \tag{9}$$

where $A = \ln (v_r^0 / v_t^0)$, $\Delta E = E_t - E_r$, $\Delta V^* = V_r^* - V_t^*$.

The v_r/v_t value indicates how many times the particle changes its orientation during the time of translational equalling its diameter (~10 Å).

An investigation of the dynamics of various probes has revealed that their rotation in polymers is much faster than the translational motion. The values of ΔE for polymers at temperature higher than T_g are 30–100 kJ/mole, and $\Delta V^* = 30$–100 cm^3/mole. In PE rotation frequencies of a TEMPO probe are two orders of magnitude higher than the frequencies of the translational motion. The particle shape and size (activation volume) and the structure and physical state of a polymer have a profound effect on the ratio of the frequencies. In addition v_r/v_t is temperature and pressure dependent. With the particle volume increasing the frequencies of two types of motion tend to approach each other. The same effect is observed with the temperature increase. An increase in pressure results in an increase in the differences in frequencies. For glassy polymers the frequency of the rotational motion greatly exceeds the frequencies of the translational motion.

For the low-frequency rotational motion (dielectrical methods) the activation volume for translational and rotational motion differ moderately (Table 1) and the frequencies of both types of motion appeared to be closely allied. During the reorientation time τ a particle has managed to cover a distance not more than 1–2 of its diameters. The ratio of the frequencies appears to be insignificantly dependent on pressure and temperature.

3   INVESTIGATION OF SEGMENTAL MOBILITY IN POLYMERS

Macromolecule mobility is changed by the structure and state of a polymer and as a result of the effect of external force fields. Some of the problems being resolved are outlined below.

3.1   Polymerization, cross-linking, destruction

The medium viscosity and correlation time of the rotational motion increases with the molecular mass. However in the range of big values of M the changes in τ cease. The threshold value of M is dependent on the length of the chain and ranges from 10^3 to 10^4.[2] For depolymerization, destruction and chain cross-linking one can observe changes in τ only at high degrees of conversion. Vulcanization of isoprene rubber results in growth of τ when the molecular mass of a chain between links appears to be less than 10^4.[1] In those cases when destruction and cross-linking processes run parallel complex dependencies of the correlation time on the conversion degree can be observed. Similar results have been obtained for the processes of chemical modification of macromolecules accompanying polymer oxidation, radiolysis and photolysis.[1,16]

3.2   Phase and Physical Transformations

The method can be employed to investigate crystallization and melting of liquids and polymers.[1,2,16] Probes are generally distributed in the amorphous region. Increasing the crystalline fraction is analogous to cross-linking with a consequent decreasing of the molecular mobility. During melting the inverse effects can be detected.

The degree of crystallinity (α) appears to be gauged through the dependence of the correlation time on α, or through the analysis of the local concentrations of spin probes. Unfortunately, both methods can give a high error in the range of small degrees of crystallinity.

The method is also successfully employed to research liquid-crystalline polymers.[78–80] In this case two parameters have to be used — microorder S_{zz} resulted from an orientation of probes in liquid-crystalline regions and macroorder $S_{z'z'}$ connected with the mutual orientation of these regions. ($S_{z'z'} = 0$ in case of mutual orientation of probes is absent and $S_{z'z'} = 1$ in case of their complete mutual orientation).

Inflection points in the temperature dependence are used for the analysis not only of transitions in liquid-crystalline polymers but transitions from one physical state to another. An inflection temperature T_1 is dependent not only on T_g but on the probe size. The difference between T_g and T_1 can be as much as 20°C. Application of small probes for investigation of rubbers is likely to give more reliable results.

Another parameter which is frequently used for T_g determination is the distance between the extremes of the ESR-spectrum.[5–7,29,33,81–83] The temperature dependence of this parameter shows a sharply necked section in the range from 65–70 to 30–40 Gs. Depending on structure and properties of a polymer the width changes from a few to tens of degrees. A mean value of this parameter, in the centre of the transitional region, is 50 Gs designated by T_{50Gs} or T_{50}. The frequency of the probe rotation is $8 \cdot 10^{-9}$ s which is dependent not only on the glass temperature but on the probe size and structure of the polymer.

Since the strict theoretical prediction of the dependence between T_g and T_{50} is absent, semiempirical equations have to be used for glass temperature calculation. Based on the results of probe-aided experiments carried out on a large number of polymeric systems the following equation was derived:[33]

$$T_{50} = T_g/(1 - 0{,}03T_g/E) \qquad (10)$$

where E is the activation energy at $T > T_g$.

Another equation which avoids the activation energy is due to:[81]

$$T_g = T_{50}/[1 + 1/\exp{(T_g/173 \text{ K})}] \qquad (11)$$

For the same purposes the relations derived from the equations of Arrhenius, WLF and VFTH may be used.[83]

3.3 Investigation of the effects of external fields

External fields cause changes of the molecular mobility of amorphous regions of a polymer, spin probes can be used for the qualitative estimation of such alterations. This method can be applied to investigate bulk compression, orientational drawing, static and dynamic stresses and physical ageing.[1,16,58,60–63,84]

Experiments on probe orientation have been made in PE.[85,86] Extension of the polymer by more than 500% leads to a drastical change in the ESR-spectrum shape.

The analysis of S_{zz} and $S_{z'z'}$ parameters shows that a fraction of nonoriented regions of a polymer does not exceed 15%.[79]

3.4 Relation between probe and macromolecular dynamics

What is the key role of the method of spin probes in the investigation of the molecular dynamics? Providing answers to this problem one should consider that the physical characteristics of a polymer are determined not only by the structural and dynamic features of macromolecules but by defects.

The first question raised is what type of segmental motion defines the frequencies of the rotational dynamics of doped molecules. The basic relaxational processes associated with the segmental dynamics in the amorphous regions consists in the α- and β-processes. These processes differ markedly in their nature and activation parameters. The first of these processes have a non-Arrhenius nature (nonlinear dependence log $\tau - 1/T$), the second one is considered to be an Arrhenius process. Distinctions between the frequencies of these processes reach the magnitude of the same order at T_g and disappear at temperatures 50–70°C above T_g.

At temperatures above T_g, the motion of the probe is unambiguously determined by the β-relaxational process. The rotational frequencies correspond to the β-process frequencies if the molecular volume of a probe equals the segment volume. This phenomenon can be applied for the estimation of the macromolecular segment size.

In the region below T_g, the activation energy of the probe and label rotation appear to be substantially lower than that of the β-process. In the early stages of the development of the method it has been proposed that in the low-temperature region the dynamics of probes are governed by small-scale relaxation processes (γ and δ). Data exemplifying the dependence of the rotation frequency on the density of polymeric glass packing have been obtained. Investigations of rigid cellular systems based on cross-linked polystyrene[11] and synthetic zeolites[87] appear to be the most convincing experiments.

In cellular systems, the probe rotation frequency is determined not by the rigidity characterized by the glass transition temperature, but by the pore size. The results of the zeolite investigation have led to similar conclusions. It has been demonstrated that the probe rotation frequencies in zeolite may amount to significant values (10^9–10^{10} s) if the "window" exceeds the probe sizes. An increase of probe size or a decrease of the "window" sizes results in a sharp reduction of the rotation frequency. These results allow the relation between the segmental motion may be caused by probes being localized in pores of polymeric glasses and their motion is dependent on the parameters of such pores.

Another cause of the low values of the activation energy has been formulated.[88] The shape of the ESR spectra remains unchanged at $\tau > 10^{-7}$ s then on cooling an increasing number of molecules cross over this border and provide identical spectra. In this case the experimental valucs of τ are larger than the mean distribution value. The magnitude of such deviations increases with the decline in temperature which results in the decreasing of the apparent activation energy. This assumption is supported by the data of the saturation transfer ESR-spectroscopy,[89] which is characterized by the low frequency behaviour ($<10^{-4}$ s).

An important question arising from examination of the relationship between probes and segmental dynamics appears to be the size of macromolecule segments defining the mobility of the defect. In experiments with polyethyleneglycol.[2] τ ceases to change when the molecular weight of the polymer exceeds 200. It follows hereforth that a kinetic element of PEG macromolecules is built up from approximately five monomeric units.

Assuming that the rotation frequencies of a probe τ_p and a segment τ_s being equal in volume are bound to coincide. One can choose a probe in such a way that the frequency of its rotation is coincident with the β-relaxation determined by dielectrical relaxation. The relaxation volume is easy to obtain. The relation between the rotation frequency and the molecular size (6) should be as follows:

$$V_s = V_p - \left(\ln \frac{\tau_p}{\tau_s} \right) / B \tag{12}$$

values of B for different polymers cover a narrow range from $(1–1.2) \cdot 10^{-2}$ Å^{-3}. Calculations have been performed for natural rubber and polyethylene.[16] The value of Van der Waals segment volume for NR and PE reached 330 and 205 Å^3 respectively, a segment of NR is built up from 5 and PE from 4 monomeric units.

Using the Buche model of free volume[82] an equation can be derived to calculate $f = V_p/V_s$ from T_{50} and T_g:

TABLE 3

Van der Waals volumes for: monomeric units V_W^m, kinetic units (segments) V_W^k and number of monomeric units in segment n

Polymer	V_W^m (cm³/mol)	Process	V^* (cm³/mol)	V_W^k (cm³/mol)	n
PVC	28.63	α	342	228	8
		β	25	125–250	4–9
PMMA	56.1	α	398	265	5
		β	21	105–210	2–4
PET	94.18	α	518	345	4
		β	31	155–310	2–3

$$T_{50} - T_g = 52\left[2.9f\left(\ln\frac{1}{f}+1\right)-1\right] \tag{13}$$

In[83] the equation of WLF was used to define f. The following relation has been obtained.

$$T_{50} - T_g = [C_2(2.303C_1 f/13.8) - 1] \tag{14}$$

The values of f calculated by equation (14) ranged from 0.24 to 0.65 depending on the density of cross-links in PP.[90]

The segment volume can be defined by the activation volumes for molecular motion obtained from high pressure experiments. The ratio of the activation volume V^* to the particle intrinsic volume is constant[62,63] and hence may be used to estimate the segmental volume by V^* value for the corresponding relaxational process. Results given in Table 3 show that a kinetic element of β-process is built up from 2 to 10 monomeric units which correlates well with the estimates made by other methods.

The ratio V_p/V_s for the relaxation is shown in Table 4. It differs radically from V_p/V_s value for β-process but the segment sizes in both processes coincide. It indicates that for the α- and β-processes the same length of chain manifests themselves as independent kinetic units. The distinctions in the character and parameters for both processes ascertained by the thermostimulated depolarization method shows that both relaxation processes vary primarily in the relaxation angle.[75]

TABLE 4

The ratio of local and total probe concentrations C_{loc}/C_{total} and the share of PE/PVC blend accessible for the probe χ[92]

The share of PE in the blend	C_{loc}/C_{total}	χ
1	2	0.5
0.2	10	0.1
0.13	15	0.07
0.06	2000	0.0005

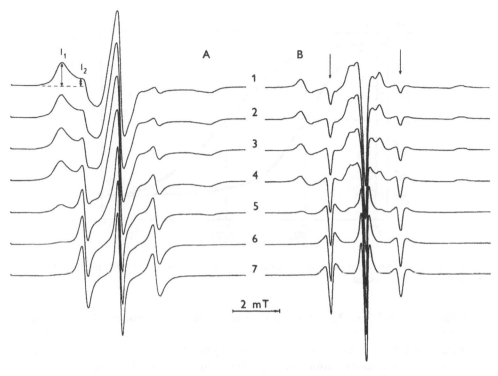

Figure 1   The first (A) and second (B) derivatives of ESR absorbtion signal of TEMPO probe in PE/PVC blends with the share of PE from 4 to 53%.[92] The arrows in Fig. 1B shows the lines of probes in PE phase of the blend.

4   INVESTIGATION OF HETEROGENEOUS SYSTEMS

Spin probes and labels utility are frequently used in the analysis of polymeric mixtures, copolymers, filled and plasticized systems, latexes, solutions and gels, interphase layers and surfaces, polymers with different supermolecule structures.[1,4,5,9,12,16]

4.1   Analysis of superpositions in ESR spectra

The linewidths and ESR spectrum shape of nitroxide radicals are dependent on the frequency of their rotation and therefore in heterogeneous systems the superposition of narrow and wide lines is observed, Fig. 1A. Having divided these spectra on their constituent parts the quantity of probes and their rotational frequencies in microphases of a system are easily determined. The temperature dependence of lines and the characteristic temperature T_{50} and T_g value in microphase system can be calculated from Eqs. 10–11.

This method provides reliable data provided the following conditions are met: a) the components of the system should have very different molecular mobilities. The differences in T_g should be at least ten degrees, b) the concentration of probe in the rigid component should be no less than 30%, and in the more mobile (soft) component — no less than

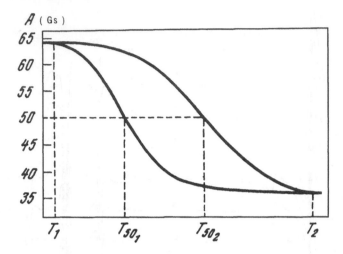

Figure 2   The temperature dependence of the distance between outer extrema of ESR spectrum of spin probe in binary polymer system.

5%. At temperatures below T_1 and above T_2 (Fig. 2) the superposition of lines in spectra is observed.

This approach appears to be conveniently used primarily for the qualitative express-analysis of a system on its uniformity. The quantitative data on the phase relationship is much more complicated to obtain because the solubility of radicals in the polymers is not known. Estimation of the coefficient of probe distribution between the components of a system k[16,91] the following equation can be used:

$$\alpha = \delta_2/(\delta_1 k + \delta_2) \qquad (15)$$

where: α is the relative quantity of probes in a single phase, δ_1 and δ_2 are volume parts of the components.

The value of k for the uncompatible system PE-PIB has been obtained equal to 4 (in PIB medium the quantity of probes is four times greater than in PE medium). With the degree of crystallinity of PE being equal to 70% this result reveals that the solubility of a probe in the amorphous region of PE is approximately the same as in PIB.

Another field of application of the superposition technique is for investigation of highly oriented PP samples containing microcracks. Fig. 3, at high concentration on intense singlet appears at the centre of the spectrum. In the starting samples containing no microcracks this effect was not observed. The appearance of a singlet is due to the localization of probes in microcracks and their adsorbtion on surface. As the emergence of a singlet is possible only at dense location of probes on a microcrack's surface their quantity can be calculated starting from the total concentration of probes and mean size of a microcrack.

4.2   Analysis of mobility and local concentrations of probes

Consider the calculational methods for the probe mobility in microsites. Correlation times

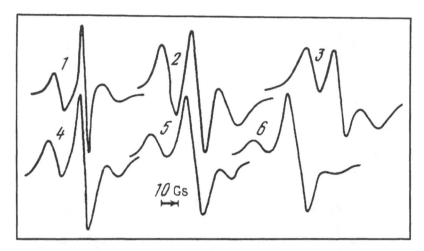

Figure 3   The concentration changes in ESR spectra of TEMPO-probe in isotropic (1–3) and highly oriented (4–6) PP.

from the spectra superpositions are unlikely to be correct because of ESR lines overlapping and distortion. The simplest technique is based on the second derivatives of the ESR spectrum (Fig. 1B). They differ from the first derivatives primarily by the imposition of the lines being observed only at the centre of a spectrum whereas at the "wings" these lines are split up and allows calculations of τ to be performed without any preliminary procedures. In the case of narrow lines the following equation has to be used:

$$\tau = 6.65 \cdot 10^{-10} \Delta H_{+1}[(I_{+1}/(I_{-1})^{1/3} - 1] \text{ (s)} \tag{16}$$

All the parameters of Eq. 16 have the same meanings as in Eq. 1.

Fig. 4 shows values of the correlation time calculated by Eq. 16 (curve 2) using the spectra shown in Fig. 1B coincide with the τ values obtained from the first derivatives of the spectrum of probes introduced only into the mobile phase of a mixture (curve 3) and drastically differ from the value of τ obtained by Eq. 1 from the spectra superposition given in Fig. 1A (curve 1).

The probe local concentrations in a system can be determined based on the analysis of ESR spectra parameters depending on the intermolecular interactions of radicals — exchange and dipole-dipole interactions. One such parameter is said to be the width of a line and more precisely the concentrational broadening. The general expression for the width of a line is of the form:[40]

$$\Delta H = \Delta H_o + \delta H = \Delta H_o + kc \tag{17}$$

where: ΔH_o is the probe concentration-independent contribution to the linewidth; $k = k_{dip} + k_{ex}$ is the coefficient of line broadening including the dipole and exchange parts; c is the concentration of probes in a sample.

In order to make the best use of Eq. 17, the mechanism of concentration broadening (exchange or dipole) should be established. In liquids the exchange interaction of radicals

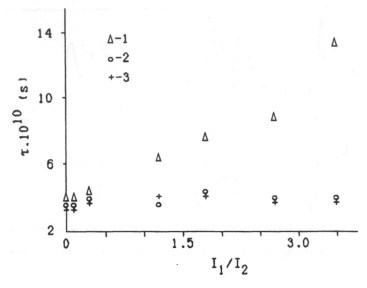

Figure 4   The dependence of correlation time for probe rotation in PE phase of PE/PV blends on the ratio of line intensities I_1/I_2. The comments are in the text.

makes itself evident at a viscosity of less than 1 cP.[40] At a viscosity of more than 50 cP, the broadening of lines is dependent only on the dipole interaction. Fig. 5 depicts the dependence of the concentration broadening of TEMPO probe lines on the diffusion coefficient. This dependence is composed of three portions. At $D < 1 \cdot 10^{-6}$ cm^2/s, the linewidth is independent of the diffusion coefficient and is defined by the dipole interaction.

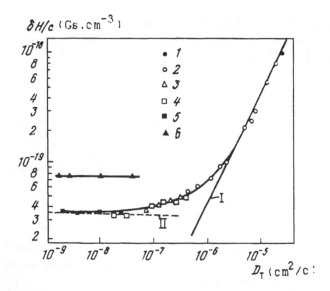

Figure 5   The dependence of ESR line concentration broadening on the diffusion constant for probe TEMPO in: heptane (1), decaline (2), polydimethylsiloxane (3), natural rubber (4), styrene rubber (5), PE (6).

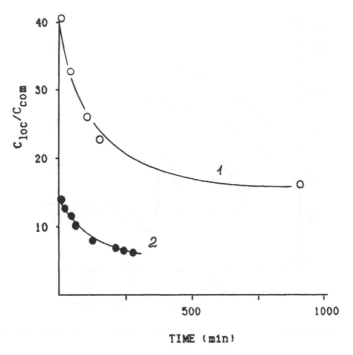

Figure 6   The kinetics of local concentration changes of TEMPO in PE/PVC blends with the PE share of 13% (1) and 20% (2).

At $D > 5 \cdot 10^{-6}$ cm^2/s the concentration broadening of lines is completely defined by the exchange interaction of radicals. It should be emphasized that the dependence given in Fig. 5 appears to be universal. It is valid for all condensed media if the uniform distribution of probes is achieved. For a nonuniform distribution of probes in a sample their local concentrations c_{loc} exceed the mean concentration over a sample c which results in the increasing of ΔH or $\Delta H/c$. The value of c_{loc} in the region of the dipole broadening can be calculated by the equation:[45,48]

$$\delta H_{dip} = c_{loc} k_{dip} \qquad (18)$$

where: $k_{dip} = 3.5 \cdot 10^{-20}$ Gs.cc.

The probe diffusion coefficient in a number of polymers.[16,63,77,88,91] The condition $D < 1 \cdot 10^{-7}$ cm^2/s is valid for vitreous polymers and for the elastic state at moderate temperatures PVC-PE blends have been investigated[92] with TEMPO probes which was introduced only into the PE-phase by vapour absorbtion for 15 min at room temperature. The change of the ESR linewidths with time was accounted for by the high concentration of radicals in surface layers, generated at the start point, being resolved. The value of c_{loc} was determined by Eq. 18 and computer-aided integration was used to define the total number of probes per volume unit of a sample c. The time dependence of c_{loc}/c ratio is given in Fig. 6. It is seen that at the particular interval the linewidth ceases to depend on the exposition and it means that probes have uniformly spread over the amorphous regions of PE. Under these conditions c_{loc}/c accounts for the blend phase ratio.

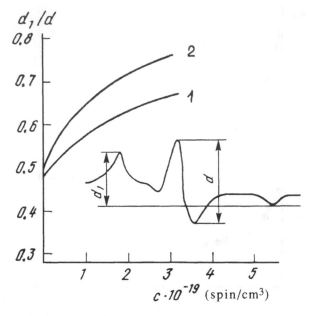

Figure 7   The concentration dependence of d_1/d for amorphous polymers (1) and crystalline polymer (PP) (2).

The dependence of the limiting values of c_{loc}/c on PE concentration is given in Table 4. The fraction of a sample available for a probe is ~2 times lower than the PE fraction in a blend. The reason of this discrepancy is the existence of a crystalline phase in PE not available to the probe. As the degree of the crystallinity of PE was ~50% the results obtained adequately reflect the blend composition. The only exception is represented by samples with a very low concentration of PE for which the parameter c_{loc}/c is an overestimate.

Another method of probe local concentration determinations is the analysis of their spectra at liquid nitrogen temperature.[16,45,93] In this case a parameter describing the dipole-dipole interaction and hence the distance between particles is the ratio of intensities for the central and end components, Fig. 7.

Here the dependence of d_1/d on the total concentration of TEMPO probe uniformly dispersed over a sample is given. If probes are nonuniformly dispersed the value of d_1/d at the same total concentration of probes should be higher, Fig. 7 for a crystalline polymer.

This method can be used to estimate the local concentrations in the range of 0.007 mol/l < c_{loc} < 0.5 mol/l. The average distance between the adjacent paramagnetic particles being randomly dispersed equals:

$$\bar{r} = c^{-1/3} \tag{19}$$

5   INVESTIGATION OF SPIN-LABELLED MACROMOLECULES

The nitroxly radicals as spin labels chemically bonded with macromolecules of synthetic polymers have been used to advantage since 1972.[2,3] Chemical modification of a polymer resulted from radical addition or preliminary reaction can change the conformation and local segmental mobility of a macromolecule. To a greater extent this is true for macromolecules with a high concentration of spin labels. However, as a set of nitroxides is rather broad, labels appear to be convenient model compounds for the investigation of the dynamical behaviour of different groups. This type of method in biological studies have received the name "the reporter groups methods".[17]

5.1 Local dynamics of spin labels

Spin labels can be considered as model fragments which give an insight into the control of the local dynamics of side groups and units of a main chain depend on many different factors. In the studies of such macromolecules with low concentration of spin labels (one label per some hundred units) are used which does not result in drastic change of properties and conformation. Analysis of the dynamic parameters is carried out using the simplest models of the isotropic rotation (Eqs. 1–3).

The dependence of the label dynamics on the nature of the connecting link and site of addition. The mobility of a spin label is primarily dependent on the flexibility of a bridge linking it with the macromolecule. The correlation times for labels rotating in dilute solutions of macromolecules decrease with increase in the length of a flexible bridge.[95] In solid polymers the rotation frequencies of such labels coincide with the rotation frequencies of probes. For this reason when studying homogenous solid polymer labels which are divorced from a chain have no advantages over the more simple method of spin probes.

The least molecular mobility is inherent in labels rigidly combined with the main chain. The correlation times which are determined in this case refer to the cooperative movement of a label with a fragment of the main chain (segment). It has been shown that the more rigid a label is associated with the chain, the closer is its movement to the frequencies identified by other methods (NMR, dielectric relaxation, polarized luminescence).

Yet another parameter responsible for the label mobility is the flexibility of the main polymeric chain. Rubbers are marked by high rotation frequencies as compared with rigid chain polymers.[16] This dependence is the most clearly defined for rigidly bound labels.

The dependence of the label rotation frequencies on its position in the chain (terminal or middle units) has been studied.[2,6] For the most part such differences in frequencies are insignificant.

The dependence on molecular weight. The correlation times of the label rotation in dilute solutions of polymers are described by the following expression:[3,7]

$$\tau^{-1} = \tau_0^{-1} + \tau_{loc}^{-1} \tag{20}$$

Here: τ_0 is the intrinsic time for whole molecule rotation, τ_{loc} is the intrinsic time of the local movement of a label.

It has been found that experimental values of τ depends on the molecular weight of macromolecule. For macromolecules with large value of M, $\tau_0 \gg \tau_{loc}$ and in accordance with Eq. 20 $\tau \approx \tau_{loc}$. For macromolecules of low molecular weight the frequency of the coil rotation appears to be limited and τ_0 depends on the solvent viscosity η_0 and the intrinsic solution viscosity $[\eta]$ in accordance with the equation $\tau_0 = 0.42\, M[\eta]\eta_0/3RT$. The dependence of τ on M can be observed only in the range of the small molecular weights; further increase of M the correlation time remains constant. The critical value of the molecular weight is determined by the polymer, solvent and temperature. For a toluene solution of polystyrene the magnitude of M varies from $\sim3\cdot10^4$ at 345 K to $\sim1\cdot10^5$ at 294 K.[3]

Influence of solvent. Change of the thermodynamic quality of a solvent results in change of a coil density and spin labels mobility. It can be exemplified by the investigation of spin-labeled solutions of polystyrene in toluene, α-chloronaphtalene (good solvents) and cyclohexane (poor solvent),[3] and polyvinylpyridine (PVP).[96]

The deterioration of the solvent quality results in an increase in correlation time. This effect has been detected on adding water to the PVP solutions in ethanol which is a poorer solvent in comparison with water. On adding a precipitant the rotation correlation times of the label increase, which can be attributed to the lamination of the solution and decreasing of the intramolecular mobility of a polymeric coil. A similar effect has been observed on cooling the cyclohexane solution of PS below θ-point (307 K).[3]

It should also be noted that spin labels allow phase transitions in solutions associated with the coil density change to be observed. The inflexion points on the Arrhenius dependences of the correlation times of spin labels in toluene solutions of PS in the temperature range of 303–343 K and in ethylacetate solutions of PMMA at 314–318 K have been observed.[97,98] These inflexions are attributed to the transition from "rigid" coil to a "flexible" one.

In[3,97,99] the analysis of modified Kramers theory has resulted in the equation connecting an apparent activation energy of the label rotation $E_{\rm rot}$ with the activation energy of the solvent viscous flow E_η and a height of the potential barrier E^* of the form:

$$E_{\rm rot} = E_\eta + E^* \tag{21}$$

Eq. (21) is limited by the condition of high viscosity[97] and hence it should be put to the test for the system being studied. Such analysis has shown that this condition is valid for PS solutions in a number of solvents. The magnitudes of E^* in good solvents — α-chloronaphtalene and toluene, appear to be 8.6 and 9.0 kJ/mole, respectively, and in poor solvent — cyclohexane — 14 kJ/mol. The high value of E^* in poor solvent,[97] can be attributed to a more dense packing of a coil.

Spin labels in solid polymers. The more extensive application of spin labels in comparison with probes is caused by the necessity to have a reporter group directly to the macromolecule. In a solid polymer the local environment of a label and probe is practically the same. The distinction lies in the fact that in the case of a label the potential barrier of the rotation increases on account of covalent bond between the radical and macromolecule. Hence, the relationship between the rotation frequencies of a label and those of a probe will depend on the mobility of the connecting bond. In case of a long and flexible bridge the rotational frequencies of a label and probe will vary moderately or coincide.[70,100] Rigid fixing of the radical to the macromolecule can make these distinctions significant. In this case, the information about the frequencies of rotation of a chain section containing a spin label, of a kinetic segment which is distorted since it contains structural defects.

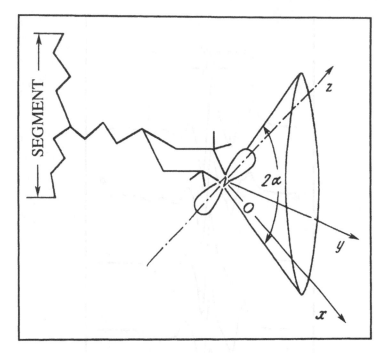

Figure 8   The diagram of motion of spin label attached to side group of macromolecule.[109]

Estimation of the segment size can be done using spin probes: selecting a probe in such a way that its frequencies will coincide with the label rotation frequency. The particle rotation frequency is basically determined by its volume, agreement between the frequencies testifies that the volumes agree as well. The values of the effective volumes of spin-labelled segments in a number of polymers[16] are 260 $Å^3$ for PVA, 226 $Å^3$ for PS and 318 $Å^3$ for PMMA. The effective energies differ little.[16,70]

One of the main distinctions between labels and probes in solid phase consists in that the probes easily migrate in loose interstructural regions and do not reach structured sections. At the same time labels rigidly bounded to macromolecules are distributed uniformly over the sample volume and enter both amorphous and crystalline regions.[101–105]

5.2   Segmental dynamics of macromolecules

In the preceding section the possibilities of the spin label method has been discussed in terms of the simplest model of isotropic rotation. In fact, for the most part the rotation of labels appears to be anisotropic. Fig. 8 depicts the schematic circuit of the rotational movements of a label situated in a side chain. It can be noticed that paramagnetic fragments can participate in two key types of the rotational movement: around the bridge connecting it with a macromolecule and together with the chain section (segment). The theoretical approaches which allow the frequencies of these two types of motion to be determined have been developed.[44,106–110] The label rotation frequencies of both types differ significantly then the characteristic splitting of the side components can be detected in ESR spectra

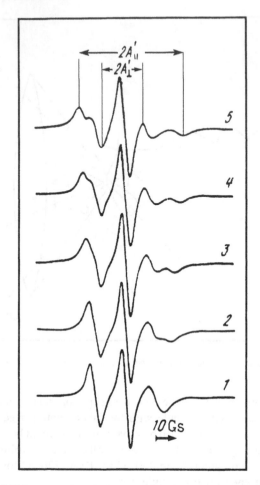

Figure 9   The calculated ESR spectra for spin-labelled macromolecules with the parameters: $\tau_s = 60$ ns, $\tau_l = 0.1$ ns, $\delta H = 2$, 4 Gs, $\Delta g = 0.052$, $a = 16.35$ Gs, $2A_{zz} = 72.4$ Gs. $S = 0.2$ (1), 0.3 (2), 0.4 (3), 0.5 (4), 0.6 (5).[110]

(Fig. 9). The frequencies of label rotation can be determined by correlation between experimental and theoretical spectra being calculated at different combinations of frequencies.[44,109,110]

To accomplish this one can use the characteristic parameters of the ESR spectra and their dependencies on the correlation time. The following relationship has been proposed[111] as such parameter:

$$S = \frac{\Delta \overline{A}}{\Delta A} = \frac{\overline{A}_{\parallel} - a}{A_{zz} - a} \tag{22}$$

here: $\Delta A = A_{zz} - 1/2\,(A_{xx} + A_{yy})$; $\Delta \overline{A} = \overline{A}_{\parallel} - \overline{A}_{\perp}$; A_{zz}, A_{xx}, A_{yy} — the hyperfine interaction tensor of nitroxide radical; $\overline{A}_{\parallel}$ and $\overline{A}_{\perp}$ — axially-symmetric partially averaged tensor; $a = 1/3(A_{zz} + A_{xx} + A_{yy})$.

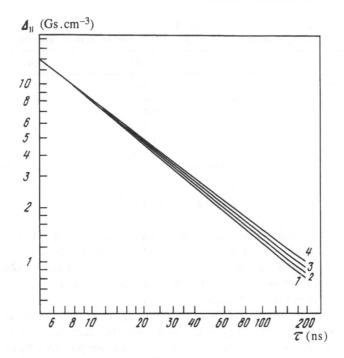

Figure 10 The calculated dependences of $\Delta = 2\overline{A}_\| - 2A_\|'$ on the correlation time of segment rotation. $S = 0.5$ (1), 0.8 (2), 0.9 (3), 1.0 (4).[110]

The parameter S is characteristic of the angle α at the top of the label fast rotation cone (Fig. 8):

$$S = 1/2(\cos^2 \alpha + \cos \alpha) \qquad (23)$$

The value $S = 1$ ($\alpha = 0$) is said to correspond to the slow rotation of radical rigidly bonded to the chain. Decreasing of S means increasing the fast rotation angle cone and at $S = 0$ ($2\alpha = 180°$) a label freely rotates about the chain without regard to the segment movement.

Yet another parameter which is characteristic of slow segmental dynamics of macromolecule is said to be the parameter $\Delta_\| = 2\overline{A}_\| - 2A_\|'$, which can be determined by the distance between the external extremes of the spectrum ($2A_\|'$). The theoretical dependence of $\Delta_\|$ on the correlation time of segmental movement τ_s at various S values which has been obtained[110] in Fig. 10. This dependence is described by the analytical equation of the type:

$$\Delta_\| \sim \tau_s^{-\beta} \qquad (24)$$

where β varies from 0.74 to 0.83 as S varies from 1 to 0.5.

The experimental evaluation of the segmental mobility and the parameter S is based on examination of the dependence of $2A_\|'$ on $(T/\eta)^\beta$. In the range of moderate values of the viscosity this dependence is linear and is described by the equation:

TABLE 5

The values of τ_s and parameter S for some polymers in solution at room temperature[110,113,114]

Macromolecule	M	solute	τ_s (ns)	S
Poly(4-vinyl-pyridine)		C_2H_5OH	3.5	0.56
Poly(vinyl-pyrrolidone)	73000	C_2H_5OH	4.4	0.81
		H_2O	2.4	0.75
Poly(vinyl-caprolactam)	41000	C_2H_5OH	8.8	0.81
		H_2O	5.5	0.83
Chitozan		CH_3COOH	14.5	0.67

$$2A'_{\parallel} = 2\overline{A}_{\parallel} - C(T/\eta)^{\beta} \tag{25}$$

The value of $2\overline{A}_{\parallel}$ results from extrapolation of the magnitude of $2A'_{\parallel}$ to $T/\eta \Rightarrow 0$ and Eq. 22 is used to calculate S value. The value of $\Delta_{\parallel}$ is determined for T/η and using Eq. 10 the magnitude of τ_s is calculated. "Double" spectra (Fig. 9) can be observed if the correlation times of the fast (local) label rotation are less than $(0.5-1) \cdot 10^{-9}$ s while those of the slow (segmental) rotation exceed $5 \cdot 10^{-9}$ s.[109]

The results of some research of the segmental motion carried out by the spin label method are given in Table 5. They demonstrate that the frequencies of the segmental movements are independent of the spin label structure and are governed by the main chain rigidity, quality of solvent and solution viscosity. The segmental mobility depends on the viscosity according to the Stokes-Einstein equation: $\tau = 4\pi\eta \ r^3/3kT$.[112] Using this equation one can calculate the effective hydrodynamic radius r of spin labelled segment.[110,113,114]

The studies of polyvinylcaprolactam solutions have shown that in proximity to the phase transformation temperature, the compression of a polymeric coil occurs accompanied by growth in τ_s.[113]

5.3   Local density and translation mobility of monomeric units in a coil

Analysis of the local concentration of labelled monomeric units in macromolecular coil in a solution have been developed.[4,115-118] As opposed to the mean value of the units density in the coil volume $\langle\rho\rangle = N/V$, (N is the total number of the units in the coil, V is its volume) the local density of the units determined over the volume appears to be much lower than V. Size of this region is controlled by the distance over which the dipole broadening of the ESR line of the label positioned at the centre of the microregion is comparable with its width (~ 1 Gs). This condition corresponds to a sphere with radius of ~ 30 Å.

Determination of the units local density is governed by analysis of the dipole and exchange contribution into the linewidth discussed in Sec. 4.2. The local concentration of labels are determined by Eq. 17 and are related with the units local density ρ_{loc} by the relationship:

$$c_{lok} = \rho_{lok} \ m/(m + n) \tag{26}$$

where m and n is a number of labelled and unlabelled units, correspondingly.

Rate constants for spin exchange are calculated by the equation:

$$k_e = \frac{\delta H_e}{c} \frac{\sqrt{3}}{2} \gamma = pfk_0 \tag{27}$$

where γ is the gyromagnetic ratio for the electron, k_0 is the rate constant of pair collisions, p is the efficiency of a collision and f is a steric factor.

The rate constant of pair collisions of a label with labels of surrounding units is equal to:

$$k_0 = 16\pi r D_{loc} c_{loc} \tag{28}$$

here: D_{loc} is the coefficient of the translational diffusion of labels in the local volume and r is the effective radius of radicals spin exchange.

From Eqs. 27 and 28 it follows:

$$\delta H_e = \frac{32\pi}{\sqrt{3}\gamma} pfr D_{loc} c_{loc} \tag{29}$$

The parameters of Eqs. 26–29 are defined experimentally as follows. At first, c_{loc} is determined in the range of the dipole broadening of the lines at low temperatures. When the temperature is increased the sample is transferred to the region where the main contribution to the linewidth is given by the exchange interaction between the labels and coefficient of the local translational diffusion calculated by Eq. 29.

It should be noted that the model used for these calculations rests on the assumption about the statistical distribution of labels over the macromolecule.

Some experimental results on the local concentrations of units in dilute polymeric solutions.[4,16]

The general conclusions resulted from the discussion above can be summarized as follows: a) local concentrations of units are only weak depending on the molecular weight of a polymer, b) the local concentration of units is 4–6 times higher than the mean value, c) the local density of units increases with the number of labels on polymeric chain. It is likely to be the result of chemical structure modification of the chain in the presence of polar groups with a consequential compression of a polymeric coil, d) the coil density is higher in a poor solvent than in a good one.

A question as to how p varies with increasing polymer concentration in solution has been tried[115] for spin labelled polyvinylpyridine (PVP) with high label content (1 label per 5–10 units). With increasing PVP content in ethanol up to 65% the magnitude of p increases from 0.3 to 0.43 mol/l. With increasing concentration insignificant compression of a polymeric coil occurs.

Another fascinating problem appears to be the study of the local density of adjacent macromolecular units in a small volume in the vicinity of a spin label. Solution of this problem results from analysis of the dipole broadening of spin labelled macromolecules. In case of their low content the dipole broadening is defined only by the interaction of the labels of the same chain. At a high concentration of spin labelled coils in solution, the interaction occurs between "own" and "foreign" labels. The distinctions between the line broadening in these two cases give the desired value. The experiments with the PVP solutions in ethanol have shown that with the increasing polymer content up to 65%, the local density of units in a coil increases not more than by 20–40%.[115,116] At the same time, the local density of the guest's units is coming to exceed the ρ-value of its own units even at a concentration of 10–15% and drastically grows with increasing concentration. It means that inside a coil the replacement of a solvent takes place — ethanol molecules are being substituted by the units of the "foreign" macromolecules.

A rigorous theory of the form of the dipole bridged ESR spectra of nitroxide radicals in magnetically diluted solid matrixes have been recently developed.[119,120] This theory offers the relationship between the line form and pair function of the distribution of units in a coil and quantitative data on conformation, rigidity and sizes of a coil. Application of this theory to spin labelled polyvinylpyridine and other polymers has demonstrated that the distribution of the units in a coil at 77 K approaches Gaussian independent of solvent.[120]

Spin labels can be used to study the permeability of a polymeric coil to low molecular weight particles.[16] Analysis of the constants of the hyperfine interaction (HFI) of radical (probe or label) depends on whether a complex between a radical and solvent molecules (or a substance added to a solution) forms or not. It has been shown that the HFI constants of spin labels and probes in PVP ethanol solutions are the same. This means that the probabilities of complex formation between the ethanol molecule and probe and between ethanol and a label are equal. Thus, all spin labels in solution are available for the ethanol molecules and the macromolecules coil is fully permeable. Modification of this method is said to be an analysis of ESR linewidths of labels on the addition to the solutions of ions of paramagnetic metal forming complexes with a radical (for example, $Cu(NO_3)_2$). This method is given the title "the method of spin exchange titration".[121–123]

The coefficients of the local translational diffusion of spin labels have been studied by exchange broadening of the ESR lines.[115,116] The results received for diluted PVP solution has shown that D_{loc} for a label is an order smaller than that of the probe. Such relationships between the diffusion coefficients are attributed to the local mobility being limited by a chemical bond. The activation energy for label translation is also much greater (36.4 kJ/mole and 13.8 kJ/mole).

With increasing of the polymer content in the solution up to 10%, D_{loc} is practically unchanged and further increase in polymer content results in a decrease in D_{loc}. This reduction can be attributed to the coils compression because of overlap in a concentrated solution.

5.4   Other applications of spin labels

When studying spin labelled adsorbed macromolecules spectral superpositions and HFI constants ($2A_{zz}$) have to be analyzed. Analysis of the first type provides data concerning the ratio of the number of units intimately attachment to a surface (SiO_2, Al_2O_3 and so on) and having no bonds with the surface of units forming loops and tails. Analysis of the second type gives data on the energy of interaction between spin labels and surface.[9,124–126]

REFERENCES

1. A.L. Buchachenko, A.L. Kovarskii and A.M. Wasserman, The Study of Polymers by Paramagnetic Probe Technique, in *Uspechi Chimii i Physiki Polimerov*, Ed. by Z.A. Rogovin, Moscow, Chimiya, p. 33 (1973); in *Advances in Polymer Sciences*, Ed. by Z.A. Rogovin, Wiley, New York, p. 26 (1974).
2. P. Tormala and J.J. Lindberg, in *Structural Studies of Macromolecules by Spectroscopy Methods*, Ed. by K.J. Ivin, Wiley, London, p. 256 (1976).
3. A.T. Bullock and G.G. Cameron, *ibid.*, p. 272.
4. A.L. Buchachenko, A.M. Wasserman, T.A. Aleksandrova and A.L. Kovarskii, in *Molecular Motions in Polymers by ESR*, Ed. by R.F. Boyer and S.E. Keinath, Harwood Academic Publishers, Chur, p. 33 (1980).
5. G.G. Cameron, *ibid.*, p. 55.
6. P. Tormala, G. Weber and J.J. Lindberg, *ibid.*, p. 81.
7. A.T. Bullock, *ibid.*, p. 115.
8. J. Sohma and K. Murakami, *ibid.*, p. 135.
9. W.G. Miller, W.T. Rudolph, Z. Veksli, D.I. Coon, C.C. Wu and T.M. Liang, *ibid.*, p. 145.
10. A.M. Bobst, *ibid.*, p. 167.
11. A.L. Kovarskii, A.M. Wasserman and A.L. Buchachenko, *ibid.*, p. 177.
12. P.L. Kumler, *ibid.*, p. 189.
13. N. Kusumoto, *ibid.*, p. 223.
14. P.M. Smith, *ibid.*, p. 255.
15. A.N. Kuznetsov, *Spin Probe Technique*, Nauka, Moskva (1974).
16. A.M. Wasserman and A.L. Kovarskii, *Spin Probes and Labels in Physical Chemistry of Polymers*, Nauka, Moskva (1985).
17. *Spin Labelling, Theory and Application*. Ed. by L.J. Berliner, Academic Press, New York, v. 1 (1976), v. 2 (1986).
18. G.I. Lichtenstein, *Spin Labelling Methods in Molecular Biology*, Wiley, New York (1976).
19. *Bioactive Spin Labels*. Ed. by R.I. Zhdanov, Springer-Verlag, Berlin (1992).
20. A.K. Hoffmman and A.T. Hendersen, *J. Am. Chem. Soc.*, **83**, 4671 (1961).
21. M.B. Neiman, E.G. Rozantsev and Yu.G. Mamedova, *Nature*, **196**, 472 (1962).
22. T.I. Stone, T. Buckman, P.L. Nordio and H.M. McConnell, *Proc. Natl. Acad. Sci., U.S.*, **54**, 1010 (1965); S. Ohnishi and H.M. McConnell, *J. Am. Chem. Soc.*, **87**, 2293 (1965).
23. E.G. Rozantsev, *Free Nitroxyl Radicals*, Plenum, New York (1970).
24. E.G. Rozantsev and V.D. Sholle, *Synthesis*, **4**, 190 (1971); **8**, 401 (1971).
25. E.G. Rozantsev and V.D. Sholle, *Organic Chemistry of Free Radicals*, Chimiya, Moskva (1979).
26. A.I. Kovarskii, *The Study of Polymers by Paramagnetic Probe Method*, Institute of Chemical Physics, Academy of Sciences of USSR, Moscow (1966).
27. A.M. Wasserman, A.L. Buchachenko, A.L. Kovarskii and M.B. Neiman, *Vysokomol, Soed.*, **10A**, 1930 (1968).
28. V.B. Strukov, Yu.S. Karimov and E.G. Rozantsev, *Vysokomol. Soed.*, **9B**, 493 (1967).
29. G.P. Rabold, *J. Polym. Sci.*, A-1, **7**, 1187 (1969), **7**, 1203 (1969).
30. N. Kusumoto and H. Mykoyama, *Rep. Progr. Polym. Phys. Jap.*, **15**, 581 (1972).
31. A.T. Bullock, J.H. Butterworth and G.G. Cameron, *European Polym. J.*, **7**, 445 (1971).
32. P. Tormala, H. Lattila and J.J. Lindberg, *Polymer*, **14**, 481 (1973).
33. P.L. Kumler and B.F. Boyer, *Macromolecules*, **9**, 903 (1976).
34. D. Kivelson, *J. Chem. Phys.*, **33**, 1094 (1960).
35. J.H. Freed and G.K. Fraenkel, *J. Chem. Phys.*, **39**, 326 (1963).
36. A.N. Kuznetsov, A.M. Wasserman, A.U. Volkov and N.N. Korst, *Chem. Phys. Lett.*, **12**, 103 (1971).
37. G.F. Goldman, G.V. Bruno and J.H. Freed, *J. Phys. Chem.*, **76**, 1858 (1972).
38. R.P. Mason and J.H. Freed, *J. Phys. Chem.*, **78**, 1321 (1974).
39. A.M. Wasserman, A.N. Kuznetsov, A.L. Kovarskii and A.L. Buchachenko, *Zh. Strukt. Chimii*, **12**, 609 (1971).
40. S.A. Goldman, G.V. Bruno, C.F. Polnaszek and J.H. Freed, *J. Chem. Phys.*, **56**, 716 (1972).
41. A.N. Kuznetsov, A.U. Volkov, V.A. Livshits and A.T. Mirzoian, *Chem. Phys. Lett.*, **26**, 369 (1974).
42. J.S. Hwang, R.P. Mason, L.P. Hwang and J.H. Freed, *J. Phys. Chem.*, **79**, 489 (1975).
43. A.N. Kuznetsov and B. Ebert, *Chem. Phys. Lett.*, **25**, 342 (1974).

44. J.H. Freed, see Ref. 22, p.
45. K.I. Zamaraev, Yu.N. Molin and K.N. Salihov, *Spin Exchange*, Nauka, Novosibirsk, 1976.
46. A.L. Kovarskii, A.M. Wasserman and A.L. Buchachenko, *J. Magn. Res.*, **7**, 225 (1972).
47. A.N. Kuznetsov and A.T. Mirzoyan, *Zh. Phiz. Chimii*, **48**, 2995 (1974).
48. A.M. Wasserman, A.L. Kovarskii, L.L. Yasina and A.L. Buchachenkjo, *Teor. i Experim. Himiya*, **13**, 30 (1977).
49. J.S. Hide and L.R. Dalton, see Ref. 17, v. 2, p. 1.
50. J.S. Hide, W. Fronzisz and C. Mottley, *Chem. Phys. Lett.*, **110**, 621 (1984).
51. V.A. Livshits, V.A. Kuznetsov, I.I. Barashkova and A.M. Wasserman, *Vysokomol. Soed.*, **25A**, 1085 (1982).
52. S.A. Dzuba and Yu.D. Tsvetkov, *Chem. Phys.*, **120**, 179 (1988).
53. S.A. Dzuba, Yu.D. Tsvetkov and A.G. Maryasov, *Chem. Phys. Lett.*, **188**, 217 (1992).
54. J.H. Freed, in *Modern Pulsed and Continuous Wave Electron Spin Resonance*, Ed. by L. Kevan and M.K. Bowman, Wiley, New York, p. 119 (1990).
55. G.G. Maresch, M. Weber, A.A., Dubinskii and H.W. Spiess, *Chem. Phys. Lett.*, **193**, 134 (1992).
56. Ya. Grinberg, A.A. Dubinskii and Ya.S. Lebedev, *Uspehi Himii*, **52**, 1490 (1983).
57. Ya.S. Lebedev, *Sov. Sci. Rev. B. Chem.*, **11**, 23 (1987).
58. A.L. Kovarskii, *Polymer Yearbook*, Ed. by R.A. Pethrick, Harwood, Switzerland, v. 9, p. 107 (1992).
59. I.I. Barashkova, A.M. Wasserman and A.L. Kovarskii, *Vysokomol. Soed.*, **24A**, 91 (1982).
60. A.L. Kovarskii, *Vysokomol. Soed.*, **28A**, 1347 (1986).
61. A.A. Dadaly, A.M. Wasserman and A.L. Kovarskii, *Doklady AN SSSR*, **237**, 130 (1977).
62. A.A. Dadaly, A.M. Wasserman, A.L. Buchachenko and V.I. Irzhak, *Europ. Polym. J.*, **17**, 525 (1981).
63. I.I. Barashkova, A.A. Dadaly, I.I. Aliev, V.A. Zhorin and A.L. Kovarskii, *Vysokomol. Soed.*, **25A**, 840 (1983).
64. A.L. Kovarskii, in *High Pressure Chemistry and Physics of Polymers*, Ed. by A.L. Kovarskii, CRC-Press, Boca Raton, 9 (1993) (in press).
65. P.B. Macedo and T.A. Litovitz, *J. Chem. Phys.*, **42**, 245 (1965).
66. P.M. Smith, *Europ. Polym. J.*, **15**, 147 (1979).
67. A.A. Dadaly, A.M. Wasserman and S.T. Kirillov, *Vysokomol. Soed.*, **22A**, 721 (1980).
68. O.G. Poluektov, A.A. Dubinskii, O.Ya. Grinberg and Ya.S. Lebedev, *Him. Phizika*, **11**, 1480 (1982).
69. L. Antsiferova and N.N. Korst, *Chem. Phys. Lett.*, **15**, 439 (1972).
70. A.M. Wasserman, T.A. Aleksandrova and A.L. Buchachenko, *Europ. Polym. J.*, **12**, 691 (1976).
71. V.A. Benderskii and N.P. Piven, *Zh. Fiz. Himii*, **59**, 1329 (1985).
72. G.P. Johari and M. Goldstain, *J. Chem. Phys.*, **53**, 2372 (1970).
73. G.P. Johari and M. Goldstain, *J. Chem. Phys.*, **55**, 4245 (1971).
74. M. Davis and A. Edwards, *Trans. Far. Soc.*, **63**, 2162 (1967).
75. A.L. Kovarskii and S.A. Mansimov, *Polym. Bull.*, **21**, 613 (1989).
76. A.M. Wasserman, L.L. Yasina, I.I. Barashkova and V.S. Pudov, *Vysokomol. Soed.*, **19B**, 820 (1977).
77. I.I. Barashkova, A.M. Wasserman and A.L. Kovarskii, *Vysokomol. Soed.*, **23B**, 436 (1981).
78. P. Meurisse, G. Friedrich, M. Dvolaitzky, *et al.*, *J. Chem. Phys.*, **77**, 3915 (1984).
79. K. Mueller, K.H. Wassmer, R.W. Lenz and G. Kothe, *J. Polym. Sci., Polym. Lett. Ed.*, **21**, 785 (1983).
80. C.L. Choy, W.P. Leung and T.L. Ma, *J. Polym. Sci., Polym. Phys. Ed.*, **23**, 557 (1985).
81. D. Brown, P. Tormala and G. Weber, *Polymer*, **19**, 598 (1978).
82. N. Kusumoto, S. Sano, N. Zaitsu and Y. Motozato, *Polymer*, **17**, 448 (1976).
83. A.T. Bullock, G.G. Kameron and I.S. Miles, *Polymer*, **23**, 1536 (1982).
84. A.A. Popov, C.G. Karpova and G.E. Zaikov, *Vysokomol. Soed.*, **24A**, 2401 (1982).
85. E. Mierovitch, *J. Phys. Chem.*, **88**, 2629 (1984).
86. N. Kusumoto and T. Ogata, *Rep. Progr. Polym. Phys. Jap.*, **23**, 599 (1980).
87. A.L. Kovarskii, J. Plachek and F. Szocs, *Polymer*, **19**, 1137 (1978).
88. A.M. Wasserman and I.I. Barashkova, *Vysokomol. Soed.*, **19B**, 820 (1977).
89. V.A. Livshits, V.A. Kuznetsov, I.I. Barashkova and A.M. Wasserman, *Vysokomol. Soed.*, **24A**, 1085 (1982).
90. Z. Hlouskova, J. Tino and E. Borsig, *Polymer Communications*, **25**, 112 (1984).
91. A.M. Wasserman, L.L. Yasina, I.I. Barashkova and V.S. Pudov, *Vysokomol. Soed.*, **19A**, 2083 (1977).
92. A.L. Kovarskii, E.I. Kulish, J. Plachek, F. Szocs and S.V. Kolesov, *Vysokomol. Soed.*, (1994) (in press).
93. A.I. Kokorin and K.I. Zamaraev, *Biofizika*, **17**, 34, (1972).
94. V.N. Parmon, A.I. Kokorin and G.M. Zhidomirov, *Stable Biradicals*, Nauka, Moskva (1980).
95. J. Labsky, J. Pilar and J. Kalal, *Macromolecules*, **10**, 1153 (1977).

96. A.I. Kokorin, Yu.E. Kirsh, K.I. Zamaraev and V.A. Kabanov, *Dokl. AN SSSR*, **208**, 1391 (1973).
97. C. Friedrich, F. Laupretre, C. Noel and L. Monnerie, *Macromolecules*, **14**, 1119 (1981).
98. A.T. Bullock, G.G. Cameron and V. Krajewski, *J. Phys. Chem.*, **80**, 1792 (1976).
99. A.T. Bullock, G.G. Cameron and P.M. Smith, *J. Chem. Soc. Far. Trans.*, Part I, **70**, 1202 (1974).
100. Z. Veksli and W.G. Miller, *Macromolecules*, **10**, 686 (1977).
101. A.T. Bullock, G.G. Cameron and P.M. Smith, *Europ. Polym. J.*, **11**, 617 (1975).
102. A.T. Bullock, G.G. Cameron and P.M. Smith, *Macromolecules*, **9**, 650 (1976).
103. P. Tormalla, H. Lattila and J.J. Lindberg, *Polymer*, **14**, 481 (1973).
104. A.M. Shapiro, V.B. Strukov and B.A. ???, *Vysokomol. Soed.*, **17B**, 265 (1975).
105. M.C. Lang and C. Noel, *J. Polym. Sci., Polym. Phys. Ed.*, **15**, 1329 (1977).
106. R.P. Mason, C.F. Polnaszek and J.H. Freed, *J. Phys. Chem.*, **78**, 1324 (1974).
107. J. Pilar, J. Labsky, J. Kalal and J.H. Freed, *J. Phys. Chem.*, **83**, 1907 (1979).
108. J. Pilar and J. Labsky, *J. Phys. Chem.*, **88**, 3659 (1984).
109. V.P. Timofeev, I.V. Dudich and M.V. Volkenstein, *Biophys. Struct. and Mech.*, **7**, 41 (1980).
110. A.M. Wasserman, T.A. Aleksandrova, I.V. Dudich and V.P. Timofeev, *Vysokomol. Soed.*, **23A**, 1441 (1981).
111. W.K. Hubbell and H.M. McConnell, *J. Amer. Chem. Soc.*, **93**, 314 (1971).
112. A.L. Buchachenko and A.M. Wasserman, *Pure and Appl. Chem.*, **54**, 507 (1982).
113. A.M. Wasserman, V.P. Timofeev and T.A. Aleksandrova, *Europ. Polym. J.*, **19**, 333 (1983).
114. T.A. Aleksandrova, A.M. Wasserman and A.I. Gamzazade, *Vysokomol. Soed.*, **12B**, 219 (1983).
115. A.W. Wasserman, T.A. Aleksandrova, Yu.E. Kirsh and A.L. Buchachenko, *Europ. Polym. J.*, **15**, 1051 (1979).
116. A.M. Wasserman, T.A. Aleksandrova and Yu.E. Kirsh, *Vysokomol. Soed.*, **22A**, 275 (1980); **22A**, 282 (1980).
117. A.I. Kokorin, Yu.E. Kirsh and K.I. Zamaraev, *Vysokomol. Soed.*, **17A**, 1618 (1975).
118. Yu.A. Shaulov, A.C. Haritonov and A.I. Kokorin, *Vysokomol. Soed.*, **19A**, 1813 (1977).
119. T.N. Khazanovich, A.D. Kolbanovsky, T.V. Medvedeva and A.M. Wasserman, *EPR Study of the Pair Distribution Function of Monomer Units in Solid State Polymers*, Abstracts of II International Workshop on Electron Magnetic Resonance in Disordered Systems. Sofia p. 26 (1991).
120. A.D. Kolbanovsky, A.M. Wasserman, T.N. Khazanovich and A.I. Kokorin, *Chim. Phizika*, **11**, 94 (1992).
121. Yu.E. Kirsh, V.Ya. Kovner and A.I. Kokorin, *Dokl. AN SSSR*, **212**, 138 (1973).
122. A.I. Kokorin, K.I. Zamaraev and V.Ya. Kovner, *Europ. Polym. J.*, **11**, 719 (1975).
123. A.I. Kokorin, S.V. Lymar and V.N. Parmon, *Vysokomol. Soed.*, **23A**, 2027 (1981).
124. I.D. Robb and R. Smith, *Polymer*, **18**, 500 (1977).
125. H. Hommel, A.P. Legrand, H. Belard and E. Papirer, *Polymer*, **24**, 959 (1983).
126. G. Thambo and W.G. Miller, *Macromolecules*, **23**, 4397 (1990).
127. N.A. Sysoeva, A.Yu. Karmilov and A.L. Buchachenko, *Chem. Phys.*, **7**, 123 (1975); **15**, 321 (1976).
128. L. Andreozzi, M. Giordano, D. Leporini and L. Pardi, *Phys. Lett.*, A, **160**, 309 (1991).
129. L. Andreozzi, M. Giordano, D. Leporini, L. Pardi and A.S. Angeloni, *Mol. Cryst. Liq. Cryst.*, **212**, 107 (1992).
130. H. Evaitar, E. van Faassen and Y.K. Levin, *Chem. Phys. Lett.*, **195**, 233 (1992).

Spatial Organization of Polymer-Metal Coils in Liquid Solutions

A.I. KOKORIN

Institute of Chemical Physics, Russian Academy of Sciences, Kosygin St., 4 117977 Moscow, B-334, Russia

INTRODUCTION

Coordination compounds with macromolecular ligands, or polymer-metal complexes (PMC) have found increasing application in catalysis, analytical chemistry, and hydrometallurgy.[1-5] A study of the structure of PMC has two essential levels. The first involves an analysis of the composition and geometry of the coordination sphere of individual metal complexes.[1-7] The second which metal ions are coordinated by the repeating units of a polymer chain.[8]

A study of spatial organization includes: a) establishment of the nature of the distribution of metal complexes in the matrix; b) measurement of their mean local concentration $<C_M>$ or the mean distance $<r_M>$ between them; c) determination of the fraction of the each structural type present and changes in the macromolecular coil during the complex-forming process. If the PMC is involved in a chemical reaction, a question of changes in coordination centres in the polymer-metal system, and its spatial structure becomes important.

Information about the structure of coordination centres of PMC can be obtained using visible, IR-, EPR, Mössbauer- and X-ray photoelectron spectroscopies, circular dichroism, and magnetic susceptibility.[7,9]

From EPR the energy of the dipole-dipole interaction (DDI) between paramagnetics depends on the $<C_M>$ and $<r_M>$.[9,10] To measure a DDI-value it is necessary to record the EPR spectra of the samples frozen at 77 K. However, such extremely low temperatures can change the composition and structure of the PMC. Comparison of the results obtained for solutions of the PMC at room temperature and glass at 77 K is of interest.

For solutions of polymer-metal coils the DDI is averaged by the molecular motions and one has to develop a new approach for analysis of the spatial organization of the PMC. One approach,[11] "the technique of spin exchange titration" (SET-technique), is used for these studies.

DIFFUSION OF SMALL MOLECULES IN MICRO-HETEROGENEOUS SYSTEMS

The method of spin exchange titration is based on observation of differences in broadening of the EPR lines of stable radical probes with free and coordinated by a polymer coil paramagnetic metal ions in solutions. The goal of SET is the quantitative measurement of the complex-forming ability of a macromolecule relative to these ions, and the effects of the PMC structure. The stable nitroxide radicals are usually used as spin probes:

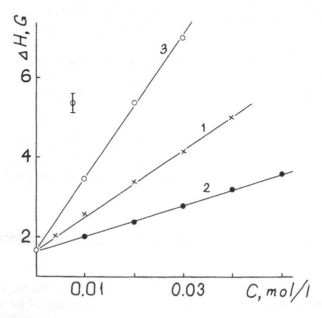

R_6H
R_6OH
R_6NH_2
$R_6CH_2CH_2OH$
R_6CH_2COOH
R_6OCOCH_3
$R_6{'}COOH$
$R_6{'}COOCH_3$
R_5COOH
R_5NH_2
R_5CONH_2

The titration of a solution of R with a paramagnetic ion gives a typical EPR line broadening δH of R, linear on concentrations used, due to intermolecular spin exchange (Fig. 1). Every radical-ion pair is characterised for a definite temperature and solvent with a value of the broadening constant K_b and connected with it the spin exchange rage constant K_e.[11]

The spin exchange rate constant K_e is usually determined in the region of exchange broadening by using[11]

$$K_e = 1.52 \cdot 10^7[(1 - \varphi)g/2]\ \delta H/C = [(1 - \varphi)g/2]\ K_b, \tag{1}$$

Figure 1   The dependence of the linewidth of the R_6OH EPR spectrum at 298 K on the concentration of R_6OH (1) in ethanol, $VOSO_4$ (2) and $Cu(en)_2(NO_3)_2$ (3) in aqueous solutions.

where $\delta H = \Delta H - \Delta H_0$ is broadening of the individual EPR line; ΔH and ΔH_0 are the peak-to-peak linewidths (in Gauss) at concentration C (in mol/l) and without spin exchange respectively; φ is the statistical weight of the EPR component for which δH is measured; K_b is the experimentally determined broadening constant. In the case of interaction between various paramagnetics, one of which is a nitroxide radical R, values of K_e and K_b are equal because of $g_R/2 \approx 1$. If one studies spin exchange between nitroxides themselves, then $(1 - \varphi) = 2/3$, i.e. $K_e = 1.5 \ K_b$.

In the case of strong exchange, the K_e value is given by the formula[11]

$$K_e = P_{max} \cdot f \cdot K_D, \tag{2}$$

TABLE 1

Rate constants of spin exchange between charged and electro neutral nitroxide radicals and metal complexes at 293 K

Radical	Complex[a]	T_1, s[12]	pH	$K_e \cdot 10^{-8}$,[b] l/mol·s
R_6OH	$Cu(H_2O)_6^{2+}$		6.9	19
$R_6CH_2CH_2OH$	"		6.9	17
R_6CH_2COOH	"		3.0	16
$R_6CH_2COO^-$	"		6.8	30
$R_5^N CH_3$	"		7.0	24.5
R_5CONH_2	"		3.0–6.8	23
R_5COO^-	"		6.9	40
R_6OH	$Ni(H_2O)_6^{2+}$		5.7	14
R_6OH	$Co(H_2O)_6^{2+}$	$7 \cdot 10^{-13}$	5.7	7
	$Cu(H_2O)_6^{2+}$	$(0.1–2) \cdot 10^{-8}$ [c]	3.0	22; 24.5 [d]
	$Ni(H_2O)_6^{2+}$	$1.3 \cdot 10^{-11}$	"	17
R_6H	$Mn(H_2O)_6^{2+}$	$4.8 \cdot 10^{-9}$	"	20
	$Cr(H_2O)_6^{3+}$	$(0.8–4) \cdot 10^{-9}$ [c]	"	11.5
	$Cu(en)_2(H_2O)_2^{2+}$	$2.7 \cdot 10^{-9}$	5.0–11	28 ± 2
	$Fe(CN)_6^{3-}$	$2.3 \cdot 10^{-11}$	3.1–11	9
	$Cu(H_2O)_6^{2+}$		3.0	12; 21 [d]
	$Ni(H_2O)_6^{2+}$		"	11
$R_6NH_3^+$	$Mn(H_2O)_6^{2+}$		"	10.6
	$Cr(H_2O)_6^{3+}$		"	6.5
	$Cu(en)_2(H_2O)_2^{2+}$		6.1	11.8
	$Fe(CN)_6^{3-}$		3.0	36
R_6NH_2	"		11.0	10.3
"	$Cu(en)_2(H_2O)_2^{2+}$		10.9	24.2
R_6CH_2COOH	$Ni(H_2O)_6^{2+}$		3.0	15
$R_6CH_2COO^-$	"		6.9	24
R_6CH_2COOH	$Fe(CN)_6^{3-}$		3.0	8.1
$R_6CH_2COO^-$	"		7.5	2

[a] anion: NO_3^-;
[b] accuracy of measurements: ± 2 at $K_e \cdot 10^{-8} > 25$; ± 1 at $10 < K_e \cdot 10^{-8} < 25$; ± 0.5 at $K_e < 10^9$ l/mol·s;
[c] measured by different methods;
[d] in the presence of 2 M $NaNO_3$.

TABLE 2

Rate constants of spin exchange between nitroxide radicals and
aquo complexes of transition metals at room temperature

Metal ion	$K_e \cdot 10^{-9}$, l/mol·s	
	$R_6 = O$ [13]	R_6H [11]
Cu^{2+}	1.4	1.6
Ni^{2+}	1.1	1.3
Co^{2+}	0.9[a]	0.9
Fe^{2+}	0.9	—
Mn^{2+}	2.0	2.2
Cr^{3+}	1.1	1.2
VO^{2+}	1.2[b]	—

[a] The same value was obtained in [14]
[b] from [15].

where f is the steric factor, K_D is the rate constant for diffusion binary collisions. P_{max} is the maximum value of the spin exchange efficiency, and was calculated for many cases.[11] If for the "broadening agent" $T_1 < \tau_c$ (τ_c is the collision time, T_1 is the electron spin-lattice relaxation time), then regardless of the spin values $P_{max} = 1$.

The parameter $f \approx 1$ in the absence of steric hindrances for collisions between particles, and one can determine the K_D from the K_e value. If there are steric hindrances in the system, then the value of f should be estimated by measuring the constant K_e. In the limiting case when by steric hindrance the paramagnetic complexes are unapproachable for collisions with the spin probes, the magnitude of $f \cdot K_D = 0$, and addition of metal ions to the solution does not lead to broadening of the EPR lines of the probe R.

Typical values of the spin exchange rate constant K_e for ions and radicals often used are listed in the Tables 1 and 2.

An example of the influence of the spatial insulation of interacting particles on the frequency of their collisions, i.e. on the K_e value are presented in Fig. 2 for spin exchange in binary solvents. Collision of particles in such systems may be far more complicated than in one-component systems. The complicating factors include, among others, specific solvation of the molecules and the possible microheterogeneous structure of the binary solvents themselves. It is known[16,17] that the water-dioxan mixtures are not ideal but may consist of microregions enriched in one or other component. If solubilities of the radical and metal complex are essentially different in these solvents, then one may expect the concentration of R and ions in various areas, and hence the decrease of a number of binary collisions between them and of the K_e value.

The radical R_6Cl as a probe is readily soluble in dioxan and practically insoluble in water,[18] while the metal complexes $Cu(en)_2(NO_3)_2$ or $VOSO_4$ are more soluble in water than in dioxan (en is the ethylenediamine). To measure the value of K_e in pure water solutions we chose the radical R_6OH equally soluble in both components.[18] For R_6Cl and R_6OH it was shown that the exchange rate constants K_e for the reaction R + R coincide in many organic solvents (Table 3)[18] what allowed us to use R_6OH instead of R_6Cl in aqueous solutions.

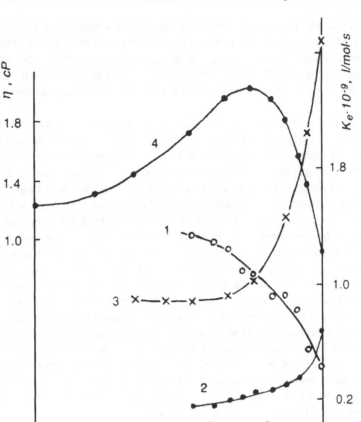

Figure 2   Rate constants K_e for the systems: (1) $VOSO_4$ + $VOSO_4$, (2) R_6Cl + $VOSO_4$, (3) R_6Cl + $Cu(en)_2^{2+}$ in water-dioxan mixtures and the viscosity η values (4) at 301 K.[17]

TABLE 3
Values of K_b and $K_e \cdot \eta$ in different solvents at 25°C

Solvent	R$_6$OH		R$_6$Cl	
	$K_b \cdot 10^{-9}$ l/mol·s	$K_e \cdot \eta \cdot 10^{-9}$ cP·l/mol·s	$K_b \cdot 10^{-9}$ l/mol·s	$K_e \cdot \eta \cdot 10^{-9}$ cP·l/mol·s
CCl$_4$	1.83	2.7	1.62	2.3
Dioxan	1.52	2.7	1.58	2.7
Ethanol	1.42	2.4	1.45	2.6
Toluene*	2.65	2.4	2.80	2.5
Benzene	2.48	2.3	—	—
Heptane	4.35	2.6	—	—
H$_2$O	1.20	1.7	—	—

* At 20°C.

Fig. 2 presents dependences of the K_e on the solvent composition for the exchange processes R + M and M + M, where M means a metal complex. The results agree qualitatively with the hypothesis that R and the metal ions are spatially separated. Indeed, when dioxan is added to water the spin exchange constants for R + VO^{2+} and for R + $Cu(en)_2^{2+}$ decrease, the scope of the change exceeding markedly the effect that could be produced by a change in viscosity.

For the system R + $Cu(en)_2^{2+}$ the condition of strong exchange was determined by the fulfilment of the viscosity criterion. Therefore curve 3 in Fig. 2 reflects directly the change in the number of binary collisions between R and M^{2+} with change in the composition of the water-dioxan mixture. In the region of dioxan mol fraction from 0.3 to 0.7, the R molecules and Cu(II) complexes are colliding approximately three times less than in pure water. The exchange rate constants between the same particles VO^{2+} + VO^{2+} behave inversely: they increase as dioxan is added to water. The data of Fig. 2 may be accounted qualitatively in the following way: the M^{2+} complexes are concentrated in the areas enriched by H_2O molecules while the radical R_6Cl prefers a dioxan surrounding.

In the case of an irregular distribution of the spin probes in a substance, and their concentrating in some part of the sample's volume the effective spin exchange rage constant should be larger than in the same pure solvent. In isotactic polypropylene with a 60% crystallinity a twice higher value of K_e was obtained compared with the atactic. This indicates that the probes R_6H are concentrated in the amorphous parts of the polymer. This was confirmed by independent experiments on measurement of the dipole-dipole broadening in frozen solutions at 77 K. Analogously, it was shown for the system butylcaoutchouc-toluene that for ratios of the polymer:solvent close to 1:1 a decrease of temperature results in formation of microphases of two components, with the free radical $R_6 = O$ concentrated in the low molecular phase.[20]

Spin exchange allows the appearance of factors embarrassing free diffusion collisions of paramagnetic particles to be studied. Steric hindrance is the strongest factors influencing on the spin exchange efficiency. These factors in the case of PMC are: the macromolecular nature of the ligand; the electrostatic interaction between the polymer chain, metal ions and the spin probe; a size of the radical R.

SPATIAL STRUCTURE OF THE Cu(II)-POLYMER COMPLEXES IN AQUEOUS SOLUTIONS

Spin exchange between the radical probe and paramagnetic metal ions in the presence of the macromolecular ligand is determined by the chemical nature and composition of the polymer. Peculiarities of the complex-forming process between a macromolecule and the Cu(II) ions are discussed in this section.

a)   Cu(II)-Polyvinylpyridine Complexes

For all polyvinylpyridines (PVP) the complex-forming nitrogen atoms are at some distance from the main chain and are able to coordinate metal ions as monodentate ligands. To make the PVP molecules soluble in water, they are usually quaternized partially by alkylating agents. Poly-4-vinylpyridine (P4VP) has been investigated.[21]

$(-CH_2-CH-)_n-(CH_2-CH-)_m$

$\beta = m/(m+n) \cdot 100\%$

$R = CH_3$ (P4VPM) ; C_2H_5 (P4VPE)
$X = Br^-$, $CH_3SO_4^-$

X^-

Here β is the content of quarternized pyridine residues.

The SET-technique demonstrates, Fig. 3, the dependence of the EPR linewidth of the probe R_6OH on the concentration of $Cu(NO_3)_2$ with and without additions of the P4VPM ($\beta = 35\%$). The titration of the aqueous solution of R with a solution of $Cu(NO_3)_2$ without polymer gives the usual linear broadening of the EPR lines of R due to spin exchange with the Cu^{2+} aqua-complexes. However the titration in the presence of P4VPM does not produce any broadening of the R lines until a definite concentration of $Cu(NO_3)_2$, which depends on the concentration of added polymer, is reached. Further addition of Cu(II) ions gives a linear broadening of the EPR lines with the same slope as at the absence of P4VP. It should be mentioned that the R molecules are not attached to a polymer and do not have any specific interaction with it. Such a dependence of ΔH on C shows that Cu(II) ions bound by the macromolecules are screened by the polymer chain from collisions with R_6OH. Previously it was shown[22,23] that the first portions of Cu(II) form with P4VPM and with P4VPE stable complexes $Cu(Py)_4^{2+}$, where Py is a free pyridine residue.

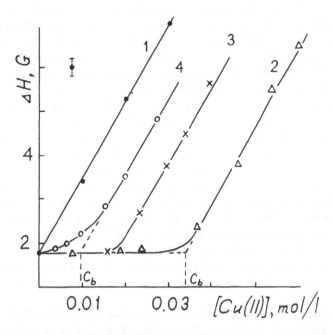

Figure 3   The dependence of the EPR linewidth of the R_6OH on the concentration of Cu(II) in the absence (1) and in the presence of P4VPM, $\beta = 35\%$ (2, 3), or of P2M5VP (4) at 300 K. [Py] = 0.36 (2); 0.21 (3); 0.4 mol/l (4).

Comparison of the concentrations of Py and of Cu(II) at the bending points of the curves $\Delta H = F(C)$, marked in Fig. 3 as C_b, allows determination of the fraction of Py fragments of polymer that can form the $Cu(Py)_4^{2+}$ complexes. This value is equal to $35 \pm 4\%$ of the total number of the Py rings in the polymer. It is of importance that independent measurements made by the spectrophotometric titration[21] and the EPR method at 77 $K^{22,23}$ yield a similar result. Small differences between the room temperature and 77 K data can be explained by the increase of the complex-forming constant for $Cu(Py)_4^{2+}$ with a decrease in temperature.

Another type of curve $\Delta H = F(C)$ was observed from the titration of aqueous solutions of poly-2-methyl-5-vinylpyridine with $\beta = 30\%$ (P2M5VP), contained R_6OH as a probe, with a $Cu(NO_3)_2$ solution. The difference is seen (Fig. 3) in broadening of the EPR spectrum lines of R even at rather low concentration of Cu(II) and in the absence of the interval with $\Delta H = $ const. The linear part of the curve is parallel to the line "1" due to interaction of R with aqua-complexes $Cu(H_2O)_6^{2+}$. Such a dependence indicates the presence in the P2M5VP coil of a large amount of coordinatively unsaturated complexes which are available for collisions with the radical.

It was shown indeed that even at 77 K at $\gamma = [Py]/[Cu(II)] \leq 25$ rather many complexes of the $Cu(Py)_3^{2+}$ composition appeared in the P2M5VP coil though at $\gamma = 40$ only $Cu(Py)_4^{2+}$ complexes could be detected.[24] At temperatures close to 300 K one should expect an increase of a fraction of the $Cu(Py)_3$ structures, as well as appearance of copper(II) bi- and monopyridinates, which are able to participate in spin exchange with higher efficiency.

The ratio of Py and Cu(II) concentrations at the C_b point (Fig. 3), as in the case of P4VP, shows the maximum fraction of the Py links in the P2M5VP molecule forming the $Cu(Py)_4$ complexes. The γ value is sufficiently lower ($\approx 10\%$) than for the P4VPM and is in a good agreement with the data obtained at 77 K.[24]

Addition of the P2VP ($\beta = 25\%$) to a water solution of R_6OH does not produce any changes in the linear dependence of ΔH on [Cu(II)] due to a low complex-forming ability of the P2VP in aqueous solution, i.e. to the absence of a noticeable amount of the PMC at ≈ 300 K.

It is known from previous studies[23] that the composition of the Cu(II)-P4VPE complexes depends on the content of quaternized Py residues (β) at 77 K. Use of the SET method at the room temperature yields titration curves of the R_6OH solutions in the presence of P4VPE with various values of β, Fig. 4. For all the samples the mean concentration of pyridine links has been held equal.

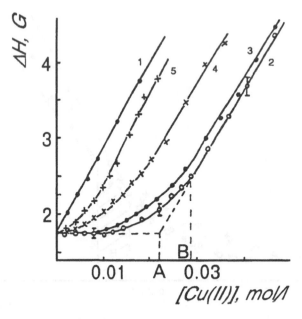

Figure 4   The concentration broadening of the R_6OH EPR line in aqueous solutions in the absence (1) and in the presence of P4VPE with β = 22 (2), 32 (3), 54 (4), 64% (5).

At small values of β the shape of the $\Delta H = F(C)$ dependence is analogous to the one, Fig. 3. The interval of independence ΔH on [Cu(II)] is evident and rather long. At $\gamma \leq 14$ an increase of ΔH is observed up to the region of linear EPR line broadening. At larger values (β = 32%) there is a shortening of this interval.

Further increase of β allows the EPR line broadening to be observed even with addition of the initial portions of Cu(II). The curves for β = 54% and 65% are similar to the SET curves obtained in the case of P2M5VP, Fig. 3. These results are in a good agreement with the data obtained in frozen solutions. It was shown[23] that even at very low Cu(II) ion content complexation in the P4VPE coils at β > 50% coordinatively unsaturated complexes $Cu(Py)_3$, and $Cu(Py)_2$, for which there are no strong steric hindrances for spin exchange as for the $Cu(Py)_4$ centres. Therefore, at β > 50% all Cu(II) complexes in the P4VPE are available for collisions with the spin probes, so the region with ΔH = const is absent.

The SET curves were obtained for P4VPE molecules at equal concentrations of free (nonalkylated) Py groups, Fig. 5. From the plots, the fraction of Py fragments forming the PMC at room temperature is close to the values measured at all β.[23] The maximum number of the links participating in coordination of Cu(II) ions with the P4VPE (γ_{lim} = [Py]$_{lim}$/[Cu(II)]) corresponds to the beginning of the linear dependence of ΔH on C (point "B" in Figs. 4 and 5), and is equal to γ_{lim} = 6.1 ± 0.4 at $20 \leq \beta \leq 65\%$.

b)   Cu(II)-Polyethyleneimine Complexes

Macromolecules of linear (PEI) and branched (BPEI) polyethyleneimine have a different type of structure compared to polyvinylpyridines. In PEI, the ligand nitrogen atoms,

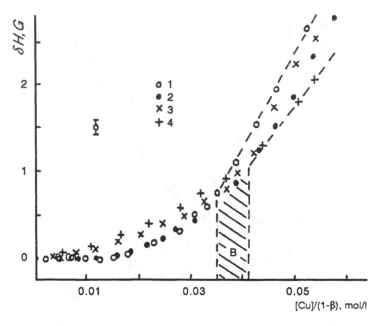

Figure 5   The data of Fig. 4 represented as the dependence of $\delta H = \Delta H - \Delta H_o$ on the parameter $[Cu(II)]/(1 - \beta)$. $\beta = 0.22$ (1); 0.32 (2); 0.54 (3); 0.64 (4).

participating in a complex formation, are localised in the main chain of the polymer. Moreover, two neighbouring N atoms act as a bidentate chelating ligand.[25] In this case the small spin probes will diffuse inside the polymer-metal coil and interact with the paramagnetic Cu(II)-PEI or Cu(II)-BPEI complexes differently from those in PVP.

The titration curve of R_6OH with copper ions in the presence of BPEI, Fig. 6, exhibit three regions.[26] At $[Cu(II)] \leq 8 \cdot 10^{-3}$ mol/l a slope of the dependence of ΔH on $[Cu(II)]$ is approximately half the slope in the case of the free copper aquoions. In the concentration range $0.009 \leq [Cu(II)] \leq 0.018$ mol/l (i.e., $9 \geq \gamma \geq 4.3$, where $\gamma = [N]/[M]$, and M is a transition metal ion) there is a region of independence of ΔH on $[Cu(II)]$. At $[Cu(II)] > 0.018$ mol/l ($\gamma < 4.3$) the usual linear EPR line broadening of the R_6OH is observed.

Initially, when the Cu(II) concentration is low, the Cu(II) complexes are randomly distributed in the macromolecular coil, which remains loose, and freely permeable for the spin probes. The value of the diffusion-controlled constant K_e is in this initial region only a half the value of Cu^{2+} aquoion, which results in a lower mutual diffusion coefficient. Indeed, Cu(II) are coordinated with a macromolecule of rather low mobility, and only

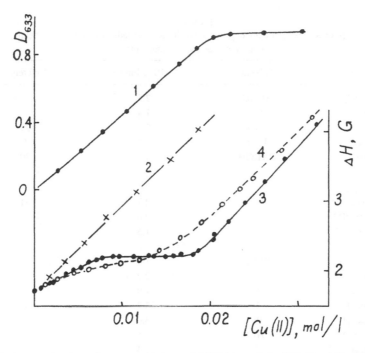

Figure 6   Optical density D_{633} of aqueous solutions of BPEI (1), and the EPR linewidth ΔH of the R_6OH radical in the 50% water-methanol mixture in the presence of BPEI (3), PEI (4) and without them (2) vs. Cu(II) concentration at 300 K. $[N]_o = 0.08$ mol/l.

the probe R diffuses freely in the solution. The half reduction in K_e is not attributable to a change in the coordination sphere of Cu(II) ions, since K_e values for the EPR line broadening of R by $Cu(H_2O)_6^{2+}$, $Cu(en)_2^{2+}$, $Cu(acac)_2$ and Cu(II)-EPP complexes coincide in the limits of experimental accuracy (Tables 1, 4). In the case of steric hindrances created by the polymer chain and impeding electron spin exchange one should expect the independence of ΔH on γ as shown for Cu(II)-P4VP complexes.

In the second region of curve 3 in Fig. 6 it is shown that at $10 > \gamma > 4.3$ part of the copper ions attached to the BPEI molecule became inaccesible for collisions with R, and only Cu(II) complexes, which are located in the polymeric coil surface, are still involved into spin exchange. Spin probes are either unable to pass through the central part of the coil, or stay there for a long time ($\tau > h/g\beta\Delta H \approx \{K_e[Cu(II)]\}^{-1} \approx 10^{-8}$ sec), and their EPR lines are broadened so much as to make their contribution to the measured linewidth ΔH. In the third region ($\gamma < 4.3$) spin exchange occurs between the R_6OH and Cu(II) aquaions.

These data has been supported by use of pyrene molecules as fluorescent probes.[26] In the presence of BPEI in water-methanol solutions at $\gamma > 4$ only a slight quenching of the fluorescent pyrene molecules was observed, while at $\gamma < 4$ the quenching was in an order more effective.

The results obtained with the SET technique indicate that the polymer-metal coils of BPEI with Cu(II) in aqueous and water-alcohol solutions are freely pervious for small uncharged molecules at $[Cu(II)] < [N]/9$, where [N] is the concentration of the imino-groups of BPEI. At higher Cu(II) concentrations a BPEI coil becomes impenetrable for

TABLE 4

Values of K_e for spin exchange between the $R_6 = O$ radical and various metal complexes at room temperature

Complex	Solvent	$K_e \cdot 10^{-9}$, l/mol·s	Reference
$CoPy_2Cl_2$	CH_3NO_2	2.5	27, 28
$CoPy_4Cl_2$	Py	1.4	"
$Cu(en)_2^{2+}$	H_2O	2.6	29
$Ni(en)_2^{2+}$	H_2O	2.1	"
$Fe(acac)_3$	C_2H_5OH	4.5	30
$Cu(Gly)_2$		2.0	
$Cu(II)$-EPP[a]		2.7	31
$Fe(III)$-EPP		2.6	
$VO(II)$-EPP		1.5	
$VO(acac)_2$	$CHCl_3$	1.6	
$Cu(acac)_2$	"	2.2	
$Cr(acac)_3$	"	1.9	
$Fe(acac)_3$	"	3.2	32
$Mn(acac)_2$	Py	3.0	
$Mn(acac)_3$	$CHCl_3$	0.5	
$Co(acac)_2$	"	0.4	
$Ni(acac)_2$	"	0.22	

[a] EPP is the Ethioporphyrin pyridinate.

the radical probes, and only copper complexes, which are localized in the external areas of the PMC, are still active in spin exchange. The maximum number of Cu(II) ions attached to the BPEI does not exceed 1/4.3 of the total concentration of the amino-groups.

The irregular spatial distribution of Cu(II) complexes inside the PEI coils has been confirmed by independent experiments on studies of dipole-dipole interaction in frozen water-methanol solutions of the Cu(II)-PEI complexes at 77 K.[34] Computer calculations of the EPR spectra established that when the PEI random coils are filled with Cu(II) ions, isolated complexes of the $Cu(N)_4^{2+}$ type are initially formed. When the number of copper ions in the solution is increased, the local concentration $<C_M>$ of these complexes increases, and regions with a very high $<C_M>$, i.e. a kind of associates, simultaneously appear in the PMC coil. The fraction of the complexes found in associates continually increases, and when $\gamma \leq 8.7$, practically all the ions appear in associates in the PEI coils. The complex-forming process is accompanied by a decrease in the volume of each macromolecular coil.[34,35] The mean distance $<r_M>$ between Cu(II) ions in the associates was estimated to be of the order of 8–9.5 Å.[34] The Cu(II) ions formed with PEI complexes have only one composition $Cu(N)_4$ for all the values of γ.[25,34] It's obvious that the spin probe molecules cannot enter into these associates because of their dimensions.

The SET method makes it possible to distinguish very small changes in conformational state of the complex-forming macromolecules, Fig. 6. Comparing the titration curves for the linear PEI and branched BPEI indicates that PEI behaves similarly to BPEI, but has a "smoother" shape. The differences are explained by a lower flexibility of the main

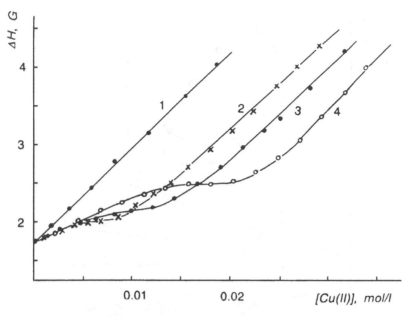

Figure 7   The EPR line broadening of the R_6OH on the Cu(II) concentration in 50% water-methanol mixtures in the presence of PEI and without it (1): $[N]_o$ = 0.045 (2); 0.08 (3); 0.12 mol/l (4) at 300 K.

polymer chain of PEI and the loose organization of the Cu(II)-PEI coil, because the structures of their copper complexes coincide at low Cu(II) concentrations ($\gamma \geq 14$).[25,26]

Such a smooth bend of the SET curve for PEI in a region of linear broadening of the R_6OH EPR lines by Cu(II) aquaions, Fig. (7), compared with BPEI, indicates at high ratios of copper ion's absorbtion into the PMC coil, the effective stability constant value for Cu(PEI) complexes is noticeably less than for Cu(BPEI) complexes. It follows from Fig. 7 that the maximum number of PEI links participating in coordination of Cu(II) ions is approximately 70–75% of all the links in the chain, and is in good agreement with a value measured at 77 K from the EPR spectra of the Cu(PEI) complexes.[25]

THE ROLE OF SPIN PROBE'S AND METAL ION'S ON THE SET RESULTS

The charge value and sign, and dimensions are the most important parameters which influence the efficiency of spin probe's diffusion inside a polymer-metal coil.

a)   Influence of Spin Probe's Charge on the Efficiency of Spin Exchange

Study of collisions of charged particles has a special interest because the rate of chemical reactions is known to be influenced by electrostatic repulsion or attraction between them.[35] The quantitative evaluation of the contribution of this interaction to the rate constant for condensed media is, however, a very complicated problem.[36] Metal ions in the PMC, as a rule, preserve their charge.

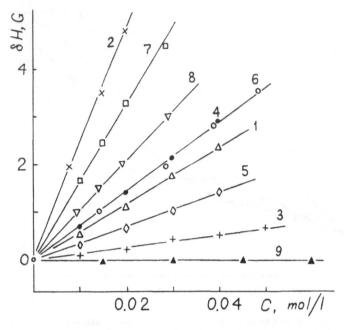

Figure 8   The dependence of the EPR line broadening δH for the R_6H (1,4,9), $R_6NH_3^+$ (2,5,6), $R_6CH_2COO^-$ (3,7,8) on the concentration of $Fe(CN)_6^{3-}$ (1–3), Cr^{3+} (4,5), Ni^{2+} (6–8), Zn^{2+} (9) in aqueous solution at pH = 6.6–7.2 (1,3,4,7,9) and 3.0 (2,5,6,8) at room temperature.

The data presented in Fig. 8 and Table 1 show the influence of the radical charge on the efficiency of spin exchange with various paramagnetic ions in aqueous solutions. The influence on the spin exchange rate constant K_e becomes stronger with increase of the ion's charge value. Similar results has been obtained[37] for the case of spin exchange between nitroxide radicals and the $KFe(CN)_6^{2-}$ and $Cr(C_6H_6)_2^+$ complexes. It is seen from Tables 1 and 5 that the value of K_e increases considerably for spin exchange between

TABLE 5
Values of K_e for spin exchange between radicals, radicals with $KFe(CN)_6^{2-}$ and $Cr(C_6H_6)^{2+}$ in 0.06 M of phosphate buffer solutions at room temperature[31,37]

Radical (R)	$K_e \cdot 10^{-9}$, l/mol·s		
	R + R	$KFe(CN)_6^{2-}$	$Cr(C_6H_6)_2^+$
$R_6'COO^-$	0.9	0.23	2.9
$R_5'COO^-$	1.4	0.18	3.1
$R_6=O$	2.4	1.22	2.8
$R_6'COOCH_3$	1.95	0.88	2.5
R_6OCOCH_3	1.95	0.91	2.5
R_5NH_2	—	0.91	—
$R_6NH_2^+$	—	0.98	—
$R_5NH_3^+$	1.9	1.68	2.0
R_6NH_3	1.6	2.5	2.0

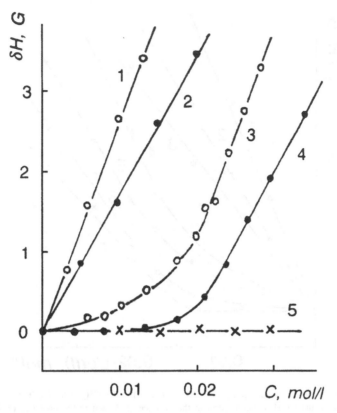

Figure 9   The dependence of the EPR line broadening of the R_6OH (1,3) and R_5COO^- (2,4,5) on the Cu(II) (1–4) and Zn(II) (5) concentration in the absence (1,2) and in the presence of P4VPM ($\beta = 35\%$) in aqueous solutions at pH 6.5. $[Py]_0 = 0.35$ mol/l.

heterocharged R and M, compared with homocharged particles. Hence, an interaction between positively charged PMC coils and the negatively charged probes R intensifies spin exchange and increases sensitivity of the SET method.

Fig. 9 presents the dependence of the linewidth ΔH on a Cu(II) concentration for the R_5COO^- radical at pH 6.5. The SET curve in the presence of the P4VPM ($\beta = 35\%$) for a negatively charged probe differs from a curve for the electro-neutral probe R_6OH: broadening of the R_5COO^- EPR lines is observed from the very beginning of the titration. This indicates that Cu(II) complexes in the P4VP coil are available for collisions with the R_5COO^- anions, but are screened for the R_6OH molecules at the same conditions of the experiment. Analogous results were obtained for the P4VP with a lower fraction of the quaternized Py groups in the case of $R_6C_2COO^-$ and R_5COO^-, Fig. 10.

Simple concentration of the R_5COO^- anions in a volume of the P4VP coil does not lead to noticeable broadening of the probe's EPR lines: substitution of Cu^{2+} cations by Zn^{2+} ones, Figs. 9 and 12 did not result in any preferentially bound with both P4VP and BPEI.[38,39] Therefore, increase of ΔH in the EPR spectra of R_5COO^- at $\gamma \geq 16$ should be explained by their more efficient interaction with Cu(II)-P4VP complexes, and not by their concentrating inside the Cu(P4VP) coil.

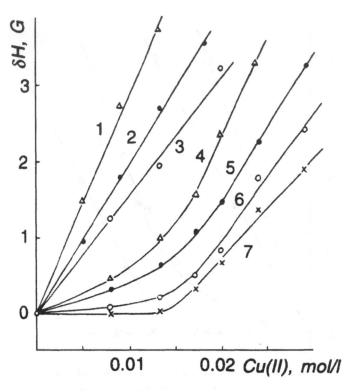

Figure 10   The SET curves for the spin probes R_5COO^- (1,4), R_5CONH_2 (3,6), $R_6CH_2COO^-$ (2,5), $R_6CH_2CH_2OH$ (7) vs. the Cu(II) concentration in the absence (1–3) and in the presence (4–7) of P4VPM ($\beta = 20\%$) at $[Py]_0 = 0.33$ mol/l and pH = 6.5–6.7.

The Coulomb electrostatic interaction facilitates penetration of the negatively charged probes into the macromolecular coil, increasing the probability of spin exchange with metal complexes. A fact that the ΔH values at $\gamma \geq 16$ are markedly smaller ΔH values defined from the equation $K_e = K_e^0 / 2$, K_e^0 is the spin exchange rate constant in the absence of a polymer indicates that there are steric hindrances for collisions of the paramagnetic centres, though less than in the case of the uncharged probe R_6OH, Fig. 9. The presence of the carboxylic fragments in the spin probes helps them to interact with Cu(II) complexes because the COO^- group is known as an effective complex-forming ligand for Cu(II) ions.

Similar results were obtained for the system Cu(II)-BPEI. Fig. 11 shows that the SET curve of BPEI in the case of $R_6CH_2COO^-$ has the same peculiarities and bending points as for the probe R_6OH. The slopes in the regions of linear broadening of ΔH on [Cu(II)] are increased for $R_6CH_2COO^-$ in accordance with a Coulomb factor. An independence of the interval with $\Delta H \approx$ const on the radical's nature shows that the areas with a high local concentration in the Cu(II)-BEPI complexes are not penetratable even for charged particles with a diameter of 6–8 Å, and confirms a hypothesis about a very compact structure for such associates.

Usage of the probes with a positive charge seems to be less of perspective for studies of the PMC structure. The dependence of ΔH on [Cu(II)] for the probe $R_6NH_3^+$ provides less information compared with the electro-neutral or negatively charged probes, Fig. 11.

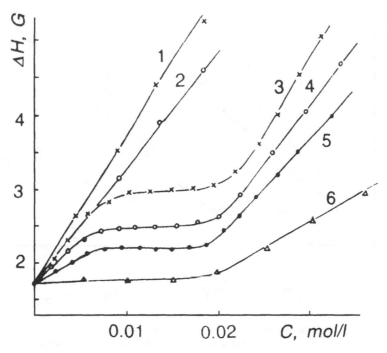

Figure 11   The EPR linewidth curves for the probes $R_6CH_2COO^-$ (1,3), $R_5^NCH_3$(2,4), R_6OH (5), $R_6NH_3^+$ (6) on the Cu(II) concentration in aqueous solutions in the absence (1,2) and in the presence (3–6) of the BPEI at $[N]_o = 0.08$ mol/l.

b) Influence of Spin Probe's Dimensions

The size of a spin probe markedly influences the efficiency of its diffusion in polymeric materials.[40,41] The analogous situation has been found for the PMC in liquid solutions. It is depicted in Figs. 10 and 11 that the replacement of the $R_6CH_2CH_2OH$ and $R_6CH_2COO^-$ radicals to the R_5CONH_2 (or $R_5^N CH_3$) and R_5COO^- leads to increase of the constant K_e value at 25–30% in the case of Cu(II) ions in aqueous solutions without additions of a complexing polymer.

The SET dependences presented in Fig. 10 show that a titration of the charged probes in the presence of P4VP broadens the EPR lines of a smaller one R_5COO^- at the same concentrations of Cu(II), than for a bigger probe $R_6CH_2COO^-$, being approximately 1.5 times more effective. One can assume that this effect occurs due to decrease of steric hindrances for diffusion of the smaller probes into a macromolecular coil.

This assumption is illustrated by comparison of the titration curves for two radicals: $R_6CH_2CH_2OH$ and R_5CONH_2. It is shown in Figs. 3 and 10 that while for the first one of the region of independency of ΔH on [Cu(II)] is well defined in the case of R_6OH, the EPR linewidth ΔH of the second probe exceeds the ΔH_0 value even at low concentrations of the copper ions. Although this broadening is not large, Fig. 10, it confirms effective collisions between R_5CONH_2 and the $Cu(Py)_4^{2+}$ complexes inside the P4VP coil.

Similar results has been obtained for polyamine macromolecules. The concentration region of Cu(II) ions, where a constant value of ΔH was observed, is the same for both

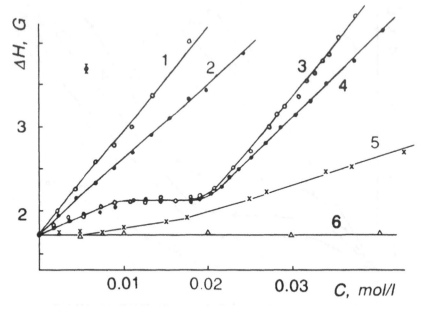

Figure 12   The EPR linewidth dependences of the R_6OH on the concentration of: 1,3–Cu(II); 2,4–Ni(II); 5–Co(II); 6–Zn(II) in aqueous solutions in the absence (1,2) and in the presence (3–6) of the BPEI at 25°C $[N]_0 = 0.08$ mol/l.

the probes R_6OH and $R_5^N CH_3$, Fig. 11. Nevertheless, the value of broadening $\delta H = \Delta H - \Delta H_0$, due to spin exchange, is higher in the case of $R_5^N CH_3$, a portion of the Cu(II) complexes available for collisions with a smaller probe is larger than for R_6OH.

It seems that a smaller probe can intercolate into areas of the metal-coordinated polymer coil, where the larger particles are not able to enter. Thus, it is better to use radical probes as small in size as possible to obtain as much information about the structure and features of the PMC.

c)   Influence of the Paramagnetic Ion

From EPR theory,[11] a broadening agent can have a great influence on the value of the spin exchange rate constant K_e: the magnitude of its electron and nuclear spin, the spin-lattice relaxation time T_1, and the type (strong or weak) of spin exchange.

Table 1 shows that shortening of the T_1 results in a decrease of the K_e value by several orders of magnitude, one has to use 3–4 times higher concentration of the ions with a short T_1 to observe the effect equal to one for an ion with a long T_1. Fig. 12 illustrates this: it presents the dependences of the EPR linewidth of R_6OH on the concentration of Cu^{2+}, Ni^{2+}, Co^{2+} at equal content of BPEI. Spin exchange of these ions with nitroxide radicals was shown to be strong.[11]

The concentration dependences of ΔH, Fig. 12 in the case of Cu(II) and Ni(II) coincide in the limits of accuracy of the experiment up to the appearance of free metal ions in solution. Using $Co(NO_3)_2$ as a broadening agent, one can observe a monotonous increase of ΔH, rather insignificant in the region of the PMC's forming. Such an unexpected dependence is explained by the small value of the spin exchange rate constant K_e for the pair R-Co(II), because the high spin complexes of Co(II) possess a very fast relaxation of the electron spin ($T_1 < 10^{-12}$ s).

Use of the SET method, one has to know to what type of spin exchange (strong or weak) the measured K_e constant corresponds to. This question is of principle importance, because in the case of weak exchange (for example, using the VO^{2+} aquoion[17]) the value of the spin exchange rate constant K_e is not a quantitative parameter of the frequency of collisions between the spin probe and ion, but also depends on several other factors: the energy of exchange interaction during collisions J_c; the collision time τ_c; the electron spin relaxation time T_1, etc.[11] Often these parameters are not precisely known.

The simple criterion to distinguish the cases of strong and weak exchange is based on measuring of K_e values at various viscosities, which may be varied in different ways: by changing the temperature T, solvent composition, or external pressure.[11] For the case of strong exchange the following equation has been found:[42]

$$K_e \sim T/\eta \tag{3}$$

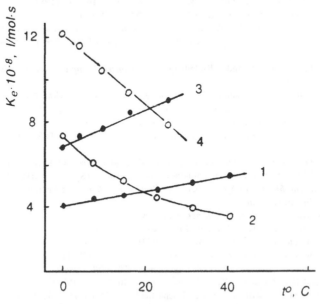

Figure 13   Temperature variation of the parameters K_e (1,3), $K_e\eta$ (2,4) and $K_e\eta/T$ (5) for the VOSO$_4$ + VOSO$_4$ (1,2), R$_6$OH + VOSO$_4$ (3,4), and R$_6$OH + Cu(en)$_2^{2+}$ (5) systems in aqueous solutions at [R] = 0.002 mol/l.

For weak exchange, depending on the relation between τ_c and T_1:

$$K_e \sim \eta/T \ (\tau_c < T_1), \ \text{or} \ K_e \sim T_1 \ (\tau_c > T_1) \tag{4}$$

At the constant temperature the criterion for strong exchange should be: $K_e \cdot \eta = \text{const.}$[11,18]

The temperature dependences of K_e and $K_e \cdot \eta/T$ for spin exchange in the systems $VOSO_4$ + $VOSO_4$ and R_6OH + $VOSO_4$ in aqueous solution are depicted in Fig. 13. With increase of temperature the K_e value increases slightly, in both cases, but the parameter $K_e \cdot \eta/T$ (which characterizes the spin exchange rate at the viscosity $\eta = 1$ cP) decreases 1.5–2 times. Strong exchange is valid, for spin exchange between R_6OH and $Cu(en)_2^{2+}$, Fig. 13 and between nitroxide radicals themselves (Table 3). In these cases diffusion collisions are the limiting stage of spin exchange.

ACKNOWLEDGEMENTS

This work was in part supported by Deutsche Forschungsgemeinschaft (436 RUS-17/108/93). The author thanks Yu.E. Kirsh, A.A. Yarslavov and N.M. Kabanov for providing the polymers used in the work, A.B. Shapiro and L.B. Volodarskii for supplying the spin probes, and A.N. Druzhinina for assistance with some of the experiments.

REFERENCES

1. M. Kaneko and E. Tsuchida, *J. Polymer Sci., Macromol. Rev.*, **16**, 397 (1981).
2. A. Skorobogaty and T.D. Smith, *Coord. Chem. Rev.*, **53**, 55 (1984).
3. F.R. Hartly, Supported *Metal Complexes: A New Generation of Catalysts* (Reidel, Dordrecht, 1985).
4. K.M. Saldadze and V.D. Kopylova-Valova, *Complex-Forming Ion Exchangers* (Khimiya, Moscow, 1980), [in Russian].
5. E.A. Bekturov, L.A. Bimendina and S. Kudaibergenov, *Polymeric Complexes and Catalysts* (Nauka, Alma-Ata, 1982), [in Russian].
6. E.A. Bekturov and S. Kudaibergenov, *Catalysis by Polymers* (Nauka, Alma-Ata, 1988), [in Russian].
7. A.D. Pomogailo, *Metal Complex Catalysts Immobilized on Polymers* (Nauka, Moscow, 1988), [in Russian].
8. A.I. Kokorin and A.A. Shubin, *Sov. J. Chem. Phys.*, **7**, 1770 (1991).
9. A.I. Kokorin, *Structure of Metal Complexes with Macromolecular Ligands*, D.Sc. Thesis, Moscow (1992), [in Russian].
10. A. Abragam, *The Principles of Nuclear Magnetism* (Clarendon Press, Oxford, 1961).
11. Yu.N. Molin, K.M. Salikhov and K.I. Zamaraev, *Spin Exchange* (Springer-Verlag, Berlin, 1980).
12. A.I. Kokorin, in *The Method of Spin Labels and Probes, Problems and Perspectives* (Nauka, Moscow, 1986). pp. 61–79. [in Russian].
13. G.I. Skubnevskaya and Yu.N. Molin, *Kinet. Katal.*, **8**, 1192 (1967).
14. D. Fiat, Z. Luz and B.L. Silver, *J. Chem. Phys.*, **49**, 1376 (1968).
15. K.I. Zamaraev, Doctor of Sci. Thesis, Moscow (1972), [in Russian].
16. Yu.I. Naberukhin and V.A. Rogov, *Russ. Chem. Rev.*, **40**, No. 3 (1971).
17. K.I. Zamaraev and A.I. Kokorin, *Russ. J. Phys. Chem.*, **50**, 1700 (1976).
18. A.I. Kokorin and K.I. Zamaraev, *Russ. J. Phys. Chem.*, **46**, No. 7 (1972).
19. A.L. Kovarsky, A.M. Wasserman and A.L. Buchachenko, *J. Magn. Res.*, **7**, 225 (1972).
20. A.M. Wasserman, G.V. Korolev, A.L. Kovarsky, *et al.*, *Izv. Akad. Nauk SSSR, Ser. Khim.*, No. 2, 322 (1973).
21. Yu.E. Kirsh, V.Ya. Kovner, A.I. Kokorin, *et al.*, *Europ. Polym. J.*, **10**, 671 (1974).
22. A.I. Kokorin, N.A. Vengerova, Yu.E. Kirsh and K.I. Zamaraev, *Dokl. Akad. Nauk SSSR*, **202**, 597 (1972).

23. A.I. Kokorin, A.A. Yaroslavov and V.A. Kabanov, *Vysokomolek. Soed.*, **B28**, 294 (1986).
24. A.I. Kokorin, K.I. Zamaraev, V.Ya. Kovner, Yu.E. Kirsh and V.A. Kabanov, *Europ. Polym. J.*, **11**, 719 (1975).
25. N.M. Kabanov, A.I. Kokorin, V.B. Rogacheva and A.B. Zezin, *Polymer Sci. USSR*, **21**, 230 (1979).
26. A.I. Kokorin, S.V. Lymar' and V.N. Parmon, *ibid.*, **23**, 2209 (1981).
27. G.I. Skubnevskaya, E.E. Zayev, R.I. Zusman and Yu.N. Molin, *Dokl. Akad. Nauk SSSR*, **170**, 386 (1966).
28. G.I. Skubnevskaya and Yu.N. Molin, *Kinet. Katal.*, **13**, 1383 (1972).
29. O.A. Anisimiv, A.T. Nikitaev, K.I. Zamaraev and Yu.N. Molin, *Teor. Eksp. Khim.*, **7**, 682 (1971).
30. Yu.B. Grebenshchikov, G.V. Ponomarev, R.P. Evstigneeva and G.I. Likhtenshtein, *Biofizika*, **17**, 910 (1972).
31. Yu.B. Grebenshchikov, Candidate of Sci. Thesis, Moscow (1973).
32. G.I. Skubnevskaya, K.M. Salikhov, L.M. Smirnova and Yu.N. Molin, *Kinet. Katal.*, **11**, 888 (1970).
33. A.I. Kokorin and A.A. Shubin, *J. Struct. Chem.*, **30**, 260 (1989).
34. A.I. Kokorin, L.S. Molochnikov, I.V. Yakovleva, A.B. Shapiro and P.A. Gembitskii, *Polymer Sci. USSR*, **31**, 597 (1989).
35. S.G. Entelis and R.P. Tiger, *The Kinetics of Reactions in Liquid State* (Khimiya, Moscow, 1973), [in Russian].
36. A.M. North, *The Collision Theory of Chemical Reaction in Liquids* (Methen Co., Ltd., New York, 1964).
37. Yu.B. Grebenshchikov, G.I. Likhtenshtein, V.P. Ivanov and E.G. Rozantsev, *Molek. Biol.*, **6**, 498 (1972).
38. N.H. Agnew, *J. Polymer Sci.*, *Pt. A-1*, **14**, 2819 (1976).
39. P.A. Gembitskii, D.S. Zhuk and V.A. Kargin, *Polyethylenimine* (Nauka, Moscow, 1971), [in Russian].
40. A.L. Buchachenko and A.M. Wasserman, *The Stable Radicals* (Khimiya, Moscow, 1973), [in Russian].
41. A.M. Wasserman and A.L. Kovarskii, *Spin Labels and Probes in Physical Chemistry of Polymers* (Nauka, Moscow, 1986).
42. G.I. Skubnevskaya, Candidate of Sci. Thesis, Novosibirsk (1972).

Kinetics and Mechanism of the Inhibited Polymerization of Vinyl Monomers

M.D. GOLDFEIN, G.P. GLADYSHEV and A.V. TRUBNIKOV

Institute of Mechanics and Physics, Laboratory of Chemical Physics, Saratov University, 112-A Bolshaia Kazachia Str., 410071, Saratov City, Russia

One of the main factors influencing the mechanism of radical polymerization is the presence of inhibitors.[1] Inhibitors can be used as a tool for studying the polymerization mechanism. Inhibitors prevent undesirable formation of high-molecular products during synthesis, purification and storage of monomers. The most widely used inhibitors for radical polymerization of vinyl monomers include oxygen, quinones, phenols, aromatic amines, nitro compounds, salts of transition metals. The polymerization inhibition is connected with interaction of the active radicals with an ineffective (weak) inhibitor which results in decrease of active centers concentration or in formation of particles which can not propagate the molecular or kinetic chains. In the presence of high-effective (strong) inhibitors the reaction is accompanied by an induction period during which there is no polymerization. After completion of the induction period stationary kinetics is established and the reaction rate coincides with the rate of the non-inhibited polymerization or is smaller. The induction period depends on the inhibitor initial concentration X, and is determined by the structure of the monomer, the mechanism of initiation and nature of the initiator (azo compound, peroxide, etc.).[2-6] Often a linear relationship between the induction period and inhibitor concentration is observed.[6]

1 EQUATION OF BAGDASARJAN-BAMFORD

The simplest scheme for radical-chain polymerization in the presence of an inhibitor may be written as follows:[3,6]

$$I \xrightarrow{K_d} 2R_0$$

$$R_0 + M \xrightarrow{K_i} R$$

$$R + M \xrightarrow{K_p} R$$

$$R + R \xrightarrow{K_t} \text{polymer}$$

$$R + X \xrightarrow{K_x} \text{polymer}$$

163

where I, M and X are the initiator, monomer and inhibitor, respectively and R_0 and R the initiation and polymer radicals, K_d, K_i K_p, K_t, K_X are the reaction constants of the corresponding elementary reactions.

Using stationary state condition during the induction period:

$$W_i = K_t[R]^2 + K_X[R][X] \tag{1}$$

where W_i is the initiation rate and rate for inhibitor consumption

$$-\frac{d[X]}{dt} = K_X[R][X] \tag{2}$$

The Bagdasarjan-Bamford equation can be derived

$$F_1 \equiv -\frac{1}{\varphi} + \ln\frac{1+\varphi}{1-\varphi} = \frac{K_X W_{max}}{K_p[M]}t + A \tag{3}$$

using the stationary polymerization rate W/W_{max} where A is the constant of integration.

Equation (3) characterizes the polymerization kinetics in the presence of the "ideal" inhibitor.[2,7] The Bagdasarjan-Bamford equation can only be used for inhibited polymerization during which total process does not change with time. The function is linear with time and the inhibition constant can be found from its slope.[6] Secondly, using the slope we can derive the ratio

$$W_i = \frac{\mu[X]_0}{\tau} \tag{4}$$

which is the theoretical basis for the inhibition method for accurate calculation of the initiation rate.

2   EQUATION DESCRIBING KINETICS OF THE INHIBITED POLYMERIZATION

Dependence of the polymerization rate on time may be influenced by the reaction of inhibitor molecules.[8,9] Interaction of inhibitor with monomers resulting in formation of active centers the stationarity relation (1) changes:

$$W_i + K_i[X][M] = K_X[R][X] + K_t[R]^2 \tag{5}$$

where K_i is the reaction constant for initiation and the inhibitor consumption follow the equation:

$$-\frac{d[X]}{dt} = K_i[X][M] = K_X[R][X] \tag{6}$$

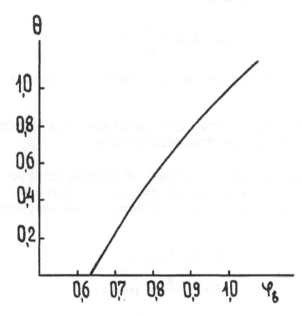

Figure 1   The dependence of parameter θ and φ_B according to Eq. (9).[9]

After completion of the induction period, the reaction proceeds with a stationary rate W_∞ and

$$W_i = K_t[R]^2 \tag{7}$$

In these cases the kinetics of inhibited polymerization are described as follows:

$$F_2 \equiv -2\theta \ln (\varphi + \theta) - (1 - \theta) \ln (1 - \varphi) + (1 + \theta) \ln (1 + \varphi)$$

$$+ \frac{1/\theta - \theta}{2} \ln \left| \frac{\varphi - \theta}{\varphi + \theta} \right| = \frac{K_x W_\infty}{K_p[M]} (1 - \theta^2)t + B \tag{8}$$

and from the inhibition constant K_X/K_p can be found if the parameter θ equalling to $K_i[M]/K_X[R]_\infty$ is known. Its value is calculated from the rate φ at the inflection point φ_B of the S-shaped relation for the inhibited polymerization with time. If φ value obeys Eq. (8) it follows from the inflection point that:

$$\theta^3 + \frac{2\varphi_B}{1 + \varphi_B^2} \theta^2 - \frac{2\varphi_B^4 - \varphi_B^2}{1 + \varphi_B^2} \theta + \frac{\varphi_B^5 + 2\varphi_B^3 - \varphi_B}{1 + \varphi_B^2} = 0 \tag{9}$$

If the inhibited polymerization kinetics are described by the Bagdasarjan-Bamford equation, then $\varphi_B = 0.643$. With addition of initiation φ_B changes from 0.643 to 1.0. In Fig. 1 the relationship between θ and φ_B calculated from Eq. (9) is shown.

Equation (8) can be arranged as:

$$F_2 \equiv K_i[M]\left(\frac{1}{\theta} - \theta\right)t + B$$

which allows determination of the additional initiation rate constant K_i.

3 KINETIC EQUATION TAKING INTO ACCOUNT THE SECONDARY INHIBITION EFFECT ON POLYMERIZATION

A high-effective inhibitor can cause not only an induction period but can reduce the reaction rate with increase of the inhibitor concentration.[6,10] Fall in the concentration of the propagating chains is usually connected with their interaction with the primary inhibition products Y:

$$R + X \xrightarrow{\ K_X\ } Y$$

$$R + Y \xrightarrow{\ K_Y\ } \text{polymer}.$$

For the stationary process during the induction period

$$W_i = K_X[R][X] + K_Y[R][Y] + K_t[R]^2$$

In this case the inhibitor consumption is characterized by Eq. (2). Bearing in mind the relation between material balance $[Y] = [X]_0 - [X]$ and the stationary state equations for the non-inhibited polymerization:

$$W_i = K_t[R]^2 \tag{10}$$

after completion of the induction period

$$W_i = K_t[R]_\infty + \alpha K_Y[R]_\infty[X]_0 \tag{11}$$

where α is the stoichiometric coefficient for secondary inhibition,

$$F_3 \equiv -\frac{1}{\varphi} - \ln(1-\varphi) + (1-\varphi_\infty^2)\ln\varphi + \varphi_\infty^2\ln\left(\varphi + \frac{1}{\varphi_\infty^2}\right) = \frac{K_X W_\infty}{K_p[M]}t + C \tag{12}$$

In the latter equation φ is the ratio of the polymerization rate at any moment in time, W_t the stationary rate of the inhibited reaction, W_∞ and φ_∞ the ratio of W_∞ to the stationary rate of the non-inhibited reaction W.

The primary inhibition constant K_X/K_p is determined from the slope of the linear dependence F_3 on time. It is evident that the inhibition product Y does not produce any inhibiting influence on the following polymerization ($W_\infty = W$ and $\varphi_\infty = 1$), then Eq. (12) transforms into Eq. (3).

Value of the secondary inhibition rate constant K_Y is derived from the ratio obtained from calculation of polymerization scheme with a low-effective inhibitor[3,6] and from analysing the non-stationary Eq. (11):

$$K_Y = (1 - \varphi^2) K_t W / \alpha \varphi K_p [M][X]_o$$

4 REFINEMENT OF THE KICE EQUATION USED FOR SOLVING THE INVERSE KINETIC PROBLEM FOR INHIBITED POLYMERIZATION

The main Kice assumptions[14] relates to two conditions: the inhibitor does not take part in the chain initiation and the rate constant of the chain termination reaction does not depend upon the presence of inhibitor. The reduced rate φ is determined by the ratio of concentration of the propagating chains with $[R]_X$ inhibitor and without $[R]$ inhibitor:[11]

$$\varphi = [R]_X / [R]$$

Usual scheme of radical polymerization with an inhibitor has been expanded with the effect of initiation reaction on the inhibitor

$$R_0 + X \xrightarrow{K_{iX}} R_X$$

chain regeneration

$$R_X + M \xrightarrow{K_r} R$$

and cross chain termination

$$R + R_X \xrightarrow{K_{tX}} \text{non-active product.}$$

with K_{iX}, K_r and K_{tX} are the rate constant of the appropriate reactions.

If $\varphi^2/(1 - \varphi^2) >> K_r K_t / K_p K_{tX}$ which is always true for the typical inhibitors (K_r/K_p $<< 1$, $\varphi \geqslant 0$, 1), then kinetics of the inhibited polymerization will be described by the following equation:

$$A\left(\chi + \frac{2\varepsilon}{\varphi}\right) = C + B \tag{13}$$

with

$$\chi = K_X[X]/\{K_d f[I](K_t)_{X=o}\}^{0,5}$$

$$C = K_r K_t K_X[M][X]/\chi K_{tX}$$

$$\varepsilon = K_{iX}[X]/(K_{iX}[X] + K_{iM}[M])$$

where K_{iM} is the reaction rate constant $R_0 + M \rightarrow R$.

$$A = \frac{\theta}{2}\left(\frac{B}{\varphi} - 1\right)$$

$$B = \varphi\left(1 + \theta \frac{1 - \gamma\varphi^2}{\varphi^2}\right)^{0.5}$$

$$\gamma = (K_t)_X / (K_t)_{X=0}$$

$$\theta = 4(K_t)_{X=0} K_{tXX} / K_{tX}^2$$

where K_{tXX} is the reaction rate constant $R_X + R_X \rightarrow 0$.

With $\varepsilon = 0$ and $\gamma = 1$ Eq. (13) transforms into the Kice equation:

$$\chi A = C + B$$

where A and B depend only on φ and θ. Under the normal conditions of inhibition the rate φ is a function of χ, C and θ. Deviation from the kinetics described by Eq. (13) is observed when the φ values are small and the rate of chain regeneration is large. The relation $\theta \ll 1$ is observed for the inhibited polymerization reaction, and $B < 1$ for any value of φ and the parameter C of chain regeneration may take any values including $C \gg B$.

When the active radical R reacts with the inhibition product we have

$$\varphi = (C/\chi)^{0.5}$$

$$W_X = K_p(K_d f K_r / K_X K_{tX})^{0.5} [M]^{1.5} [I]^{0.5} [X]^{-0.5}$$

If chain termination is the result of inhibitor radicals interaction with each other, then

$$\varphi = 2C/\chi\theta^{0.5}$$

$$W_X = \{K_p K_r (K_d f / K_{tXX})^{0.5} / K_X\} [M]^2 [I]^{0.5} [X]^{-1}$$

Both expressions show that the inhibition itself does not increase the kinetic order in terms of the initiator.

The inhibition mechanism changes when stable radical and ions of transition metals are presented which do not take part in the initiation reaction and do not contribute to the chain regeneration. The inhibition stage is expressed as:

$$R + X \xrightarrow{\ K_X\ } \text{non-active products}$$

$$R + M^{2+} L_2^- \xrightarrow{\ K_X\ } M^{3+} L_2^- R^-$$

$$R + M^{3+} L_3^- \xrightarrow{\ K_X\ } M^{2+} L_2^- RL \quad (L\text{-ligand})$$

For these conditions the rate constant of the inhibition reaction may be found from:

$$K_X = \frac{2(1 - \varphi^2)(K_d f [I] K_t)^{0.5}}{\varphi[X]}$$

This equation determines the rate constant of reaction of inhibitor interaction with polymer radical K_Y under conditions of weak inhibition in which there is no induction period and only a decrease of the polymerization rate is observed. However the stable radicals and salts of transition metals are usually the effective inhibitors of the radical polymerization. When they are used the polymerization rate and consequently φ depends upon time and during the induction period they are equal to zero.

5   INHIBITORS AND MECHANISM OF THE INHIBITION REACTIONS UNDER THE RADICAL POLYMERIZATION CONDITIONS

First systematic review of data[2–5] and interesting recent surveys[15,16] have appeared. The problem of control by the gel-effect has been reviewed.[17] To inhibit the polymerization of acrylates and methyl acrylates, divinyl acetylene,[18] hexamethylene tetramine,[19] salts of dibutyl dithiocarbamic acid,[20] composition containing phosphoric acids and/or phosphorous oxide, cresol, benzaldehyde,[21] aromatic amines with different structure,[22–27] n-methoxyphenol[28] and other phenols,[29] combinations of the inhibitors of amine and phenol types,[30] compositions including nitroso group bonded with the carbon atom of the aromatic ring[31] etc. are proposed.

There are a small number of works devoted to the study of the relation between the structure of polycondensated aromatic hydrocarbons and their ability to inhibit the polymerization of vinyl monomers.[32–36] Aromatic hydrocarbons do not influence chemically initiated polymerization of methyl acrylate but they inhibit the photosensitised reaction reducing its rate but they do not produce an induction period. The inhibiting activity increases from: naphthalene, phenanthrene, anthracene, chrysene. The inhibiting activity of these substances is the result of their transition into the triplet state upon the influence of light.

Polymerization of vinyl acetate initiated by azo-bis-isobutyronitrile with aniline and its derivatives[37] the substituent's inhibition stoichiometry does not change but the rate constant of the inhibition reaction decreases for aniline, N-methyl aniline and N,N'-dimethyl aniline K_X equals to 17.6 and 0.7, respectively. When the substituent is introduced into the benzene ring (para- and metha positions) the inhibitor's activity is characterized by the Hammet equation. Ortho- and polysubstituted of aniline have lower reactivity.

Many quinoid compounds and their derivatives effectively inhibit radical polymerization. Inhibiting activity of quinones is determined by their chemical structure and redox potential as well as by the electron donors ability of the macro radicals.[38] Depending on all these factors recombination and disproportionation are possible:[3]

or formation of non-active radicals by means of an intermediate charge-transfer complex.[39,40]

The limiting stage of the active radical interaction with quinone[41] is transformation of the intermediate charge-transfer complex into the inhibiting radical. In the case with methacrylates this reaction proceeds slowly and can be accelerated by adding electron-donor compounds, for example, phenothiazine.

Taking the methyl acrylate polymerization with chloroanil as an example, the mechanism of inhibition by means of quinone can be described as:

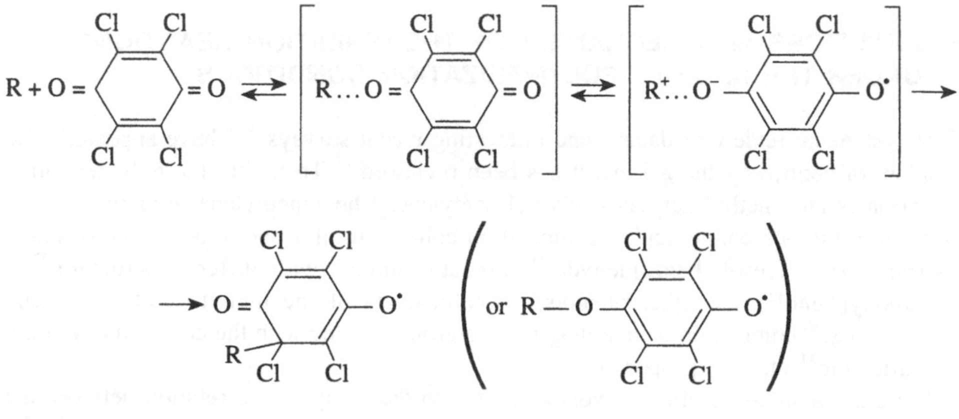

In all cases regardless of the mechanism of quinone interaction with the propagating radicals the stoichiometric inhibition coefficient is approaching 2.

Polymerization with nitro- and nitroso derivatives is unusually accompanied by an induction period and reaction inhibition.[16] Comparing quinone to nitro compounds the decrease in the reaction rate is less because their interaction with macro radicals is difficult. Their inhibiting ability is determined by monomer structure and the nitro groups growth increases the inhibiting ability. The mechanism of polymerization indicates that with these substances $\mu \approx 2$ [16,42,45]

$$R + R'NO_2 \longrightarrow R' - O - R + NO^{\bullet}$$

or

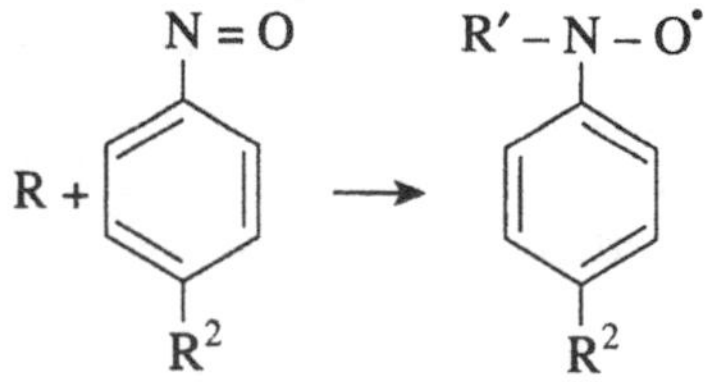

and further

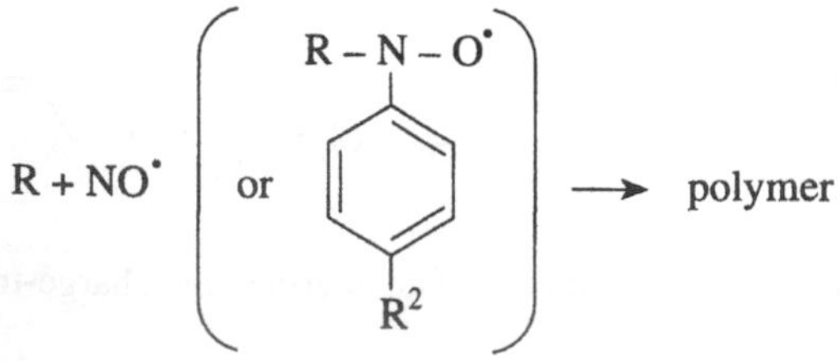

Nitro compounds can react with the propagating chains, as well as with initiator radicals.[46] The nitro compounds may decompose with formation of nitric oxide[47]

The inhibition of n-nitroso dimethyl aniline and n-nitroso dimethylamine in the process of styrene, acrylonitrile and methyl acrylate depends on the monomer concentration.[48] In acrylonitrile-DMF-n-nitrosodiphenylamine it increases from 0.86 to 1.8 with decreasing monomer mole fraction from 1 to 0.3 μM and in methyl acrylate-DMF-n-nitroso diphenylamine it increases from 1.54 to 1.94 with decreasing monomer mole fraction from 1 to 0.5 μM. Tüdes explains changing of the stoichiometric coefficient based on the hot radicals theory.[48]

Inhibition of nitroso compounds are closely connected with the problems of monomer releasing stabilizer before being processed into polymer. Optimal solution of the problem is the stabilizer in which the inhibitor remains active during storage but does not influence the polymerization. As an example of such an inhibitor is the radical-anion complexed TCNQ.[49,50]

Another possible variant is use of inhibitors with poor solubility in water. When water is stored with small quantities of dissolved inhibitor effective stabilization may be achieved. However the same inhibitor concentration does not produce significant influence on the polymerization and properties of the polymers formated. The results achieved in the process of polymerization with cupferronate addition (ammonium salt, N-nitroso-N-phenyl-hydroxylamine and its derivatives)[51] corresponds exactly to this situation. Cupferronate of sodium, potassium and copper effectively inhibits the polymerization of acrylate and methyl acrylate in the dimethylformamide solution. The duration of the induction period increases linearly with the additive concentration. Unlike the styrene polymerization which has a stoichiometric inhibition coefficient equal to 2,[16] the methacrylate polymerization has $\mu = 1.5$. In these cases, interaction of cupferronates with monomer is found which results in free radicals formation and μ decreasing.

Polymerization of methyl acrylate, butyl acrylate and methyl methacrylate, the induction period depends non-linearly on the cupferronate concentration, the rate of inhibition is lesser than for the non-inhibited reaction. These salts inhibit polymerization in air as well as under oxygen-free conditions. Differences in the influences of different cupferronates on polymerization is caused by their different solubility. Copper cupferronate has the highest solubility among acrylic monomers, sodium salt has the lowest solubility.[51]

The interaction of cupferronate with methacrylates result in their deactivation and this is one of the causes of the non-linearity of the concentration dependence of the induction period. Gradual decomposition of cupferronates takes place with time. Cupferronates may be arranged in the following row according to their stability: copper cupferronate, sodium, potassium, ammonium. Their degree of deactivation depends on the monomer structure, the highest is in methyl methacrylate.

At some stage of the monomer synthesis, the reaction mixtures contain quite a large amount of water. Existence of water in monomers influences the cupferronate solubility and consequently their inhibition activity. Water content of a monomer influences much more the induction period for polymerization, the higher the cupferronate solubility in monomer, the higher is the water solubility in it.

Metal salts have a different influence on the kinetics and mechanism of radical polymerization of vinyl monomers.[2-4] Concentration inversion of the catalytic properties is typical for ferric and copper chloride.[52,53] The inhibiting action of the fatty acid salts depends on the characteristics of the metals included in them, the reaction medium and the anion size.[54] Inhibition by zinc, cobalt and lead stearates on the styrene and methyl acrylate polymerization initiated by benzoyl peroxide has been studied.[55,56] The inhibition mechanism consists of interaction of free radicals with the intermediate complex formed between styrene and peroxide. Chlorides of transition metals and fatty acid salts are not widely used in practice.

Ferric and copper chlorides have a medium influence on the inhibition of vinyl monomers.[57] Inhibition of acrylic nitrile polymerization in water solution by rhodanide sodium, the inhibition constant increases by two-three orders compared to polymerization in solution of dimethylformamide.[58]

Effective inhibitors for acrylic monomers are metal salts of dialkyl dithiocarbamic acid. These compounds easily accept both peroxides and alkyl radicals. The inhibiting mechanism is complicated in the presence of hydroperoxides because the salt interacts with the active radical or hydroperoxide resulting in the formation of a paramagnetic compound having an increased inhibiting property.[59] The inhibition activity increases in the row[60]

$$Zn(S_2CNR_2) < Ni(S_2CNR_2)_2 << Cu(S_2CNR_2)$$

Some authors[61] consider that the weakest place for the attack of the salt molecules by the active radicals is the sulphur atoms. Others suppose[62,63] that interaction with metal ions provides electron transfer from the ligand to the radical due to existence of covalent bond between metal and sulphur atoms.[64]

Polymerization of acrylic acid in water, acetic acid and isopropylacetate solution indicate copper diethyl dithiocarbamate has good inhibiting activity.[65] Dependence of the induction period on the salt concentration is linear and values of the stoichiometric inhibition coefficient exceed 1 and depend on solvent type. The highest inhibiting activity of copper diethyl dithiocarbamate is observed in water and is lowest in acetic acid solution. Values of μ obtained in acrylic acid polymerization in air and oxygen-free conditions coincide perfectly. This indicates a similar nature for the stoichiometry of copper diethyl dithiocarbamate interaction with alkyl and peroxide radicals. The inhibition mechanism is evidently similar to the mechanism of dithiophosphate oxidation[66] and it is accomplished by electron transfer to the active radicals. The increased value of μ (>2) may be the result of radicals interaction with two atoms of sulphur.

The acetic acid polymerization rate in organic mediums decreases drastically in comparison to polymerization in water. It is dictated by complexation of the acetic acid molecules and isopropylacetate with the polymer radicals.[67,68] Interaction of these solvents with the propagating chains is the cause of decreasing μ value in the polymerization of acetic acid in acetic acid and isopropylacetate. For this systems K_X/K_p and the secondary $\alpha K_Y/K_p$ inhibitions were determined from Eqs. (14) and (15), respectively.

Mixtures of chloroanil with phenothiazine have proved to be good inhibition systems.[69] Unlike the styrene polymerization in which chloroanil is a strong inhibitor[70] a slight inhibition is observed in methyl acrylate and chloroanil does not influence the methyl acrylate polymerization. At the same time addition of phenothiazine or oligoester product of its oxidative condensation gives a synergistic effect with a clearly defined induction period.[71,72] The limiting stage of inhibition is interaction of the propagating chain not with the phenothiazine. . . chloroanil complex but with the chloroanil molecule.[69] Drastic increase of K_X approximately by two orders in the presence of phenothiazine or oligoester product having oxidative condensation is explained by the mechanism described above.[41]

The interaction of phenothiazine with the charge-transfer complex results in formation of a phenothiazine radical:

$$R...CA + \text{(phenothiazine)} \longrightarrow R-O-\text{(tetrachloro ring)}-OH + \text{(phenothiazine radical)}$$

The phenothiazine radicals can interact with R, M and XA with formation of a new radical which can terminate one or more chains. However the stoichiometric inhibition coefficient is less than 2 caused by chain regeneration with phenothiazine interaction with methyl acrylate. In methyl acrylate polymerization, the inhibiting activity of the chloranil-phenothiazine system is noticeably less. In this case chloranil reacts not only with the active radicals but also with the charge-transfer complex. So the decisive role in the inhibition mechanism belongs to the monomer structure and donor-acceptor properties of the monomers and propagating radicals.

Studying the inhibition kinetics of vinyl monomers in the presence of a binary system[73,74] shows that mixtures of chloroanil with triethylamine and mixtures of anthracene with phenothiazine and oligophenothiazine inhibit radical polymerization of methyl acrylate. Existence of a critical concentration of inhibitor and initiator is the result of the inhibiting activity of N,N-diethylamine vinyltrichrorine-n-benzoquinone formed in the process of charge-transfer complex transformation into a radical pair. The synergistic effects are explained using radical chain reactions and inhibitor consumption.

Stable radicals cannot recombine with each other and are effective inhibitors of the vinyl monomer polymerization. Studies on hydraziles, veraziles and imidazolidine and some radical-anions have been shown these to be spin traps.[50]

Nitroxyl radicals of imidazoline and imidazolidine row are very interesting from theoretical and practical point of view.[75,76] They belong to the five-unit compound class of the heterocyclic free radicals containing nitrogen:

Compared to the above nitroxyles based on imidazoline and imidazolidine have a number of advantages. The additional atom of nitrogen or of N-oxide group in combination with the functional group allows complexation and cyclic metallizing without the radical center being used. Existence of imines or imine-N-oxide groups gives a much higher stability in acid medium and is an important factor when they are used as inhibitors of polymerization of methyl acrylate. Imidazoline-N-oxides have the largest inhibiting activity as the agents terminating the oxidizing chains and traps for short-lived radicals.

$$R^2 \equiv (CH_3)_2$$

According to the above scheme the stoichiometric inhibition coefficient should exceed 1.

The radical 2,2,5,5,-tetramethyl-4-phenyl imidazoline-3-oxide-1-oxyl (TMPHIZ) inhibits polymerization of styrene and methyl acrylates.[77,78] Dependence of the induction period on the initial concentration of the stable radical is linear and values of the stoichiometric inhibition coefficient for styrene $\mu = 1.15$ and methyl acrylate $\mu = 1.25$, do not depend on the initiator type azo or peroxide compounds, for iminoxyles and arylnitroxyles.[50]

TMPHIZ strongly inhibits methyl acrylate polymerization in the presence of both concentrated and diluted sulfuric acid. The relationship between the stationary rate and TMPHIZ concentration is determined by the concentration of sulfuric acid in water.

The inhibiting activity of imidazoline nitroxyl in acid medium allows use of TMPHIZ for determination of the initiation rate by sulfuric acid and then determination of the rate constants for reactions of chain propagation and termination corresponding to different acid content in water.[78]

Inhibition of methyl acrylate thermal polymerization by sulfuric acid has a more non-linear relationship between the induction period and TMPHIZ concentration. Some authors[79,80] suppose that thermal polymerization is initiated by triplet radicals and all radicals react with the inhibitor. Other authors pay attention to possibility of inhibitor interaction with monomers[81–83] and reaction between inhibitor and initiation rate.[84] A detailed analysis for styrene[85] divides the inhibitors into two groups; reacting and non-reacting with the intermediate product having biradical nature in thermal initiation.

6 POLYMER STABILIZERS AS EFFECTIVE INHIBITORS OF RADICAL POLYMERIZATION

6.1 Inhibition of Thermal-Oxidative Polymerization

Aromatic amines and phenols are often united into a single group of inhibitors in spite

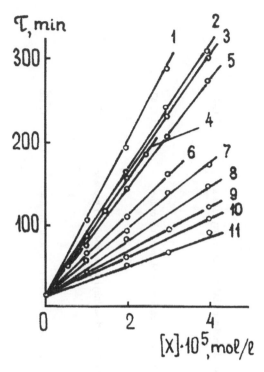

Figure 2   Induction period of methylmethacrylate polymerization as a function of inhibitors concentration (60°C, [AIBN] = 10^{-3} mol/l): 1–C-5, 2–C-1, 3–C-47, 4–C-49, 5–C-41, 6–C-789, 7–C-933, 8–C-875, 9–ON, 10–HQ, 11–PT.[94]

of their different structures. If there is oxygen in these compounds they effectively inhibit radical polymerization, however if there is no oxygen these compounds demonstrate properties of weak inhibitors.[17] Amines as well as phenols are inclined to hydrogen bond with peroxide radicals and contribute to the increased inhibition rate.

In many cases the inhibition is hydrogen atom detachment from the active radical

$$RO_2 + R^1 - NH - R^2 \longrightarrow ROOH + R^1 - N^{\cdot} - R^2,$$

$$R(or\ RO_2) + R^1 - N^{\cdot} - R^2 \longrightarrow polymer$$

$$R^1 - N^{\cdot} - R^2 + R^1 - N^{\cdot} - R^2 \longrightarrow non\text{-}active\ product$$

In this case the high rate of RO_2 interaction with amine is caused by formation of the polar activated complexes. $ROO^-\ldots H^+\ldots N^- <$ than by the low bond N—H energy. The inhibiting activity of the aromatic amines depends on the substituents' polarity in the benzene ring; electropositive substituents increases the inhibiting effect and electronegative substituents decrease it.

New high-effective inhibitors for vinyl monomers in the presence of oxygen belonging to aromatic amines;[87–94] have been found. These compounds in Table 1 are easily dissolved in many organic solvents, acrylate and methacrylate monomers and concentrated sulfuric acid.

TABLE 1

New effective inhibitors of vinyl monomers polymerization[87-94]

Structural formula	Chemical name	Technical name
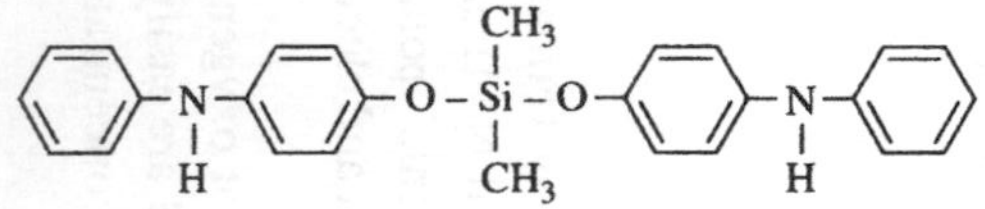	dimethyldi-(n-phenylamino phenoxy) silane	C-1
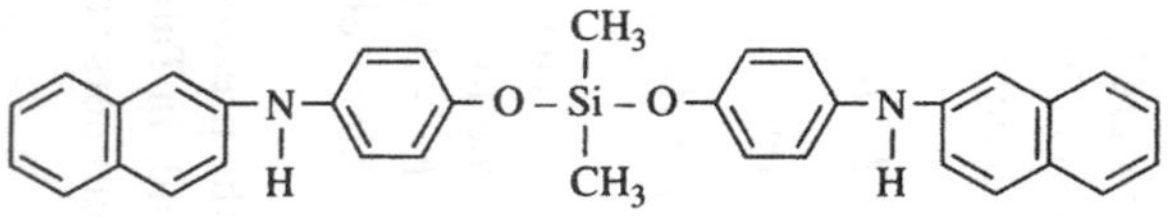	dimethyldi-(n,β-naphthyl aminophenoxy) silane	C-41
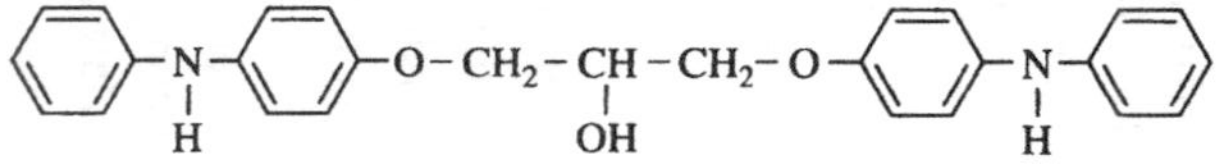	2-oxy-1,3-di(n-phenylamine phenoxy) propane	C-47
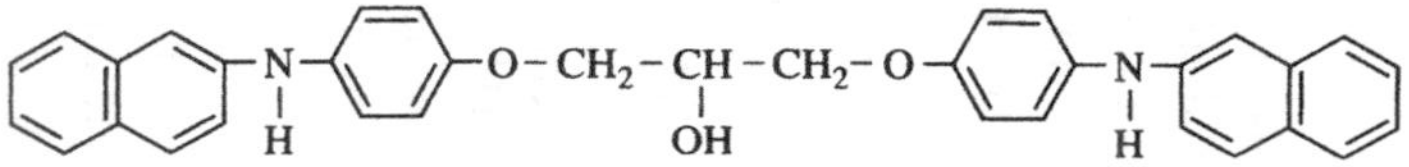	2-oxy-1,3-di-(n,β-naphthyl aminophenoxy) propane	C-49
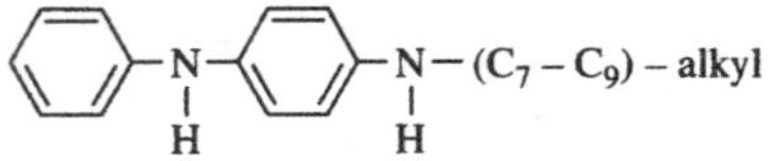	N-(C7-C9)alkyl-N-phenyl-n-phenylene diamine	C-789
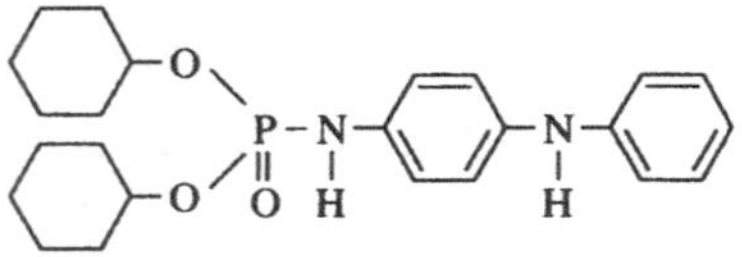	(n-phenylaminophenyl) amide diphenyl of phosphoric acid	C-875

TABLE 1 *Continued*

Structural formula	Chemical name	Technical name
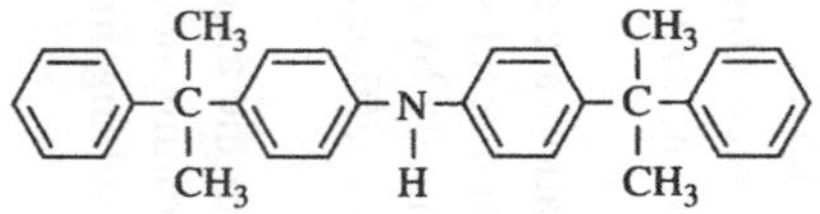	4,4'bis-(α,α'-dimethyl benzyl) diphenylamine	C-933
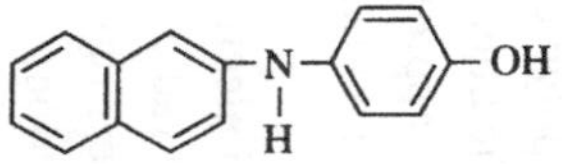	n-oxyphenyl-β-naphthylamine (oxyneozone)	ON
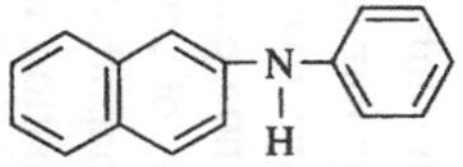	phenyl-β-naphthylamine	neozone-D
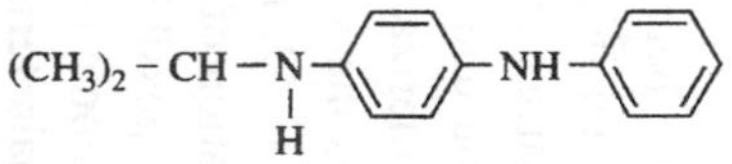	phenyl-n-isopropylamine	diaphenophenylamine
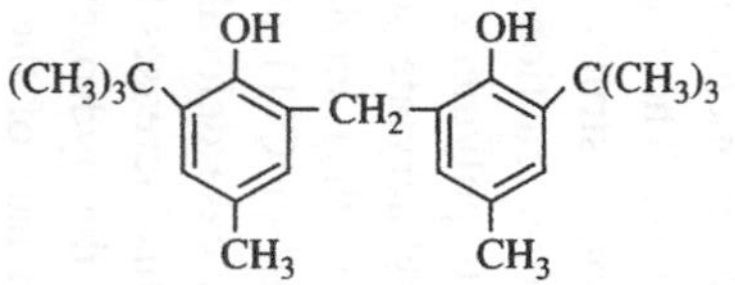	2,2'-methylen-bis-(4-methyl-6-tret butyl) phenol	22-46

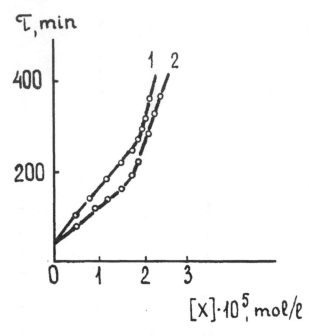

Figure 3   The dependence of induction period of methylmethacrylate thermopolymerization in the presence of air oxygen on C-I (1) and C-47 (2) concentrations at 100°C.[94]

The induction period increases linearly with concentration of amine for the oxidative polymerization of styrene, acrylic and methacrylic acids and their esters initiated by dinitrile of azo-isobutyric acid. All the substances exceed hydroquinone (the mostly widely used stabilizers at present) if the induction period is taken as the criterion of inhibition activity, Fig. 2. The transition region width from non-stationary to stationary kinetics is slightly dependent on the inhibitor structure. Under these oxygen-free conditions these substances do not practically influence methyl acrylate polymerization in the initial stages.

An important characteristic of inhibitor effectiveness is the stoichiometric coefficient μ. For bifunctional inhibitors such as C-1, C-41, C-47, C-49, $\mu = 4$ is expected. However in all cases, μ determined from Eq. 1 exceeds 4. Such differences in μ from the theoretical value is caused by amine regeneration due to interaction with the non-active inhibitor radical and the propagating chain not only by means of recombination but also by disproportionation. The relation between the two types of chain termination depends on both the monomer structure and inhibitor.

Recombination reactions during butyl acrylate polymerization are smaller than during methyl acrylate polymerization due to steric difficulties. This results in increase in the μ value in butyl acrylate polymerization inhibition approximately by two times. For both acrylates and methyl acrylates the stoichiometric inhibition coefficient is larger for C-1 and C-47 than for C-41 and C-49, respectively.

In the latter case delocalization of a non-paired electron in the aminol radical is more significant. This increases the possibility of active radicals joining not to a nitrogen atom but to one of the hydrogen atoms of naphthalene rings, resulting in an increase of the recombination part of the chain termination reaction.

Other polyfunctional aromatic amines type (C-789; C-875; C-933, etc.) have an effect on the polymerization, the μ value does not depend on the structure of inhibiting additives but it is determined by the monomer structure and initiator properties.[92] In these cases the active radicals interact with the NH-group located between the phenyl rings and a $\mu \geqslant 2$ indicates inhibitor regeneration.

Aromatic amines may be arranged according to their inhibiting activity. For the inhibited bulk oxidative polymerization of methyl acrylate hydroquinone and phenothiazine look like

C-5 > C-I > C-4I > C-47 > C-49 > C-789 > C-875 > C-933 > ON > HQ > PT

The inhibition activity of the above compounds are noticeably reduced for the polymerization of vinyl monomers by benzoyl peroxide. Interaction of the peroxide initiator with the amine molecule is indicated by a change of benzoyl peroxide solution colour. The benzoyl peroxide interaction mechanism with arylnitroxyl[92,95] is proposed to occur as follows:

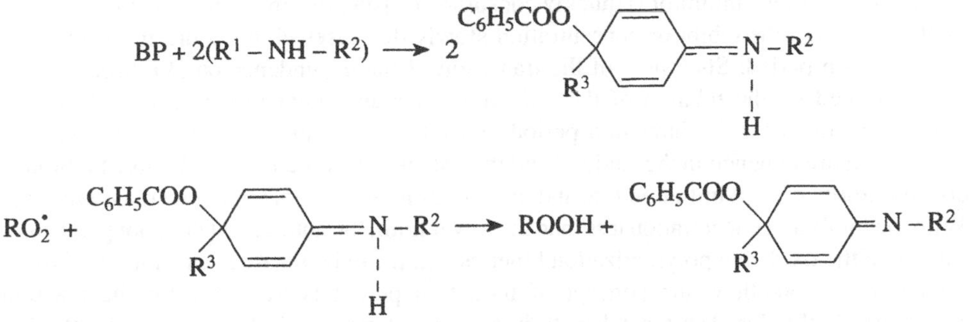

Under the conditions of synthesis, purification and storage of methacrylate monomers, spontaneous polymerization often results in formation of high-molecular products. Such polymerization usually proceeds at higher temperatures (80–150°C) in air and it has a thermal oxidative nature.

The main difference between the inhibited thermal polymerization of methacrylates as distinct from initiated one lies in the fact that the dependence of induction period at high temperatures on the inhibitor concentration has a break, Fig. 3. It follows that there is critical inhibitor concentration $[X]_{cr}$ above which a drastic increase of the induction period duration is observed. Values of $[X]_{cr}$ is greatly dependent not only on the temperature but also on the structure of monomer and inhibitor. For branching chain reactions especially for organic compounds many critical phenomena have been found, including a critical inhibitor concentration.[96,97]

The analogous effects may be observed in the process of inhibition of thermal-oxidative polymerization of vinyl monomers. The polymer peroxide decay giving nucleation of new reactive chains occurs during the induction period and it can be considered as the degenerated branch of the chain. It should be noted that the degenerated branches of the chain[98] have been ignored in all recent works on studying the thermal-oxidative polymerization inhibition.[99–103]

The first time these questions have been considered is[91,92] in the inhibiting activity of aromatic amines.

Kinetic analysis of the radical-chain scheme for oxidative polymerization in the presence of an inhibitor results in the equation describing the polyperoxide concentration change (P)

$$\frac{d[P]}{dt} = vW_i + \varphi[P] \tag{14}$$

where $v = K_p[M]/\mu K_X[X]$ is the length of the kinetic chain with the inhibitor, $\varphi = K(v\alpha - 1)$ is the autocorrelation factor; α is the effectiveness of the polyperoxide initiation.

The condition $\varphi = 0$ is achieved with the critical inhibitor concentration $[X]_{cr}$ and determined using the reaction rates of chain propagation, polymer peroxide decay and secondary reactions, for example, inhibitor regeneration

$$[X]_{cr} = \alpha K_p[M]/\mu K_X \tag{15}$$

With $[X] \leqslant [X]_{cr}$ the inhibitor is quickly consumed causing insignificant induction periods. With $[X] > [X]_{cr}$ the inhibitor concentration slowly decreases giving a drastic increase in the induction period. Sharpness of the transition of the dependence on $[X]$ through $[X]_{cr}$ is determined by the relation of the induction rates and polyperoxide decay (K_i/K_t).

The equations for the induction period and rate of oxidative polymerization with an antioxidant are obtained in the study[103] and the values of lower and upper limits of inhibitor concentration, critical initiation rate and critical monomer's concentration are determined. With the inhibitor concentration less than the lower limit, the antioxidant does not practically influence the oxidative polymerization kinetics and it can be regarded as an inert mixture. Under these conditions the concept of induction period is true only for the reaction proceeding in the closed system when its duration is determined by the oxygen concentration alone and rate of its consumption. With the high inhibitor concentration exceeding the upper limit, the antioxidant acts as the monomer oxidant and polymerization inhibitor and the concept of an induction period is true both for the closed and open systems. The mode of non-inhibited oxidative polymerization on the inhibited polymerization occurs between the lower and upper limits of antioxidant concentrations.

The process of oxidative styrene polymerization inhibited by 2,6-ditetra-butyl phenols[104,105] has a stepped character accompanied by primary τ_1 and secondary τ_i induction periods. The observed critical laws are interrelated with period formation of secondary inhibitor X^2 which concentration oscillates in time, i.e. it is caused by auto oscillation mode. As the studies have been conducted in a closed system oscillation of the secondary inhibitor concentration is non-attenuated in the observed period of time. Polymerization rate in the secondary induction periods drastically falls with the antioxidant concentration growth. The equation for the induction period of polymerization with oxygen[106,107] we have:

$$\tau = \frac{\mu K_X[O_2]_0[X]_0}{(K_p[M] + K_X[X]_0)W_i}$$

the latter should be absent at the end of the primary induction period. Then the oxygen molecules role in X^2 formation should be indirect and it should be achieved by means of the products of oxidative transformation of phenol. In general, the mechanism of the auto oscillating mode for styrene radical polymerization inhibition includes three main reactions resulting in formation of secondary products Y, the intermediate product Q and to the regeneration of the inhibitor X^2:

$$RO_2^{\cdot} + X^1 \longrightarrow Y$$

$$R + Y \longrightarrow Q \text{ — non-active product,}$$

$$Q + X^1 \longrightarrow X^2$$

An attempt has been made to carry out a mathematical simulation of the auto oscillating polymerization with phenols.[108] The main prerequisite for this is the idea that polymerization is a non-branching radical-chain process, that occurs after the primary induction period is finished.

6.2 Inhibited Polymerization of Methacrylates with Sulfuric Acid

Sulfuric acid is one of the main components in the process of synthesis of acrylate and methyl acrylate monomers. The problem of polymerization inhibition with sulfuric acid is interesting because the acid can intensify polymer formation and can deactivate many well-known stabilizers.

To understand the inhibition mechanism with sulfuric acid, it is necessary to understand specific features of the influence of acid on the elementary stages of polymerization. When studying sulfuric acid influence on methyl acrylate photochemical polymerization under oxygen-free conditions[109-111] complexation between acid and monomer has been found. Growth of the polymerization rate and degree is explained by the monomer's reactivity increase (or macro radicals) and by increase of the macroscopic viscosity and change of conformation and effective stiffness of the propagating chains. The influence of sulfuric acid on the kinetics and polymerization mechanism of the acrylic monomers at higher temperatures with oxygen has been studied.[112] The sulfuric acid addition drastically decreases the induction period caused by the oxygen inhibiting action and increases the polymerization rate. The relation of the rate and induction period with the sulfuric acid concentration has a quite complicated nature. The data obtained show the complexation between monomers and acid with a different quantitative composition. The complexation accelerates the peroxides' decay due to its protonation with acid with homolysis of the bond —O—O— by analogy with mechanism of induction decay of peroxides in the presence of some acids.[113]

The compounds given in Table 1 effectively inhibit the acrylic monomers polymerization with sulfuric acid.[51,114] Sulfuric acid growth in solution leads to drastic reduction of the induction period caused by both initiation rate increase and inhibitor interaction with acid decreasing its efficiency. Unlike the polymerization in bulk where critical phenomena exhibit themselves at sufficiently high temperatures during polymerization with sulfuric acid they are observed at relatively low temperatures, Figs. 4 and 5.

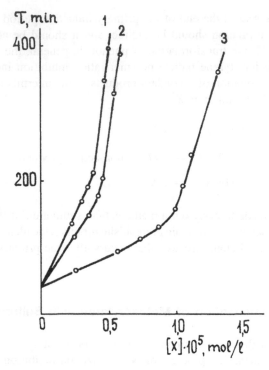

Figure 4   Induction period of methylacrylate polymerization in the presence of air oxygen and 5% vol. H_2SO_4 as a function of C-1 (1), C-41 (2) and ON (3) concentration at 75°C.[51]

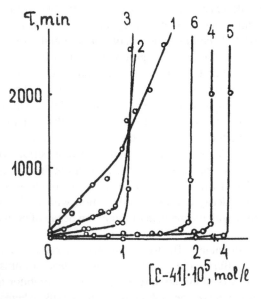

Figure 5   The dependence of induction period of methylmethacrylate polymerization in the presence of air oxygen on C-41 concentration at 60°C and various contents of sulphuric acid: 1–2, 2–5, 3–10, 4–20, 5–30, 6–40% vol.[51]

TABLE 2

Values of the inhibitor's critical concentration for the methacrylate polymerization with sulfuric acid[51]

Inhibitor	$[X]_{cr} \cdot 10^5$ mol/l, 100°C	
	5 volume % H_2SO_4	30 volume % H_2SO_4
C-1	0.7	0.8
C-41	0.9	1.3
C-47	—	1.3
C-49	—	1.1
ON	3.0	2.0
HQ	90.0	8.0

Relations (14) and (15) are used for inhibited polymerization of methacrylates with sulfuric acid. As the latter accelerates the polyperoxide decays. In Table 2 values of $[X]_{cr}$ for some new inhibitors of polymerization in comparison to hydroquinone are given.

The critical inhibitor concentration, the transition sharpness of relation of τ with $[X]$ and stoichiometric inhibition coefficient depends on the acid content in the system. The μ fall with acid concentration growth is possible if in the process of interaction with it, the substances C-1, C-41, etc. transform into compounds do not inhibit polymerization or have much lower μ value. Both substances form dynamic equilibrium and increase of sulfuric acid content shifts it to the side of a less effective inhibitor that results in the stoichiometric coefficient falling.

Electron paramagnetic resonance spectra of the systems shows the interaction of the non-paired electron with nitrogen and hydrogen atoms which gives amine transformation X into the radical cation:[51]

$$2X + 3H_2SO_4 \longrightarrow 2X^{\cdot+} + 2HSO_4^- + H_2O + H_2SO_3$$

Such reaction have been recently described[115] in which intensification of the phenothiazine action in the presence of sulfuric acid is found. The increased value of μ was explained by the following inhibition mechanism:

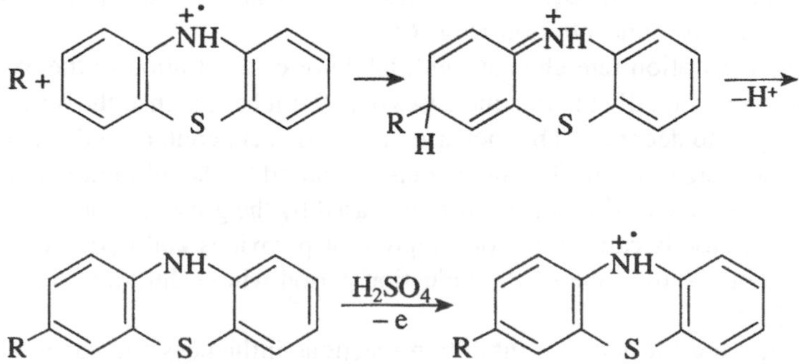

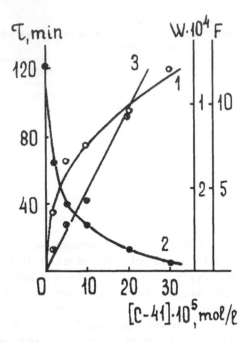

Figure 6   The relationship between induction period (1), stationary rate of methylmethacrylate polymerization (2), $(1 - \varphi^2)/\varphi$ ratio (3) and C-41 concentration in oxygen-free conditions and in the presence of 20% vol. H_2SO_4 (60°C, [AIBN] = $2 \cdot 10^{-4}$ mol/l).[51]

The electron paramagnetic resonance signal and consequently a number of radical-cations change depending on the reaction medium composition. In sulfuric acid the electron paramagnetic resonance signal is less than in the solution of methyl acrylate-sulfuric acid.

Phenols and aromatic amines are used as a rule as antioxidants.[86] With sulfuric acid transformation of the amines into the stable radical-cations give them the possibility to inhibit polymerization under oxygen-free conditions. After completion of the induction period the polymerization under oxygen-free conditions proceeds with a stationary rate which decreases with increase in inhibitor concentration. This is the result of reaction inhibition by the products of interaction of active radicals with the amine radical-cations. The kinetics obtained correspond to Eq. (15) indicated by a linear relationship of the ratio $(1 - \varphi^2)/\varphi$ with the inhibitor concentration (Fig. 6).

The stationary polymerization rate changes with inhibitor concentration in the oxygen presence has another form. Initially the rate increases but after it has reached the maximum value with $[X]_{cr}$ it begins to decrease. The increase of sulfuric acid content in the solution makes these efforts more significant. The rate increase is caused by the inhibition products — hydroperoxides whose rate of decay is in turn increased by the growth in acid content. The stationary rate decrease is caused by consumption of peroxides and hydroperoxides initiating polymerization for the time of the induction period whose duration drastically increases in the $[X]_{cr}$ area.

If initiating products are formed, the inhibition reactions influences the shape of the curve induction period with the inhibitor concentration.

7 INHIBITION FEATURES OF HETEROGENEOUS POLYMERIZATION AND POLYMERIZATION WITH A HIGH DEGREE OF TRANSFORMATION

The process of polymerization in highly viscous media with a high degree of transformation and in the systems containing polymer, the quadratic chain termination is diffusion controlled and the constant K_t drastically decreases. In this case termination, even by a weak inhibitor, can become a determining factor and the derived expression for the reduced rate is:

$$\varphi = \frac{W_X}{W} = \frac{(K_t W_i)^{0.5}}{K_X[X]}$$

it follows that with K_O and W_R fall (or with increase of the degree of transformation), the inhibiting effect grows. The relation of K_O and K_X with temperature we have:

$$\varphi = A \exp\left[(E_X - 0.5\, E_t)/RT\right], \quad A = A_t W_i^{0.5}/A_X[X]$$

Usually $E_X > E_O$ and φ value should decrease with temperature increase that results in the gel-effect decreasing and it should be kept in mind that φ decrease can compensate for the inhibition effectiveness falling with K_t growth caused by the medium viscosity decreasing with temperature increase. However this is observed only in the case when sufficiently inert products are formed during inhibition.

The gel-effect decrease and consequently decrease of the polymerization with a high degree of transformation influences the inhibition effect of the introduction of oxygen into the reactive system. Simultaneous actions of oxygen and weak inhibitor can be very effective in inhibiting a heterogeneous reaction. Calculations of the corresponding radical-chain schemes W_{O_2} it follows that:

$$\varphi = W_{O_2}/W = 10^{-3} - 10^{-2}$$
$$\varphi = W_{O_2,X}/W_X = 10^{-6} - 10^{-6}$$

Comparison of these data show that oxygen increases the weak inhibitor activity. In the presence of oxygen phenol, amines and nitro compounds inhibiting action is caused not only by their oxides formation (of quinone, quinonimine, etc. type) but also by the increased reactivity of these compounds in reactions with peroxy-radicals in comparison to reactions with alkyl radicals.[116] The weak inhibition method was used to show high activity of some monomers in copolymerization reactions.[117]

The heterogeneous polymerization features are the result of the macro radicals low mobility in the solid phase and sharp deacceleration of their mutual termination rate. Under these conditions reaction of the chain termination at a weak inhibitor may be important in the initial stages of transformation. That is why in some cases the heterogeneous polymerization is a more convenient model for researching the weak inhibitor reactions than polymerization with a high degree of transformation. The efficiency of weak inhibitors

TABLE 3

Values of the apparent reaction rates of alkyl (K_X) and peroxide (K_{XO}) radicals with the inhibitors determined while testing (60°C)[124,126,127]

Inhibitor	Structural formula	$K_X \cdot 10^2$; $K_{XO} \cdot 10^4$ l/mol·sec
hydroquinone	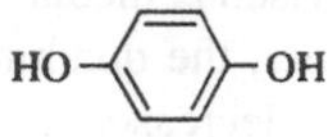 HO—⟨benzene⟩—OH	1.8; 21.4
α-naphthol	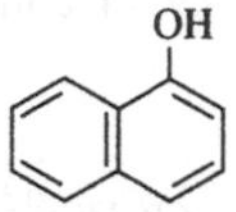 ⟨naphthalene with OH⟩	1.6; 18.6
22-46	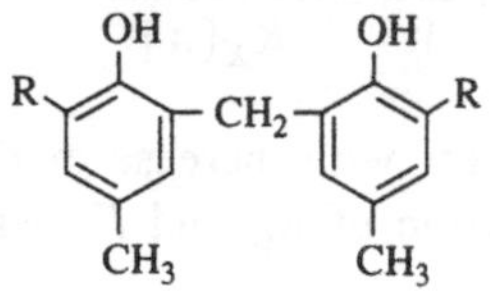 ⟨structure: two phenol rings with R, CH$_2$ bridge, CH$_3$⟩	2.1; 39.8
phenozane-1	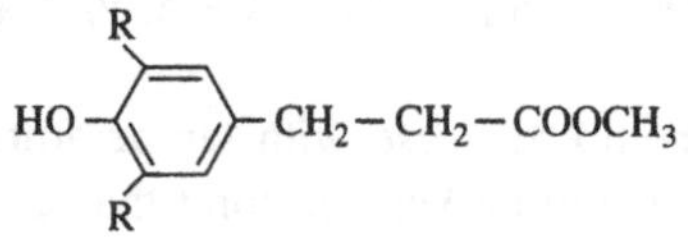 HO—⟨ring with R⟩—CH$_2$—CH$_2$—COOCH$_3$	0.7; 2.1
phenozane-28	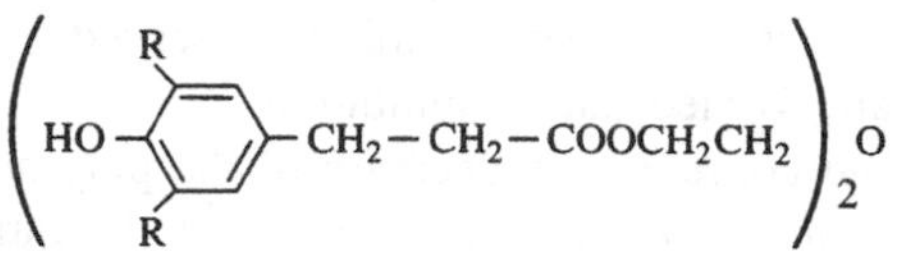 (HO—⟨ring with R⟩—CH$_2$—CH$_2$—COOCH$_2$CH$_2$)$_2$ O	0.46; 1.7
ionol	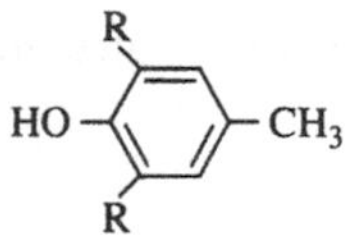 HO—⟨ring with R⟩—CH$_3$	0.7; 2.2
β-naphthol	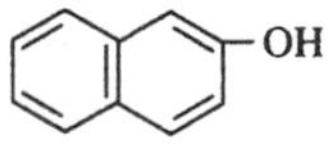 ⟨naphthalene with OH⟩	0.8; 2.4
neozone-D	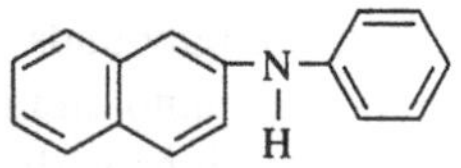 ⟨naphthalene—N(H)—phenyl⟩	0.75; 6.7

Note: R=C(CH$_3$)$_3$

belonging to the aromatic amines class, nitro compounds and phenols drastically increases in the acrylonitrile polymerization in bulk compared to the reaction in the dimethylformamide solution.[118]

The same effect is observed in the case of diphenylamine, σ-dinitro benzene and other inhibitors on the polymerization of vinyl chloride and methyl acrylic acid in bulk and on methyl acrylate polymerization in the percipitant mixture "methanol-water".[119-121] The examination of the heterogeneous polymerization of different monomers in the precipitant media allows information to be obtained about the reactivity of a large number of inhibitors.[17]

Weak inhibitors efficiency increases with polymerization in emulsion and depends upon their solubility in the reaction zone, topochemistry of elementary events, free-radicals, generation rate and number of polymer and monomer particles.[122,123]

Recently[124–127] a number of test methods for the weak inhibitors of radical polymerization which are strong inhibitors for the oxidative polymerization and polymer destruction have been proposed. One of these methods is based on the use of the analog reaction of the heterogeneous radical polymerization of acrylonitrile. It is easy to demonstrate that for evaluation of apparent inhibition constant K_X in the reaction

$$R + X \quad RH + \dot{X}$$

the following criterion should be met

$$\frac{W_X}{W} = \frac{K_t W_i}{K_X [X]} = \varphi \ll 1$$

Comparison of data on the acrylonitrile polymerization inhibition to data on the hydrocarbon oxidation allow correlation between the inhibition constants in the polymerization (K_X) and oxidation (K_{XO}) reactions for a number of the substituted phenols (Table 3). The correlation discovered allows the prediction of reactivity of antioxidants on the basis of the analog heterogeneous polymerization.

REFERENCES

1. *The Polymer's Encyclopedia.* (Soviet Encyclopedia, Moscow). v. 1, p. 836 (1972).
2. K. Bamford, W. Bard, Jenkins A. and P. Onjon, *Kinetics of the Vinyl Compounds Radical Polymerization.* (Inostrannaja Literatura, Moscow, 1961).
3. H.S. Bagdasarjan, *The Radical Polymerization Theory.* (Nauka, Moscow, 1966).
4. G.P. Gladyshev, *The Vinyl Monomers' Polymerization.* (AN KazSSR Alma-Ata, 1964).
5. F. Tüdes, *Examination of the Radical Polymerization Kinetics on the Basis of the Hot Radical Theory.* (Mir, Moscow, 1966).
6. M.D. Goldfein, *Kinetics and Mechanism of the Vinyl Monomers' Radical Polymerization.* (SGU Publishing House, Saratov, 1986).
7. C.H. Bamford, A.D. Jenkins and R. Jonston, *Proc. Roy. Soc.,* **239**, 214 (1957).
8. A.D. Stepukhovich, N.V. Kozhevnikov and L.T. Leonteva, *Vysokomoleculjarnye Soedinenija,* **16A**, 1522 (1974).
9. N.V. Kozhevnikov, *Master's Thesis on Chemistry.* (SGU Publishing House, Saratov, 1977).
10. M.D. Goldfein, E.A. Rafikov, N.V. Kozhevnikov, *et al., Vysokomoleculjarnye Soedinenija,* **17A**, 1671 (1975).
11. B.R. Smirnov and Z.A. Karapetjan, *The Carbo-Chain Polymers.* (Nauka, Moscow). p. 10 (1977).
12. B.R. Smirnov, *Vysokomoleculjarnye Soedinenija,* **24A**, 787 (1982).
13. B.R. Smirnov, *ibid.,* **24A**, 877 (1982).
14. J.L. Kice, *J. Amer. Chem. Soc.,* v. **76**, 6274 (1954).
15. E.V. Lazareva and V.A. Sidorov, *The monomers' stabilization.* (TSNIITENEFTECHIM, Moscow, 1973).
16. R.A. Barabashkina, V.A. Fomin, E.A. Sivenkov and A.A. Tupitsyna, *Inhibition of the Acrylic Monomers.* (NIITECHIM, Moscow, 1981).

17. G.P. Gladyshev and V.A. Popov, *The Radical Polymerization with High Degrees of Transformation.* (Nauka, Moscow, 1974).
18. USA Patent 4010082. *Izobretenija za Rubezhomn*, **24**, N 9, 44, (1977).
19. USA Patent 3876686. *Izobretenija za Rubezhom*, **18**, N 13, 58 (1975).
20. Application 49-3514 Japan. *Izobr. za Rubezhom*, **15**, N 14, 149 (1974).
21. Application 49-3974 Japan. *Izobr. za Rubezhom*, **15**, N 15(II), **20** (1974).
22. USA Patent 3857878. *Izobr. za Rubezhom*, **18**, N 1, 18 (1975).
23. USA Patent 3855281. *Izobr. za Rubezhom*, **15**, N 24, 16 (1974).
24. Application 2244749 France. *Izobr. za Rubezhom*, **18**, N 10, 63 (1975).
25. USA Patent 3896162. *Izobr. za Rubezhom*, **18**, N 17, 29 (1975).
26. GFR Application 2028183. *Izobr. za Rubezhom*, **18**, N 17, 130 (1975).
27. Application 50-1005 Japan. *Izobr. za Rubezhom*, **18**, N 20, 76 (1975).
28. GFR Application 1468118. *Izobr. za Rubezhom*, **24**, N 14, 151 (1976).
29. Application 1425788 Great Britain. *Izobr. za Rubezhom*, **24**, N 5, 10 (1976).
30. USA Patent 3959358. *Izobr. za Rubezhom*, **24**, N 18, 78 (1976).
31. Application 1477298 Great Britain. *Izobr. za Rubezhom*, **55**, N 3, 13 (1978).
32. M. Magat and R. Boneme. *Compt. Rend.*, **232**, 1657 (1951).
33. E.V. Anufrieva, M.V. Volkenshtein and M.M. Kotton, *Zhurnal of Phys. Chem.*, **31**, 1532 (1957).
34. F. Tüdes, T. Berezhnykh and B. Turchani, *Vysokomolec. Soed.*, **4**, 1584 (1962).
35. B.T. Foldeshe and F. Tüdes, *Magyar kem. foluoirat*, **70**, 500 (1964).
36. T.Ja. Smirnova, S.P. Rafikov and G.P. Gladyshev, *Vysokomolec. Soed.*, **10**, 562 (1968).
37. M. Simonyi and F. Tüdes, *Kinetics and Mech. Polyreacts.*, **3**, Preprs. Budapest, p. 119 (1969).
38. A.A. Ivanov, N.I. Vorobeva, G.M. Lysenko and G.V. Kaljakina, *Chemistry and Chemical Technology.* (High School Publ. House), v. 23, p. 1315 (1980).
39. A.A. Yassin and N.A. Rizk, *Polym. J.*, **10**, N 1, 77 (1978).
40. A.A. Yassin and N.A. Rizk, *Eur. Polym. J.*, **13**, 441 (1977).
41. A.A. Ivanov, T.M. Lysenko and I.N. Zhulina, *Vysokomolec. Soed.*, **22B**, 515 (1980).
42. G.P. Gladyshev and V.A. Sechkovskaja (AN KazSSR Publ. House), *Chemistry series*, **3**, 89 (1967).
43. G.V. Lepljalnin, S.R. Rafikov, I.I. Furlei, *et al.*, Papers of AN USSR, **207**, 905 (1972).
44. F. Tüdes, I. Kende, T. Berezhnykh and O.O. Solodovnikov, *Kinetics and Catalysis*, **6**, 203, (1965).
45. V.E. Zubarev, V.N. Belevski and L.T. Bugaenko, *Uspekhi Chimii*, **8**, 1361 (1979).
46. I. Kende, L. Sumegi and F. Tüdes, *Kinetic and Mech. Polyreacts Prepts. Budapest, Akadem. Tiadc.*, **3**, 109 (1969).
47. S.P. Ljakina and B.A. Dogadkin, *Vysokomolec. Soed.*, **15A**, 2773 (1973).
48. I. Tanczos, T. Foldes-Berezscnich and F. Tüdes, *Eur. Polym. J.*, **19**, 225 (1983).
49. N.V. Kozhevnikov, *A.s. 1175932 USSR. B.I.*, **32** (1985).
50. E.G. Rozantsev, M.D. Goldfein and A.V. Trubnikov, *Uspekhi Chimii.*, **55**, 1881 (1986).
51. M.D. Goldfein, N.V. Kozhevnikov and A.V. Trubnikov, *Kinetics and Mechanism of the vinyl monomer polymerization control.* Dep. ONIITECHIM. Cherkassy: N 1007-chp. VINITI, **12**, 151 (1986).
52. W.I. Bengough and W.H. Fairservice, *Trans. Faraday Soc.*, **61**, 1206 (1965).
53. W.I. Bengough and W.H. Fairservice, *ibid.*, **63**, 382 (1967).
54. A.I. Jurzhenko and S.S. Ivanchev, *Kolloid. Zhurn.*, **22**, N 1, 120 (1960).
55. M.D. Goldfein and A.D. Stepukhovich, *Heads of Reports of the International Symposium on Macromolecular Chemistry*, Budapest, p. 34 (1969).
56. M.D. Goldfein, *The Author's Summary of the Master's Thesis on Chemistry and Physics.* (SGU Publ. House, Saratov, 1970).
57. E.A. Rafikov, M.D. Goldfein and B.A. Zjubin, *Heads of Papers of the XIV All-Union Chugaev Meeting on the Complex Compounds Chemistry.* Ivanovo. Part 2, p. 446 (1981).
58. E.A. Rafikov, M.D. Goldfein, N.V. Kozhevnikov, *et al.*, *Vysokomol. Soed.*, **18A**, 2424 (1976).
59. V.G. Vinogradova and Z.K. Maizus, *Kinetics and Catalysis*, **13**, 298 (1972).
60. A.I. Zverev and Z.K. Maizus, (An USSR Publ. House) *Chemistry series*, **11**, 2437 (1973).
61. A.I. Bern, *Tetrahedron*, **22**, 2653 (1966).
62. V.G. Vinogradova, Z.K. Maizus and N.M. Emanuel, *Papers of AN USSR*, **188**, 616 (1969).
63. V.F. Anufrienko, T.M. Kogan, E.G. Rukhadze and V.V. Dudina, *Teor. and experim. chimija.*, **3**, 370 (1967).

64. S.K. Ivanov and V.S. Juritsyn, *Neftechimija.*, **11**, N 1, 99 (1971).
65. M.D. Goldfein, A.V. Trubnikov, L.V. Aizenberg and N.V. Simontseva, *Chemistry and Chemical Technology.* (High School Publ. House). v. 29, N 7, p. 91 (1986).
66. S.K. Ivanov and I. Kateva, *Comp. Rend. Acad. Bulg. Sci.*, **21**, N7, 681 (1968).
67. O.A. Juzhakova, V.M. Furman, G.N. Gerasimov and A.D. Abkin, *Vysokomolec. Soed.*, **25B**, 454 (1983).
68. R.Ju. Makushkin, G.I. Baeras, Ju.K. Schulskus, *et al.*, *ibid.*, **27A**, 567 (1985).
69. A.A. Ivanov, T.M. Lysenko and I.N. Zhjulina, *Kinetics and Catalysis*, **24**, 275 (1985).
70. A.A. Ivanov, N.I. Vorobeva, T.M. Lysenko and G.V. Kaljagina, *Chemistry and Chemical Technology.* (High School Publ. House). v. 23, p. 1315 (1980).
71. A.A. Ivanov, T.M. Lysenko, A.I. Kadantseva, *et al.*, *A.s. 734216 USSR. B.I.*, **18** (1980).
72. A.A. Ivanov, A.I. Kadantseva, N.I. Vorobeva, *et al.*, *A.s. 737406 USSR. B.I.*, **20** (1980).
73. A.A. Ivanov and Ju.K. Romanovitch, *Vysokomolec. Soed.*, **28A**, 2272 (1986).
74. A.A. Ivanov, *ibid.*, **28B**, 873 (1986).
75. L.B. Volodarski and G.A. Kutikova, *Chemistry series*, N 5, p. 937 (AN Publishing House, 1971).
76. A.A. Shevyrev, G.S. Belikova, L.B. Volodarski and V.I. Simonov, *Kristallografija*, **24**, 787 (1979).
77. M.D. Goldfein, L.B. Volodarski, E.G. Rosantsev, *et al.*, *A.s. 1147708 USSR. B.I.*, **12** (1985).
78. M.D. Goldfein, L.B. Volodarski and E.E. Kochetkov, *Chemistry and Chemical Technology*, **29**, N 4, 93 (High School Publishing House, 1986).
79. M.D. Goldfein, E.A. Rafikov, A.D. Stepukhovich and L.A. Skripko, *Vysokomolec. Soed.*, **16A**, 672 (1974).
80. A.V. Trubnikov, M.D. Goldfein, N.V. Kozhevnikov, *et al.*, *ibid.*, **20A**, 2448 (1978).
81. F. Tüdes, V. Fürst and M. Azory, *Khimija i Tekhnologija Polimerov.*, **1**, 78 (1960).
82. W.A. Pryor and L.D. Laswell, *Polymer Preprints*, **11**, N 2, 73 (1970).
83. A.L. Buchachenko and A.M. Wasserman, *The Stable Radicals.* (Khimija, Moscow, 1973).
84. R.R. Hiatt and P.D. Bartlett, *J. Amer. Chem. Soc.*, **84**, 1189 (1959).
85. V.A. Kurbatov, *Vysokomolec. Soed.*, **24A**, 1347 (1982).
86. P.I. Levin and V.V. Mikhailov, *Uspekhi Khimii.*, **39**, 1678 (1970).
87. A.D. Stepukhovich, M.D. Goldfein, N.V. Kozhevnikov, *et al.*, *A.S. 743988 USSR. B.I.*, **24** (1980).
88. A.D. Stepukhovich, M.D. Goldfein, N.V. Kozhevnikov, *et al.*, *A.S. 793997 USSR. B.I.*, **1** (1981).
89. M.D. Goldfein, N.V. Kozhevnikov, L.A. Skripko, *et al.*, *A.S. 895979 USSR. B.I.*, **1** (1982).
90. M.D. Goldfein, N.V. Kozhevnikov, E.A. Rafikov and L.A. Skripko, *A.s. 981313 USSR. B.I.*, **46** (1982).
91. M.D. Goldfein, N.V. Kozhevnikov, A.V. Trubnikov, *et al.*, *Vysokomolec. Soed.*, **25B**, 268 (1983).
92. M.D. Goldfein, L.A. Skripko, R.V. Kosyreva and L.V. Konkova, *Chemistry and Chemical Technology*, **27**, 1065 (1984).
93. M.D. Goldfein, N.B. Chumaevski and L.A. Skripko, *A.S. 1065405 USSR. B.I.*, **1** (1984).
94. M.D. Goldfein, N.V. Kozhevnikov, A.V. Trubnikov and N.B. Chumaevski, *Khimicheskaja Promyshlenmost.*, **6**, 13 (1987).
95. A.V. Trubnikov, M.D. Goldfein, N.V. Kozhevnikov and A.D. Stepukhovich, *Vysokomolec. Soed.*, **25A**, 2150 (1983).
96. N.N. Semenov, *On Some Problems of Chemical Kinetics and Reactivity.* (AN USSR Publishing House, Moscow, 1968).
97. N.M. Emanuel, G.E. Zaikov and Z.K. Maizus, *The Medium Role in the Radical-Chain Reactions of the Organic Compounds Oxidation.* (Nauka, Moscow, 1973).
98. G.P. Gladyshev and D.Kh. Kitaeva, *AN USSR Papers. Chemistry Series*, **271**, 889 (1973).
99. N.N. Tvorogov, I.A. Matveeva, P.K. Tarasenko, *et al.*, *A.s. 516700 USSR. B.I.*, **21** (1974).
100. N.N. Tvorogov, *Vysokomolec. Soed.*, **17A**, 1464 (1975).
101. N.N. Tvorogov, I.A. Matveeva, A.A. Volodkin, *et al.*, *ibid.*, **18A**, 347 (1976).
102. M.M. Mogilevich, *The Oxidative Polymerization during Filming.* (Khimija, Leningrad, 1977).
103. N.N. Tvorogov, *Vysokomolec. Soed.*, **25A**, 248 (1983).
104. V.A. Kurbatov, N.P. Boreiko, A.G. Liakumovich and P.A. Kirpichnikov, *AN USSR Papers*, **264**, 1428 (1982).
105. V.A. Kurbatov, N.P. Boreiko and A.G. Liakumovich, *Vysokomolec. Soed.*, **26A**, N 3, 541 (1984).
106. N.N. Tvorogov, *ibid.*, **17A**, 1461 (1975).
107. L.L. Hervits, N.V. Zolotova, E.T. Denisov, *ibid.*, **18B**, 524 (1976).
108. V.A. Kurbatov, A.N. Ivanova, G.A. Furman and E.T. Denisov, *Chem. Phys.*, **1316** (1984).
109. N.A. Vengerova, V.R. Georgieva, V.P. Zubov and V.A. Kabanov, *Vysokomolec. Soed.*, **12B**, 46 (1970).

110. P.P. Nechaev, V.P. Zubov and V.A. Kabanov, *ibid.*, **9B**, 7 (1967).
111. B.M. Abu-el-Hair, M.B. Lichinov, V.P. Zubov and V.A. Kabanov, *ibid.*, **17A**, 831 (1975).
112. N.V. Kozhevnikov, A.V. Trubnikov, A.D. Stepukhovich and N.M. Larina, *ibid.*, **26A**, 687 (1984).
113. E. Rizzardo and D.H. Solomon, *J. Macromolec. Sci.*, **14A**, N 1, 33 (1980).
114. M.D. Goldfein, N.V. Kozhevnikov, A.V. Trubnikov, *et al.*, *Thesis of Reports of VIII All-Union Scientific and Technical Conf. "Synthesis and Research of Effectiveness of Chemicals for the Polymer Materials".* Cherkassy: (ONIITECHEM Publishing House), p. 66 (1986).
115. A.A. Ivanov, T.M. Lysenko, A.I. Kadantseva, *et al.*, *Vysokomolec. Soed.*, **23A**, 689 (1981).
116. G.P. Gladyshev, V.A. Popov, D.Kh. Kitaeva and E.I. Penkov, *AN USSR Papers*, **215**, 898 (1974).
117. R.G. Karzhubaeva, G.P. Gladyshev and S.R. Rafikov, (AN KazSSR Publishing House). N 3, p. 54 (1967).
118. V.A. Popov and G.P. Gladyshev, *Uspekhi Khimii*, **42**, 273 (1973).
119. V.A. Popov and G.P. Gladyshev, *Vysokomolec. Soed.*, **15B**, 102 (1973).
120. V.A. Popov, G.P. Gladyshev and E.I. Penkov, *ibid.*, **16A**, 2196 (1974).
121. V.A. Popov, E.P. Shvarev, Ju.A. Zvereva, *et al.*, *ibid.*, **17A**, 1226 (1975).
122. V.A. Popov and G.P. Gladyshev, *Plastmassy.*, **5**, 10 (1972).
123. G.P. Gladyshev, V.A. Popov and E.I. Penkov, *Vysokomolec. Soed.*, **16A**, 1945 (1974).
124. G.P. Gladyshev, D.Kh. Kitaeva and E.G. Gladysheva, *On Use of the Model Reactions of Radical Polymerization in the Kinetic analysis of the Complex Compositions.* Preprint Paper. (Bash. F. AN USSR Publ. House, Ufa, 1983).
125. T.M. Turechanov, Zh.Kh. Ibrasheva and L.B. Iriskine, *The 1st All-Union Symposium on the Macroscopic Kinetics and Chemical Gas Dynamics. Thesis of Reports.* Chernogolovka, **1**, Part 1, 131 (1984).
126. T.M. Turekhanov, Zh.Kh. Ibrasheva and L.B. Iriskina, *Vysokomolec. Soed.*, **26B**, 386 (1984).
127. T.M. Turekhanov, Zh.Kh. Ibrasheva and L.B. Iriskina, *Chemistry series*, N 5, p. 49 (AN KazSSR Publ. House, 1986).

Main Factors Influencing Flame Spread Velocity Over Polymer Surfaces

NIKOLAI N. BAKHMAN

Institute of Chemical Physics, Russian Academy of Sciences, Moscow

PREFACE

Flame spread velocity over a polymer surface, w, is an important characteristic of polymer flammability. Indeed, the time elapsed from the ignition to full-scale fire is determined to a considerable extent by the flame velocity. The significance of this parameter has been emphasized in a number of books.[1,2] Many parameters influence the value of w and can be grouped into three categories:

(1) chemical and physical properties of the polymeric materials,
(2) dimensions and orientation of a specimen,
(3) properties of the gaseous environments.

To understand burning it is assumed that the flame velocity is affected primarily by processes taking place near the flame tip, the "flame velocity controlling zone", FVC-zone.[3]

Since the mixing of gaseous oxidiser with the products of polymer thermodestruction begins prior to burning, the chemical reactions in the flame tip proceed in a homogeneous mixture. Hence, the flame velocity depends to a considerable extent on kinetic factors even in the cases when the dominant part of a polymer specimen is consumed in the diffusion flame zone propagating behind the flame tip.

1 INFLUENCE OF CHEMICAL AND PHYSICAL PROPERTIES OF POLYMERS AND FILLERS

Investigations in this field deals with the critical conditions of burning, particularly, Oxygen Index, OI. It is better to consider the effect of fire retardants (FR) on the flame velocity together with the critical conditions for burning.

1.1 Influence of Char Layer on the Burning Surface

The flammability of polymers depends strongly on the ability of a polymer to form

TABLE 1

Burning rate (in air) of char-forming and char-free polymer films of various thickness, Δ, at various orientation[2], φ, of specimens.[4]

| | w, mm/s | | | |
| | $\Delta = 60\ \mu$ | | $\Delta = 500\ \mu$ | |
Polymer	$\varphi = -90°C$	$\varphi = +90°C$	$\varphi = -90°C$	$\varphi = +90°C$
Polysulphone	2.5	9.4	0.5	1.6
Polycarbonate	3.6	14.0	0.6	2.0
Polyethylene	7.0	21.0	—	—
Polyamide	9.0	28.0	1.2	3.5

[2] φ is the angle between burning direction along specimen surface and horizontal plane ($\varphi = -90°$ correspond to vertical downward burning, $\varphi = +90°$ to vertical upward burning, and $\varphi = 0°$ — to horizontal burning).

a char layer on the burning surface. A char layer, firstly, diminishes heat transfer from the flame to polymer surface and, secondly, hinders the diffusion of volatile products of thermo-destruction of a polymer to the flame. Char-forming polymers (polysulphone and polycarbonate) burned more slowly than char-free polymers (polyethylene and polyamide). Other factors (e.g. heat of combustion) can also influence the burning velocity of polymers.

1.2   Influence of Melting of Polymers

Melting of polymers and dripping affects the value of w. Burning of PMMA and PE coatings on copper wires[5] at $\varphi = -90°$ (vertical downward burning) the PE coatings burned 2 to 4 times faster than PMMA coatings. In contrast, at $\varphi = +90°$ (vertical upward burning) the value w_{PE} was nearly twice as small as w_{PMMA}. No significant amounts of melt were observed in PMMA coatings in the course of burning, whereas intense melting and dripping occurred on the PE coatings. At $\varphi = -90°$ the flow of PE melt increases heat transfer in the pre-heating zone. In contrary, at $\varphi = +90°$ the dripping of PE melt means heat losses diminishing the flame velocity.

1.3   Influence of Fillers

1.3.1   Powdered fillers

The influence of nine fillers on the burning rate of horizontal specimens of an epoxy resin were studied.[6] All the fillers decreased significantly (~ 1.4 to 4,1 times) the burning rate, Table 2 where w_o and w are the flame velocity of basic and filled specimens, respectively. Relation exists between the effect of fillers on w and that on OI.

It is clear that the decrease in the ratio w/w_o is due to heat losses connected with heating and for some fillers endothermic decomposition of the fillers.

TABLE 2

Effect of powdered fillers (43.5% by mass) on flame velocity along the surface of horizontal specimens of an epoxy resin composition burning in 35% O_2 + 65% N_2 mixture.[6]

Filler		$Ca(OH)_2$	CaO	MgO	CuO	Al_2O_3	$CaCO_3$	$Mg(OH)_2$	Sb_2O_3	$Ba(OH)_2$
w, mm/s	0.41	0.30	0.25	0.25	0.25	0.24	0.23	0.17	0.13	0.10
w/w_o	1.00	0.73	0.61	0.61	0.61	0.59	0.56	0.41	0.32	0.24
OI	19.3	20.1	21.1	20.6	19.8	20.1	20.5	22.1	20.6	22.1

1.3.2 Fiber-reinforced polymers

A characteristic feature of polymers reinforced with continuous fibers is the anisotropy of physical properties including heat conductivity. As a consequence, the value w for oriented composites depends on the flame propagation direction. Three cases may be considered:

(1) fibers are parallel to specimen surface and flame spreads along the fibers;
(2) the same as 1, but flame spreads across the fibers;
(3) fibers are perpendicular to the specimen surface (the direction of flame spread is arbitrary).

If the heat conductivity of fibers is much higher than that of the polymer matrix, the flame velocity is maximum for case 1 ($w_{max} = w_1$) and minimum for case 3 ($w_{min} = w_3$). In the first case fibers provide paths of high thermal conductivity to transmit heat forward along the flame spread direction. In the third case fibers increase heat losses deep into the specimen.

With a graphite fibers — epoxy resin composites burning in air:[7]

$\varphi°$	−30	−10	0	+10	+20	+30
w_1/w_2	1.96	1.52	1.19	1.21	1.51	1.14

The ratio w_1/w_2 decreases with φ increasing probably due to the rise of contributions from convective heat transfer through the gas phase.

1.4 Effect of Porosity of Polymer Specimen

Flame spread over horizontal layer of powdered PMMA was studied[8] at various levels of porosity, P. In Fig. 1 the linear burning rate, w, increases monotonously with P, whereas the mass burning rate, $\dot{m} = \rho w$, diminishes with P increasing.

These results seem to be analogous to those obtained for some explosives, such as RDX. If the rate-controlling chemical reaction proceeds in the gas phase and no heat losses occur in the burning zone mass burning rate is independent of P, whereas linear burning rate increases with P:[9]

$$u = \dot{m}/\rho = \dot{m}/\rho_{max}(1-P) \tag{1}$$

where ρ_{max} is the highest possible density of specimen.

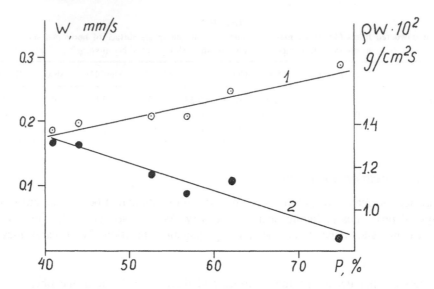

Figure 1 Effect of porosity, P, of a layer of powdered PMMA on the linear (1) and mass (2) flame velocity (at $\varphi = 0°$).[8]

However, if heat losses are significant, $\dot{m}$ diminishes with increasing P. In this case a rise of P and corresponding reduction of density of layer $\rho = \rho_{max} (1 - P)$, results in a lowering of the heat release per unit of volume of the specimen. In contrast, heat losses diminish only slightly with ρ decreasing. So, the temperature in the burning zone and mass burning rate decline with P increasing. Linear burning rate in the presence of heat losses increased with P less rapidly than in the absence of heat losses or even can diminish with P increasing if heat losses are too high.

Of course, this analogy between the dependence $\dot{m}(P)$ and $w(P)$ for flame spread over polymer surface and that for burning of explosives is rather approximate. Indeed, in the first case the problem is two-dimensional and the oxidizer is delivered to the diffusion flame from the environment. In the second case the problem is one-dimensional and the flame is either kinetic (for homogeneous systems) or microdiffusional (for heterogeneous systems such as composite solid propellants).

2 INFLUENCE OF SPECIMEN DIMENSIONS AND ORIENTATION

Experiments performed by many authors[8,10–13] showed that dependences of flame velocity on specimen dimensions and orientation may be strong and therefore are of great importance for fire-safety studies.

2.1 Effect of Specimen Dimensions

2.1.1 Effect of specimen thickness[14]

The following cases were examined as typical:

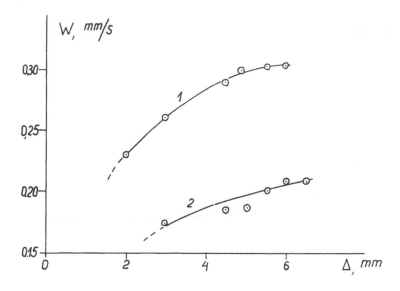

Figure 2   Dependence w (Δ) for horizontal layers of powdered PMMA burning in still air (at normal conditions) on thick underlay (aluminium alloy).[8]

(1)   One-side burning of flat specimens on a thick underlay with high heat conductivity, λ_{un};
(2)   The same as 1, but underlay is thin;
(3)   The same as 1, but λ_{un} is very small;
(4)   Two-side burning of flat specimen without any underlay;
(5)   Candle-like burning of bare polymer rods.

The dependence $w(\Delta)$ (where Δ is specimen thickness) for the case 1 is quite different from that for the cases 3–5. Namely, in the case 1 the flame velocity decreases with Δ (until extinction occurs at some $\Delta = \Delta_{cr}$). In contrast, in the cases 3–5, w increases with Δ (or rod diameter, d) decreasing. Fig. 2 illustrates case 1. The decline of w with Δ is due to heat losses to the metallic underlay.

Fig. 3 is relevant to case 3 and Fig. 4 to case 4. The rise of w with Δ decreasing is connected with the interaction of heat wave propagation from the burning surface with the asbestos underlay, (Fig. 3) or interaction of two heat waves propagating from both burning sides of specimen, (Fig. 4). The smaller is Δ, the more intense is the heating of the sheet and the higher is the temperature of the burning surface, T_s, and w. Only at very small Δ does the burning rate begins to decrease until extinction occurs due to heat losses in the air.

For two-side burning of PP films[15] w increases nearly monotonously with Δ decreasing:

Δ, μ	307	254	201.5	181.5	166	140	100	75.5	45.5
$w_{\varphi = -90°}$, mm/s	0.77	0.92	1.20	1.11	1.42	1.69	4.17	3.91	9.10

The dependence $w(\Delta)$ or $w(d)$ for two-side burning may be approximated by the power function:

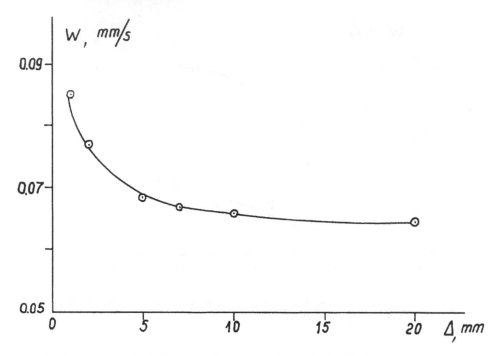

Figure 3   Dependence $w(\Delta)$ for horizontal sheets of PMMA burning in still air (at normal conditions) on thick asbestos underlay.[10]

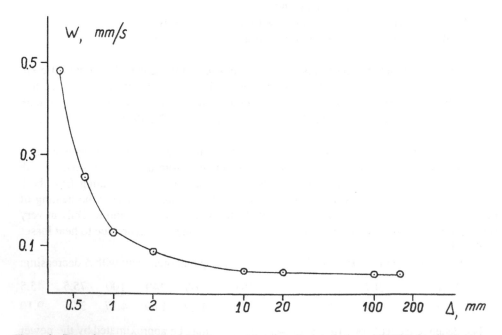

Figure 4   Dependence $w(\Delta)$ for two-side downward burning of vertical sheet of PMMA in still air at normal conditions.[10]

TABLE 3
The exponent n in Eq. 2 for polymeric films burning
in air at various orientations of specimens.[4]

Polymer	$\varphi = -90°$	$\varphi = +90°$
	n	
Polyamide	0.95	0.98
Polycarbonate	0.85	0.92
Polysulphone	0.76	0.84

$$w = A/\Delta^n, \qquad \text{or} \qquad w = A/d^n \qquad (2)$$

The following values of exponent n (calculated simply as $n = \log (w_{60\mu}/w_{500\mu})/\log (500/60)$) may be obtained, Table 3.[4]

The dependence $w(d)$ for PMMA rods is shown in Fig. 5. In the range $d \simeq 20\text{–}7$ mm the value of w rises with d decreasing. It is worth noting that w for PMMA rods[11],[12] and PS rods[13] is nearly inversely proportional to d, Table 4.

A thick metallic underlay decreases the burning rate of polymer specimens. In contrast, a thin metallic underlay with high thermal conductivity may increase flame spread velocity, Fig. 6. Such an underlay augments the heat flux along the flame direction. Incorporation of metallic wires or foils in solid propellant strands is a well-known method of augmenting burning rate.[16]

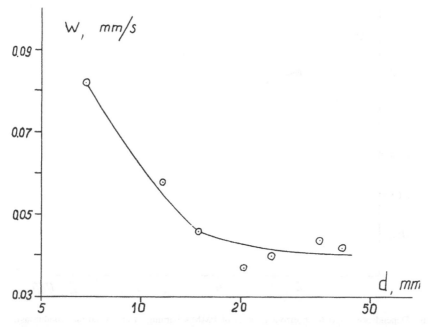

Figure 5   Dependence $w(d)$ for vertical downward burning of bare PMMA rods in still air at normal conditions.[11]

TABLE 4
The exponent n in Eq. 2 for bare polymer rods burning in air vertically downward

Polymer	Range of d, mm	n	Reference
PMMA	1.5–6	1.0	[12]
	7–20	0.8	[11]
PS	2–8	1.08	[13]

Thus, the shape of $w(\Delta)$ or $w(d)$ curves for small Δ or d may be quite different depending on boundary conditions. In contrast, at large Δ or d the dependence $w(\Delta)$ or $w(d)$ is just the same for all the cases studied: at some $\Delta \gtrsim \Delta_*$ or $d \gtrsim d_*$ the flame velocity ceases to vary with increasing Δ or d. For PMMA sheets Figs. (3 and 4) at $\Delta \gtrsim 10$ mm, as for PMMA rods[11] (burning in air):

d, mm	6.8	11.7	15.0	20.4	25.0	35.2	40.7
$w_{\varphi = -90°}$, mm/s	0.082	0.058	0.046	0.037	0.040	0.044	0.042

Taking into account the scatter of experimental data, the flame velocity in this case is independent of the rod diameter at $d \gtrsim 20$ mm in accord with the theoretical model of de Ris,[17] for "thin" specimens $w \sim 1/\Delta$, whereas for "thick" specimens $w \neq f(\Delta)$.

The latter theoretical prediction is consistent with experimental data. As for "thin" specimens, it was shown above that the exponent n Eq. 2 is equal ~1.3 for PP films

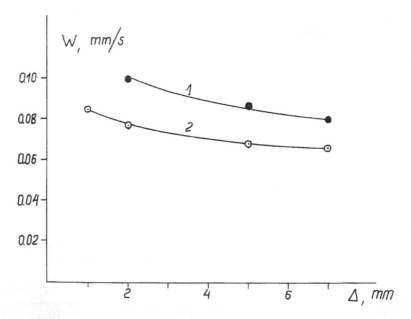

Figure 6   Dependence $w(\Delta)$ for horizontal sheets of PMMA burning in air (at normal conditions).[10]
Underlays: 1. Copper foil (25 or 100 μ thick) + asbestos (25.4 mm thick).
          2. Asbestos (25.4 mm thick).

$(\Delta \leqslant 300~\mu)^{15}$ and $n = 0.76 + 0.98$ for polyamide, polycarbonate, and polysulphone films $(\Delta \leqslant 500~\mu)^4$ instead of theoretical value $n = 1.0$. Finally, the value $n = 1$ was obtained for downward burning of paper.[18] In index cards were used (each 0.22 mm thick) pressed together to obtain specimens of various thickness $\Delta \simeq 0.22 - 1.96$ mm.[18] The agreement between experimental values n and theoretical value $n = 1$ for "thin" specimens seems rather satisfactory. However, it is not clear how to determine the value Δ_{**} such as at $\Delta < \Delta_{**}$ the specimen is "thin enough".

It is useful to note that the concept of FVC-zone the independence w of Δ or d at great enough Δ or d may be explained as follows. At big enough Δ or d the dimensions of FVC-zone become small as compared with the depth of heat wave penetration into the specimen. Hence, further increase of Δ or d has no effect on w.

2.1.2 Dependence of w on width, b, of flat specimen

The dependence $w(b)$ is important for modelling burning of real constructions. Two quite different cases are described in the literature:

(1)   the flame velocity diminishes with b decreasing;
(2)   the flame velocity rises with b decreasing.

The first case was observed for sheets of paper in metallic frames.[19,20] Heat losses to the frames retard the flame velocity for small b and at some b_{cr} the burning fails.

For PMMA sheets without any frames:[15]

b, mm	35	30	25	20	15	10	5
$w_{\varphi = -90°}$, mm/s	0.062	0.063	0.072	0.088	0.089	0.090	0.090

An increase of w with b decreasing may be connected with the interaction of heat waves propagating from the burning side surfaces of a specimen. This interaction is intense for small b.

It is natural that in the first case the flame front is convex in the flame propagation direction, whereas in the second case it is concave.

2.2 Effect of Specimen Orientation

Flame spread velocity over polymer surface can depend significantly on specimen orientation. This dependence is connected with the contribution of natural convection to the heat exchange between the flame and specimen surface.

2.2.1 Maximum and minimum flame rate

For all polymers studied the maximum flame rate, w_{max}, was observed for vertical upward burning ($\varphi = +90°$). In this case the direction of flow of hot gaseous combustion products coincides with the direction of flame propagation and the convection heat transfer from the hot gases to the virgin polymer specimen is the highest possible.

The minimum flame velocity, w_{min}, for polymer specimens not melting in the course of burning is observed for vertical downward flame propagation ($\varphi = -90°$) because the

TABLE 5
The ratio $w_{+90°}/w_{-90°}$ for polymer films burning in air.[4]

Polymer	PE		Polyamide			Polysulphone			Polycarbonate		
Δ, μ	60	120	60	200	500	30	60	500	25	60	500
$w_{+90°}/w_{-90°}$	3.0	5.1	3.1	3.2	2.9	4.3	3.8	3.2	3.3	3.9	3.3

direction of hot gases flow is opposite to that of flame spread. In this case $w_{max}/w_{min} = w_{+90°}/w_{-90°}$.

However, for specimens melting in the course of burning w_{min} can be observed for horizontal specimens ($\varphi = 0°$) because at $\varphi < 0°$ the melt flows in the direction of flame spread thus increasing the heat transfer to virgin polymer. For melting specimens w_{max}/w_{min} may be equal to $w_{+90°}/w_{0°}$.

Note that the burning rate is steady only at small φ. In contrary, for large φ and especially at $\varphi = +90°$, the burning rate is frequently non-stationary and reproducibility of w is not satisfactory.[19,21-23]

Few data relevant to the ratio w_{max}/w_{min} can be found in the literature. The following data were obtained for PMMA sheets[24] of various thickness:

Δ, mm	2.5	3.0	4.0	5.0	6.0	8.0
$w_{max}/w_{min} = w_{+90°}/w_{-90°}$	20	24	26	30	25	27

Somewhat lower values $w_{+90°}/w_{-90°} = 15–17$ were obtained for thick (25 mm) PMMA sheets.[22]

For thick PMMA sheets the $w(\varphi)$ dependence is very strong. For thin specimens, the available data are rather contradictory. For blank computer card (0.16 mm thick) the ratio $w_{+90°}/w_{-90°}$ was very high (~30).[21] In contrast, for polymer films studied[4] the ratio $w_{+90°}/w_{-90°}$ was much smaller, Table (5).

The ratio $w_{+90°}/w_{-90°}$ for polycarbonate and polysulphone films (60 μ thick) increases slightly with an increase in C_{O_2} from 0.2 to 0.3 but remains nearly constant with further rise in C_{O_2} in the range $C_{O_2} = 0.3–0.5$. For example, for polycarbonate films the following data were obtained:

C_{O_2}	0.2	0.3	0.4	0.5
$w_{+90°}/w_{-90°}$	3.9	4.4	4.3	4.4

For PE coatings on copper wires intense melting was observed in the course of burning.[5] In this case the ratio $w_{max}/w_{min} = w_{+90°}/w_{0°}$ was equal to 3–5 depending on coating thickness.

2.2.2 Floor and ceiling flames

The role of natural convection in the process of flame propagation is to compare the value w_f for floor and ceiling flames, w_c.[4] At small $|\varphi|$ the buoyancy can increase w_c but decrease w_f. In the ceiling configuration, the Archimedean force press the hot gases to bottom surface of specimens. In the floor configuration this force pushes the hot gases away from the surface of specimens. One may expect that $w_c > w_f$, all other conditions being equal.

TABLE 6
Ratio of w_c/w_f for one-side burning (in still air) of flat PMMA
specimens glued to copper underlay heated to 80°C[24]

Δ, mm	w_c/w_f		
	$\varphi = -45°$	$\varphi = 0°$	$\varphi = +45°$
3	1.6	2.6	1.4
4	1.4	3.0	1.5
5	1.3	3.4	1.2
6	1.5	3.6	1.2
8	1.5	4.0	1.4

In all the range studied ($-45° \leq \varphi \leq +45°$) $w_c > w_f$, Table (6).[24] The ratio w_c/w_f reached a peak for horizontal specimens ($\varphi = 0$) and decreased with $|\varphi|$ increasing. At $\varphi = 0$ the ratio w_c/w_f increased with specimen thickness, Table (6).

The flame spread was investigated over the surface of horizontal flat PMMA specimens 12.7 mm thick in a wind tunnel in a concurrent air flow at $v = 0.25$–4.5 m/s and various turbulence intensity, v'/v.[25] Both w_c and w_f increased monotonously with v. The ceiling flames were faster than floor flames only at $v \gtrsim 1$ m/s and the reverse was observed at $v \lesssim 1$ m/s. The ratio w_c/w_f increased first with v but then reaches a maximum and begins to diminish:

v, m/s		0.25	2	3	4.5
w_c/w_f	$v'/v = 1\%$	0.25	1.08	1.14	1.0
	$v'/v = 15\%$	0.83	2.35	2.33	1.15

The values w_c/w_f were sufficiently smaller[25] than those in (at $\varphi = 0$) still air,[24] Table (6).

A gas stream along the burning surface decreases the role of natural convection and the ratio w_c/w_f must diminish monotonously with v. However, the form of $w_c/w_f = f(v)$ curve in[25] is contradictory to this suggestion.

2.2.3   Form of w(φ) curve

The typical form of $w(\varphi)$ curve is as follows. In the range $\varphi = -90° + \varphi_*$ (where φ_* depends on specimen properties) the flame velocity is nearly constant or increases very slightly with φ. However, at $\varphi > \varphi_*$ the burning rate begins to grow with increasing φ. The following data were obtained for 100% cotton fabric:[23]

$\varphi°$	−90	−60	−45	−30	−20	0	+20	+30
w, mm/s	0.7	0.7	0.7	0.75	0.85	1.1	3.0	>5

For sheets of paper (240 μ thick) this dependence was somewhat stronger:[26]

$\varphi°$	−90	−70	−50	−30	−10	0	+10	+30
w, mm/s	1.05	1.1	1.15	1.3	1.6	2.8	6.0	7.4

TABLE 7

The slope of $w(\varphi)$ curve in various intervals of φ

Range of φ, grad		$-90 + -30$	$-30 + 0$	$0 ++ 20$	$0 ++ 30$
K, %/grad	cotton fabric[23]	0.12	1.6	8.6	—
	paper[26]	0.40	3.8	—	5.5

For blank computer card (160 μ thick)[21] the flame velocity was constant in the range $\varphi = -90° + -40°$ and began to grow at $\varphi > -40°$. The slope of $w(\varphi)$ curve may be characterized by the value $K = (w_{\varphi_2}-w_{\varphi_1})/w_{\varphi_1} (\varphi_2-\varphi_1)$. The values K for a cotton fabric and paper are listed in Table 7. A simple theoretical model[27] for $w(\varphi)$ dependence was obtained:

$$w_\varphi/w_{0°} = 1 + A_1 \sin^g \varphi \tag{3}$$

or more simply:

$$w_\varphi/w_{O°} = 1 + A_2 \sin \varphi \tag{4}$$

In some φ ranges the agreement between Eqs. 3 and 4 and experimental data for PMMA, fabrics, paper etc. is reasonable,[26,27] but for some other experimental data this agreement is not satisfactory. Data for filter paper grouped rather adequate near a straight line on log w, log sin φ plot.[19]

3   INFLUENCE OF PARAMETERS OF ENVIRONMENT

Effect of the following parameters on flame spread velocity was studied: oxygen content, C_{O_2}; pressure, P; temperature, T_o; gas flow velocity over polymer surface, v.

3.1   Effect of Oxygen Content

In all the cases studied w increased monotonously with C_{O_2}. This increase is due to the rise of heat production rate both in the homogeneous mixture in the flame tip and in the diffusion flame zone.

TABLE 8

Dependence of $w_{-90°}$ on C_{O_2} for polymeric films (60 μ thick) at atmospheric pressure[4]

Polymer	$C_{O_2} = 0.2$	0.3	0.4	0.5	0.6	0.7	0.8	1.0
				w, mm/s				
Polyamide	–	10.3	16.5	23.4	25.6	26.6	38.4	57.5
Polyethylene	7.0	14.3	19.2	22.5	25.6	31.3	35.4	50.5
Polycarbonate	4.0	6.7	9.9	13.1	15.2	17.4	19.4	22.7
Polysulphone	2.3	4.2	6.8	8.0	10.6	12.6	13.5	16.8

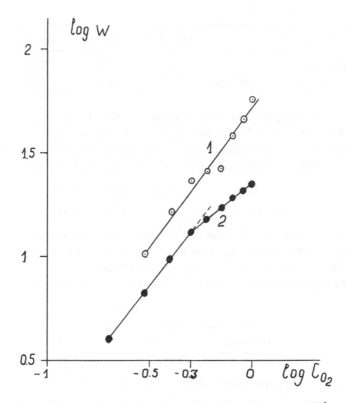

Figure 7   Dependence of log w on log C_{O_2} for polymer films 60 μ thick at $\varphi = -90°$.[4]

A common approximation for $w(p, C_{O_2})$ dependence is:

$$w = Ap^\nu \cdot C_{O_2}^m \tag{5}$$

It is seen in Fig. 7 that for polyamide films the data fall on a straight line over the range C_{O_2} studied. In contrast, for polycarbonate films two straight lines (the first at $C_{O_2} = 0.2 - 0.5$ and the second at $C_{O_2} = 0.5 - 1.0$) are needed to approximate the experimental data (Fig. 7). The values of exponent m in Eq. 5 are presented in Table 9.

The effect of inert diluents on $w(C_{O_2})$ dependences was investigated[28] for vertical downward burning of PS and PMMA specimens (3.18 mm thick) at $p = 0.03 - 3$ MPa and $C_{O_2} = 0.46$; 0.62, and 1.0. The dependences $w(C_{O_2})$ in $O_2 - $He mixtures was weaker than that in $O_2 - $Ar and $O_2 - N_2$ mixtures:

TABLE 9
Exponent m in Eq. 5 for vertical downward burning of polymeric film (60 μ thick) calculated on the base of data[4]

Polymer	PA	PE	Polycarbonate		Polysulphone	
C_{O_2} range	0.3–1.0	0.2–1.0	0.2–0.5	0.5–1.0	0.2–0.6	0.6–1.0
m	1.40	1.15	1.28	0.81	1.37	0.86

Atmosphere		$O_2 - He$	$O_2 - Ar$	$O_2 - N_2$
Exponent m	PS	1.52	2.16	2.28
in Eq. 5	PMMA	1.48	2.03	2.46

The downward burning of paper (0.8 mm thick) in $O_2 - He$ and $O_2 - N_2$ mixtures at $C_{O_2} = 0.18$–0.30 and $p \leqslant 1$ MPa was investigated.[29] In all the range C_{O_2} and p studied the flame velocity in $O_2 - He_2$ mixtures, $w_{O_2 - He}$, was greater than that in $O_2 - N_2$ mixtures, $w_{O_2 - N_2}$. At $p = 1$ MPa the following values of ratio of these velocities were obtained:

C_{O_2}	0.18	0.21	0.30
$w_{O_2 - He}/w_{O_2 - N_2}$	2.12	1.79	1.47

This result is quite natural as the heat conductivity of $O_2 - He$ mixtures is higher than that of $O_2 - N_2$ mixtures. Hence, the heat flux through the gas phase from the flame to polymer surface must be higher in the $O_2 - He$ atmosphere.

However, the dependence $w(C_{O_2})$ in $O_2 - He$ mixtures was weaker than that in $O_2 - N_2$ mixtures:[28]

Atmosphere	$O_2 - He$	$O_2 - N_2$
m	0.88	1.51
Pressure range, MPa	0.4–1.0	0.2–1.0

The exponent m in Eq. 5 for paper sheets burning vertical downward depends strongly on sheet thickness, Δ:[18]

Δ, mm	0.22	1.92
m	0.9	2.1

Consider now the effect of gaseous oxidiser flow velocity, v, over specimen surface on the exponent m in Eq. 5.

The downward burning of PMMA specimens 3.18 mm thick in an opposed flow of $O_2 - N_2$ mixture was studied.[18] The flame velocity, w, in an opposed flow at high enough C_{O_2} increases primarily with v but then begins to decrease (at $v > v_*$).

The exponent m calculated[18] is nearly independent of v at $v < v_*$ (and equal 1.6–1.7 at $v \simeq 0.2$–2 m/s). However, at $v > v_*$ the value m is much higher (and equal nearly 2.6 at $v \simeq 6$ m/s).

Thick (12.7 mm) horizontal PMMA specimens burning in co-current stream of $O_2 - N_2$ mixture were studied.[30] In the interval of v where the flame spread velocity augmented with v, the exponent m calculated[30] was nearly independent of v:

v, m/s	1	2	3
m	1.9	1.8	1.8
C_{O_2} range	0.2–1.0	0.2–1.0	0.3–1.0

The data[4,18,28-30] show that the exponent m in Eq. 5 varies in the interval $m \simeq 0.8 + 2.5$ depending on chemical nature of polymers, specimen thickness, oxygen content, properties of gaseous diluents, pressure, and gas flow velocity.

However, it is necessary to note that Eq. 5 is not an universal equation capable to approximate experimental data for all the polymers studied. The data[31] for PS specimens (0.75 mm thick) do not group satisfactory near straight lines on log w, log C_{O_2} plot.

3.2 Effect of Pressure

The vertical downward burning of paper sheets 0.8 mm thick was investigated[29] at $p \leqslant 1$ MPa. At first the flame velocity increased slightly with pressure but at $p > 0.1-0.2$ MPa became constant.

The following values of exponent v in Eq. 5 were obtained[18] for paper specimens (index cards, single or pressed together) burning downward at $p = 0.06-2$ MPa:

Δ, mm	0.22	1.92
v	$0.05 + 0.1$	0.63

Thus, the exponent v increased considerably with specimen thickness.

The data[28] relevant to $w(C_{O_2})$ dependence were presented in the preceding Sect. 3.1. Consider now the data of this paper relevant to $w(p)$ dependence. The exponent v was nearly the same for PMMA and PS specimens and nearly independent of chemical nature of diluents in the environment:[28]

Atmosphere	$O_2 - He$	$O_2 - N_2$	$O_2 - Ar$
v_{PMMA}	0.78	0.82	0.78
v_{PS}	0.80	0.76	0.83

Return now to the theoretical model of de Ris.[17] It is assumed that for "thin" specimens the only mass transfer mechanism is molecular diffusion. It is natural that for "thin" specimens the flame velocity is pressure-independent ($v = 0$). Indeed, the rate of heat release in the diffusion flame is proportional to ρD where $\rho \sim p$ is the density of gaseous oxidiser and $D \sim 1/p$ is the diffusion coefficient.

For "thick" specimens the heat losses due to natural convection were taken into account in[17] and the dependence $w \sim p^{2/3}$ was deduced.

It does not seem reasonable to ignore the role of the kinetic zone on the flame tip. Starting from the concept of FVC-zone it is possible to demonstrate that $w(p)$ dependence for thick specimens must be stronger than that for thin specimens.

The value w both for thin and thick specimens is controlled to a considerable extent by kinetic factors. However, for thick specimens heat losses deep into specimens take place resulting in a decrease of flame velocity as compared with that for thin specimens.[5] This decrease is more pronounced as p is lowered because the depth of penetration of heat wave into the specimen increases with w decreasing. Heat losses intensify the $w(p)$ dependence for thick specimens.

In the concept of FVC-zone the exponent v for thin specimens depends on kinetics of chemical reactions on the flame tip and, in general, is not equal to zero.

3.3   Effect of Temperature

The literature on $w(T_o)$ dependence (where T_o is the initial temperature of a specimen and environment) is even more limited than that on the $w(p)$ dependence.

In all the cases studied the flame velocity rises monotonously with T_o. To describe the $w(T_o)$ dependence the so called "temperature coefficient of flame velocity", $\beta = d \ln w/d T_o$, as well the ratio w_T/w_{T_o} may be used.

Burning of filter paper (specimens consisted of N sheets, each 0.15 mm thick glued together) was investigated.[32] The value $w_{-90°}$ diminished with N increasing. In contrary, the $w(T_o)$ dependence was the more pronounced, the greater was N. The following values of ratio $w_{200°C}/w_{20°C}$ and β may be obtained from the data:[32]

N	2	4	8	12
$w_{200°C}/w_{20°C}$	2.2	2.9	3.4	3.6
$\beta_{20+200°C}\cdot10^3$, 1/grad	4.5	5.9	6.9	7.1

The rise of $w_{200°C}/w_{20°C}$ and β with N can be interpreted starting from the postulate of Belyaev.[33] The temperature coefficient β increases with decrease of temperature in the FVC-zone, T_*, which, in turn, diminishes with an increase of specimen thickness. For paper specimens the decrease of T_*, with N increase is more pronounced than that for char-free burning polymers (like PMMA etc.) because only outer sheets of paper burn completely in the process of flame spreading.

Burning of PS specimens 0.75 mm thick was studied[31] at $T_o = 21 + 130°C$ and $C_{O_2} = 0.20–0.35$. The following values of β may be obtained:[31]

C_{O_2}	0.20	0.30	0.35
$\beta_{50+130°C}\cdot10^3$, 1/grad	9.0	5.9	4.2

It is reasonable to suppose that T_* increases with C_{O_2}. Thus, the reduction of β with C_{O_2} increasing is consistent with Belyaev's postulate.

It is ascertained[14] that two regimes are possible for flame spread over the surface of liquid fuels:

(1)   slow-velocity regime (up to 1 cm/s) at low enough T_o;
(2)   fast-velocity regime (up to 1 m/s) at high enough T_o.

Rather narrow interval ∂T_o exists where the transition from slow-to fast-velocity regime occurs. In this interval the flame velocity increases extremely rapidly. The value β for n-butanol (burning in air) calculated on the base of data[34] in the range $T_o = 26 + 38°C$ is equal 0.25 1/grad. This is nearly two orders of magnitude higher than common values β for flame spread over the surface of various solid and liquid fuels burning in air.

The mechanism of the fast-velocity regime[14] at high enough $T_o = (T_o)_{ll}$ the tension of saturated vapour, p_s, of a liquid fuel becomes equal to the partial pressure of fuel in gaseous fuel-air mixture on the lean concentration limit, p_{ll}. For n-butanol p_s is equal to 6.9 Torr

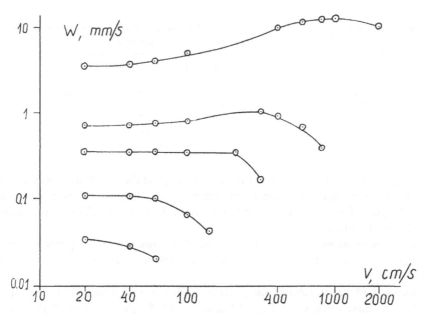

Figure 8   Effect of opposite gas flow velocity, v, of $O_2 - N_2$ mixture on flame spread rate over the surface of thick sheets of PMMA ($\varphi = -90°$).[35] 1–C_{O_2} = 1.0; 2–C_{O_2} = 0.432; 3–C_{O_2} = 0.329; 4–C_{O_2} = 0.247; 5–C_{O_2} = 0.211.

at 25°C and 33.1 Torr at 50°C, where p_{ll} is equal to 12.9 Torr under normal conditions. The partial pressure of fuel vapour in the stoichiometric mixture n-butanol — air (under normal conditions) is equal to $p_{st} = 25.6$ Torr.

It is reasonable to propose that at $T_o - (T_o)_{ll}$ the flame can propagate through the homogeneous gaseous mixture over the surface of liquid fuel. The common values of normal flame velocity of laminar flames in near-stoichiometric hydrocarbon-air mixtures are 30–60 cm/s, whereas velocity of turbulent flames can be much higher (up to 10 m/s).

The transition from slow- to fast-velocity regime means the transition from the propagation of a diffusion flame over liquid fuel — air interface to the propagation of a kinetic flame through premixed gaseous mixtures: fuel vapour — air over the surface of liquid fuel.

It is natural to suppose that for solid polymers homogeneous mixtures of air and gaseous products for thermodestruction of polymers also can be formed at high enough T_o. Certainly, the values $(T_o)_{ll}$ in this case must be much higher than those for common liquid fuels. It seems quite possible that $(T_o)_{ll}$ can be acquired in real fire conditions and then w may increase rather drastically.

3.4   Effect of Gas Stream along the Specimen Surface

Experimental data was obtained for the opposite gas flow,[35] for PMMA sheets are presented in Fig. 8. If the oxygen content in the gas stream is low enough the flame rate declines monotonously with gas velocity, v, increasing due to reduction of heat transfer from the

TABLE 10
Dependence of burning rate of one sort of polyimide films on the velocity of an opposite stream of $0.6 \, O_2 + 0.4 \, N_2$ mixture at various film thickness[36]

Δ, μ	$w_{-90°}$, mm/s					
	$v \approx 0.1$ m/s	0.5 m/s	1.0 m/s	1.2 m/s	1.4 m/s	1.8 m/s
60	2.1	2.2	2.1	2.2	2.1	1.8
240	0.33	0.45	0.59	0.61	0.54	—

flame to virgin polymer. In contrast, at high enough C_{O_2} the flame rate increases first with gas velocity but then reaches a maximum and begins to decrease.[6] A rise of w with v on the initial part of $w(v)$ curve is connected with a decrease of thickness of boundary layer on burning surface.[35]

A paradoxical result was obtained for horizontal specimens of filter paper in counter-stream of air: w decreased at first with v increasing but then became constant (in the range $v = 15$–25 cm/s).[19]

Experiments with paper, fabrics, and polyimide films were conducted[36] in the opposite flow of $O_2 - N_2$ mixtures at $C_{O_2} = 0.21$–1.0 and $v = 0.1$–5 m/s. In contrast[19] the value of w for paper and one sort of fabric remained nearly constant at $v \leqslant v_*$ and decreased with v increasing at $v > v_*$. For paper 140 μ thick at $C_{O_2} = 0.6$ the following data were obtained:

v, m/s	0.5	1	2	3	4
$w_{-90°}$, mm/s	8.3	8.4	8.5	8.4	6.2

It is interesting to note the results for polyimide films. Namely, at $v < v_*$ the value of w for thin enough films is independent of v but for thick enough films w increases significantly with v, however, at $v > v_*$ in both cases w decreases with v increasing, Table 10.

Experiments in wind tunnel were conducted[37] with flat horizontal specimens of PMMA (12.7 mm thick) and Whatman (0.33 mm thick) in the opposed air stream at various turbulence intensity, v'/v, where v' is the pulse velocity.

At $v > 0.5$ m/s the flame spread velocity declined with v increasing at all v'/v studied. At $v < 0.5$ m/s and $v'/v = 1$ and 5% an increase in w with v was observed. This initial rise of w may be connected with underventilation of wind tunnel at small enough v.[37]

The dependence $w(v'/v)$ was not monotonic. An increase in v'/v from 1 to 5% resulted in rise of w (1.15–1.3 times depending on v). However, further increase of v'/v diminished w, the value of w at $v'/v = 25\%$ was 1.5–2 times smaller than that at $v'/v = 5\%$.

For Whatman specimens the flame velocity decreased monotonously with an increase in v and v'/v. At some $v = v_{cr}$ the burning failed, v_{cr} was as smaller as greater was v'/v.

As we have seen above, in Sect. 2.2, the burning of specimens in still air at high enough φ (and especially at $\varphi = +90°$, i.e. vertical upward) may be unsteady and flame front may be non-one dimensional.

It is natural to expect that just the same character of burning can be observed for flow-assisted flame spread in a non-stationary air stream (wind, convection flows in the course of real fires etc.).

In contrast, according to Refs. 25, 30, the burning in a well-organized steady-state concurrent stream of air may be quite stationary.

The burning rate in a concurrent gas stream, w_{cc}, is much higher than that in an opposed stream, w_{op}. Thus, at $v = 1$ m/s and $v'/v = 5$ and 15% w_{cc} for PMMA specimens (12.7 mm thick)[25] was nearly by one order of magnitude higher than w_{op}.[37]

The $w_{cc}(v)$ dependence was nearly linear:[30] $w_{cc} = a + bv$, where a and b depended on C_{O_2}. The value w_{cc}[25] also increases monotonously with v in the range 0.25–4,5 m/s. However, the dependence $v_{cc}(v'/v)$[25] was not monotonic: the rise in turbulence intensity from 1 to 15% results in an increase in w_{cc} at $v < 1$ m/s but in a decrease in w_{cc} at $v > 1$ m/s.

4 SUPPLEMENT, NORMAL BURNING RATE (NBR) OF POLYMERS

Normal burning rate, u, (i.e. normal regression rate of polymer surface) is also a significant characteristic of polymer flammability. Really, the greater is u, the higher is the heat evolution rate, J/cm^2 s.

4.1 NBR in the Course of Flame Spreading

NBR may be influenced by the same parameters as flame spread velocity, w, but in addition, it depends significantly on the distance, x, from the flame tip. It is a principal feature of diffusion flames. Indeed, the path of diffusion of oxygen to the flame through a layer of combustion products increases with x.

In the course of burning a cavity appears in the specimen behind the flame tip. According to theoretical model,[3] the profile of this cavity far enough from the flame tip is parabolic:

$$x = ay^k \tag{6}$$

where $k = 2$, x axis is turned along $\vec{w}$ vector, but in the opposed direction, and y axis is normal to it and directed inside the specimen.

The form of cavities and NBR was measured[38,39] for a special case of PMMA and PS slabs contacting with $KClO_4$ and NH_4ClO_4 slabs, Table 11.

TABLE 11
Flame spread velocity, w, and NBR, u, of PMMA slabs contacting with $KClO_4$ slabs

p, MPa		1.0	1.5	2.5	3.0
w, mm/s		1.72	2.03	3.08	3.60
u, mm/s	$y = 0.2$ mm	0.60	—	0.75	0.81
	$y = 0.3$ mm	0.50	0.53	0.58	0.69
	$y = 0.4$ mm	0.41	0.44	0.50	—
	$y = 0.5$ mm	0.35	0.38	—	0.39
	$y = 0.9$ mm	0.16	0.18	—	0.15
	$y = 1.0$ mm	0.15	0.15	0.17	—

TABLE 12

Flame spread velocity and NBR (in air at normal conditions) for horizontal flat specimens of PMMA
on various underlays at near-critical specimens thickness[40]

Underlay	Δ_{cr}, mm	W_{cr}, mm/s	$\bar{u}_{cr}$, mm/s	$\bar{u}_{cr}/w_{cr}$
Dural, 15 mm thick	7.4	0.031	0.0015	1/20.7
Copper, 1 mm thick	5.2	0.039	0.0020	1/19.5
Copper, 0.3 mm thick	2.5	0.068	0.0038	1/17.9
Copper, 0.05 mm thick	1.5	0.080	0.0047	1/17.0
Textolite, 12 mm thick	2.5	0.052	0.0031	1/16.8

It follows that the NBR of PMMA slabs decreases very fast with y (and thus with x) increasing. The dependence $u(p)$ is much more weak than $w(p)$ dependence. Moreover, at great enough y NBR is pressure-independent. As a consequence, the ratio u/w is the smaller, the greater is y (and x) and pressure. Thus, the following values u/w can be obtained from Table 11 at 1 MPa:

y, mm	0.2	0.3	0.4	0.5	0.9	1.0
u/w	1/2.9	1/3.4	1/4.2	1/4.9	1/10.8	1/11.5

and at $y = 0.2$ mm:

p, MPa	1.0	2.5	3.0
u/w	1/2.9	1/4.1	1/4.4

Experimental values of exponent k in Eq. 6 for $PMMA_0$-$KClO_4$ system ranged from 1.4 to 2.4 depending on pressure and x.[38] Such a departure from the theoretical value $k = 2$ may be partly explained by the presence of random unevenness on quenched polymer surface (soot, frozen bubbles etc.).

Consider now the case of specimens burning in air at atmospheric conditions. NBR (averaged over the interval $x = 0$–3 mm) was measured[40] for PMMA specimens whose thickness was close to critical (Table 12). It is seen from Table 12 that in these conditions NBR ranges from 0.0015 to 0.0047 mm/s. This is 100–30 times smaller than that for PMMA slabs contacting with $KClO_4$ slabs at $p = 1$ MPa and $y = 1$ mm, Table 11. It seems probable that this difference is connected mostly with two factors. Firstly, the data of Table 12 are relevant to the burning in air, whereas those of Table 11 to that in pure oxygen. Secondly, the flow of hot combustion products inside the cavities in fuel and oxidizer slabs may increase the NBR considerably. This effect is absent for specimens burning in open atmosphere.

Another result worth noting in Table 12 is as follows: NBR for PMMA specimens burning in air is 17–21 times smaller than w.

The downward burning of vertical cylindrical PMMA rods was investigated.[11] The values w and u (averaged over nearly conical burning surface) are listed in Table 13.

It is seen in Table 13 that u diminishes approximately twice with rod diameter, d, increasing from 6.8 to 20.4 mm but then becomes nearly independent of d. Further, in all the range studied NBR is 4–5 times smaller than w.

TABLE 13
Flame spread velocity and mean value NBR for PMMA rods burning in air at normal conditions
$(\varphi = -90°)$[11]

Rod diameter, mm	6.8	11.7	15.0	20.4	25.0	35.2	40.7
w, mm/s	0.082	0.058	0.046	0.037	0.040	0.044	0.042
u, mm/s	0.020	0.015	0.011	0.0091	0.0091	0.0090	0.0086
u/w	1/4.1	1/3.9	1/4.2	1/4.1	1/4.4	1/4.9	1/4.9

4.2 NBR on the Entirely Burning Surface

The burning of spherical particles of polymers was studied.[41,42] The well-known equation[43,44] for the diffusion burning of spherical drops of liquid fuels, was used to describe the $d(t)$ dependence:

$$d^2 = d_0^2 - Kt \qquad (7)$$

where d_o is the initial diameter of a particle, t is the time, and K is a constant. It follows from Eq. 7:

$$u = \frac{1}{2}\left|\frac{\partial d}{\partial t}\right| = K/4d \qquad (8)$$

So, NBR of spherical particles (or drops) is proportional to K and inversely proportional to particle diameter.

The following results were obtained[41] for spherical particles ($d_o = 2$ mm) of PMMA and co-polymers of MMA and tri(ethylene glycol) methacrylate (TEGMA — a cross-linking agent):

Polymer	MMA (monomer)	PMMA	98% MMA + 2% TEGMA	90% MMA + 10% TEGMA
$K \cdot 10^3$, cm^2/s	7.8	6.9	8.3	7.7

No significant difference was observed between the values K for MMA drops and particles of PMMA, as well as between those for particles of PMMA and PMMA + TEGMA. It is necessary to note that the values K obtained[41] lie in the interval characteristic for most part of organic liquids,[45] the constant K for 32 organic liquids ranges from $6.6 \cdot 10^{-3}$ to $11.4 \cdot 10^{-3}$ cm^2/s.

In contrast,[42] the values K for particles ($d_o = 1$–2 mm) of several polymers burning in still air were much higher than those of organic liquids:

Polymer	PMMA	PP	PE	PS
$K \cdot 10^3$, cm^2/s	28	25	25	17

It is hard to explain such a drastic difference between the results.[41,42]

TABLE 14

NBR at nearly atmospheric pressure of PMMA blocks (internal burning of the concentric hole) at various distance, x, from inlet and various t[47]

t, s	u, mm/s		
	$x = 2-4"$	$x = 10-12"$	$x = 22-24"$
10	0.13	0.15	0.22
30	0.084	0.088	0.11
60	0.066	0.064	0.078
120	0.053	0.048	0.055
150	0.050	0.044	0.050

NBR of horizontal rotating PMMA cylindrical specimens burning in air was measured.[46] The mean value u, averaged over interval from $d_o = 15.0$ to $d = 5.4$ mm, was measured equal to 0.032–0.035 mm/s and was surprisingly independent of air temperature in the interval $T_o = 24 + 170°C$, air velocity (from ~0 to 30 mm/s), and rotation velocity (from 0.033 to 100 rps). Besides, some time after ignition NBR ceased to depend on d. This result seems to be rather doubtful since the path of oxygen diffusion depends on d.

In the 1960's efforts were undertaken to develop a hybrid rocket motor: a strand of solid polymeric fuel burning in a stream of liquid or gaseous oxidizer.

In the frame of this problem the internal burning of PMMA specimens with cylindrical hole was investigated.[47,48] Specimens studied[47] consisted of 12 blocks (each 2" long) pressed together. A stream of gaseous O_2 was delivered in the concentric hole ($d_o = 1"$). The mean value u was measured on the basis of mass-loss of each block during the time, t, of a run. The data obtained at oxygen flux 22 g/s are listed in Table 14.

It follows from Table 14 that NBR decreases significantly with an increase of t but augments with x. The first of these effects may be connected with an increase of hole diameter, d, in the course of the experiment and hence a decrease of oxygen flux per unit of cross-sectional area, g/cm² s. The second effect may be due to an increase of temperature of gas stream along the length of the hole.

Cylindrical specimens of PMMA (8 mm i.d., 30 mm o.d., and 70 mm long) were investigated[48] at an oxygen flux $(\rho v)_{O_2} = 11-40$ g/cm² s (at the first moment of run, $t = 0$) and $p = 0.5-7$ MPa. The following expression was used for NBR:

$$u = A(\rho v)^{\gamma}_{O_2} p^v d^{\eta} \qquad (9)$$

At $(\rho v)_{O_2} = 11-30$ g/cm² s and $p = 0.5-5$ MPa the exponents γ and v were equal to 0.40 + 0.42 and 0.10, respectively. The NBR data are presented in Table 15.

TABLE 15

NBR of PMMA specimens (internal burning of a concentric hole 8 mm dia in axial stream of gaseous O_2; $t \leq 5$ s)[48]

p, MPa		1.0	2.0	3.0	4.0	5.0
u	$(\rho v)_{O_2} = 11$ g/cm² s	0.36	0.38	0.40	0.41	0.42
mm/s	$(\rho v)_{O_2} = 30$ g/cm² s	0.54	0.58	0.60	0.62	0.63

TABLE 16

NBR and exponent γ_1 in Eq. 10 for several polymers burning in an impinging flow of air (at volume flux 100 cm^3/cm^2 s)[49]

Polymer	PMMA		PP	PE	
	nozzle type A	nozzle type B		low ρ	high ρ
u, mm/s	0.036	0.028	0.022	0.018	0.016
γ_1	0.47	0.50	0.50	0.52	0.46

The burning of the flat top of vertical rods (12.7 mm dia) of several polymers were studied.[49] Air (at normal conditions) flowed out a nozzle normal to specimen surface. A common approximation was used for $u(v)$ dependence:

$$u = Av^{\gamma_1} \tag{10}$$

The data relevant to NBR and exponent γ_1 are listed in Table 16.

4.3 Comparison of Values u Measured at Various Conditions

In conclusion of Section 4, it is of interest to compare the values u obtained for specimens of various configurations at various conditions.[7]

The data, Table 17 show that the difference between the values u measured at various conditions may be as much as two order of magnitude and even more. NBR depends strongly on the velocity of gas stream over the burning surface and oxygen content in this stream. Such dependences are typical for diffusion burning of solid fuels in a gaseous oxidizing atmosphere. The maximum NBR values were obtained[48] for internal burning of PMMA specimens with cylindrical hole at high enough p and $(\rho v)_{O_2}$ as well as[38] for PMMA blocks contacting with $KClO_4$ blocks (also at high pressure).

TABLE 17

Normal burning rate of PMMA specimens of various configurations at various conditions

Specimens configuration and burning conditions	Interval of $u \cdot 10^2$, mm/s	Refs.
1. NBR in the Course of Flame Spreading		
1.1. Horizontal sheets on various underlays in still air	0.15–0.47	40
1.2. Downward burning of vertical rods in still air	0.86–2.0	11
1.3. PMMA slabs contacting with $KClO_4$ slabs at $p = 1$–3 MPa	15–81	38
2. NBR on Entirely Burning Surface		
2.1. Axially rotating rods burning in air at $v \leq 30$ mm/s	3.2–3.5	46
2.2. Internal burning of specimens with hole at various d_o, p, $(\rho v)_{O_2}$, and t	4.4–63	47,48
2.3. Top end of rods burning in air flow (100 cm^3/cm^2 s)	2.8–3.6	49

CONCLUSIONS

1. Flame spread velocity over polymer surface, w, is one of the most significant characteristics of polymer flammability because the time of development of a full-scale fire depends in a considerable extent on w.
2. The value w depends on chemical and physical properties of polymers and fillers, dimensions and orientation of a polymer specimen, and properties of gaseous atmosphere (oxygen content, pressure, temperature, gas velocity over specimen).
3. The formation of a char layer on the burning surface can diminish considerably the flame velocity (or even prevent burning). The melting of polymer in the course of burning and dripping the melt can decrease or, in contrast, augment the burning rate depending on orientation of a specimen. For oriented composites the flame rate depends on flame propagation direction.
4. Flame spread velocity can depend rather strongly on specimen thickness, Δ. For thick enough specimens $w \neq f(\Delta)$ in all the cases studied. However, for thin enough specimens $w(\Delta)$ dependence may be quite different for various specimen types. For one-side burning of flat specimens on an underlay with high enough thermal conductivity, λ_{un}, the flame velocity at some Δ_* begins to decrease with Δ due to heat losses in the underlay until extinction at Δ_{cr}. In contrast, in the case of one-side burning of a specimen on an underlay with small enough λ_{un}, as well as in the case of two-side burning of a flat specimen without underlay (or candle-like burning of bare polymer rods) the flame velocity augments with Δ (or rod diameter, d) decreasing.
5. The flame velocity can depend considerably on specimen orientation (commonly characterized by the angle, φ, between burning direction and horizontal plane). This dependence is connected with the contribution of natural convection to the heat exchange between the flame and the specimen. For all polymers studied w_{max} is observed for vertical upward burning ($\varphi = +90°$). The value w_{min} is usually observed for vertical downward burning ($\varphi = -90°$). However, for specimens melting in the course of burning w_{min} may be observed for horizontal specimens ($\varphi = 0°$). For thick enough PMMA specimens the ratio w_{max}/w_{min} may be as high as 15–30.
6. Flame spread velocity increases monotonously with oxygen content, C_{O_2}, in the environment. The $w(C_{O_2})$ dependence may be approximated by power function $w = AC_{O_2}^m$ where $m \simeq 0.8 + 2.5$ depending on composition of polymer and environment, specimen thickness, pressure, and gas stream velocity over specimen.
7. The value w is pressure-dependent. The $w(p)$ dependence is usually approximated by power function $w = B_p^v$ where $v = 0 + 1$. According to some theoretical predictions and experimental data the $w(p)$ dependence for thick specimens is more pronounced than that for thin specimens.
8. In all the cases studied w rises monotonously with initial temperature, T_o, of specimens and environment. An abrupt increase in the flame velocity may be forecasted at high enough T_o due to formation near specimen surface homogeneous mixtures cf gaseous products of polymer thermodestruction with air.
9. The flame velocity depends on gas stream velocity, v, over specimen surface. If gas flow direction is opposite to flame spread direction, the value w_{op} at small enough C_{O_2} diminishes with v increasing. However, at high enough C_{O_2} the burning rate augments first with v, but then begins to drop until extinction. For the burning in

concurrent stream the value w_{cc} increases monotonously with v. The value w_{cc} may be at least by one order of magnitude higher than w_{op}.

10. Normal burning rate, u, i.e. normal regression rate of polymer surface behind the flame front, is also an important characteristic of polymer flammability. Indeed, the heat production rate, J/cm^2 s, on burning surface is proportional to u. The measured values u lie in a very broad interval from 0.0015 to ~1 mm/s depending on specimen configuration and environmental conditions.

REFERENCES

1. C.F. Cullis and M.M. Hirschler, *The Combustion of Organic Polymers* (Clarendon Press, Oxford, 1981).
2. R.M. Asseva and G.E. Zaikov, *Gorenie Polimernykh Materialov* (Nauka, Moscow, 1981).
3. N.N. Bakhman and V.B. Librovich, *Combust. Flame*, **15**, 143 (1970).
4. V.N. Vorobjev, L.V. Bychikhina, E.N. Basenkova and B.V. Perov, *Vysokomolekularnye Soedineniya*, (A) **26**, 2181 (1984).
5. N.N. Bakhman, L.I. Aidabaev, B.N. Kondrikov and V.A. Filippov, *Combust. Flame*, **41**, 17 (1981).
6. V.A. Ushkov, V.M. Lalayan, S.E. Malashkin, D.Kh. Kulev, L.G. Filin, M.S. Skralivetskaya and N.A. Khalturinskii, *Plastmassy*, No. 1, **66** (1989).
7. N.N. Bakhman, V.I. Kodolov, K.I. Larionov and I.N. Lobanov, *Fizika Goreniya i Vzryva*, **22**, 22 (1986).
8. L.I. Ildabaev, N.N. Bakhman, B.N. Kondrikov and L.A. Shutova, *Fizika Goreniya i Vzryva*, **17**, 82 (1981).
9. N.N. Bakhman and A.F. Belyaev, *Gorenie Geterogennykh Kondensirovannykh Sistem* (Nauka, Moscow, 1967; English translation: Combustion of Heterogeneous Condensed Systems, RPE Trans. -19, 1967).
10. A.C. Fernandez-Pello and F.A. Williams, *15th Symposium (Int.) on Combustion* (The Combustion Institute), p. 217 (1974).
11. N.N. Bakhman and L.I. Aldabaev, in: *Gorenie Geterogennykh i Gasovykh Sistem* (Chernogolovka), p. 7 (1977).
12. M. Sibulkin and C.K. Lee, *Combust. Sci. Technology*, **9**, 137 (1974).
13. K. Kishore, K. Mohandas, and I. Spilda, *Combust. Flame*, **52**, 215 (1983).
14. N.N. Bakhman, *Fizika Goreniya i Vzryva*, **23**, 41 (1987).
15. R.Mc Ilhagger and B.J. Hill, *Fire & Materials*, **5**, 123 (1981).
16. N.N. Bakhman and I.N. Lobanov, *Fizika Goreniya i Vzryva*, **19**, 46 (1983).
17. J.N. de Ris, *12th Symposium (Int.) on Combustion* (The Combustion Institute), p. 241 (1969).
18. F.A. Lastrina, R.S. Magee and R.F. Mc Alevy, *13th Symposium (Int.) on Combustion* (The Combustion Institute), p. 935 (1971).
19. M. Sibulkin, M. Ketelhut and S. Feldman, *Combust. Sci. Technology*, **9**, 75 (1974).
20. A.D. Margolin and V.G. Krupkin, *Fizika Goreniya i Vzryva*, **14**, 56 (1978).
21. T. Hirano, S.E. Noreikis and T.E. Waterman, *Combust. Flame*, **22**, 353 (1974).
22. A. Ito and T. Kashiwagi, *Combust. Flame*, **71**, 189 (1988).
23. N.N. Bakhman, I.N. Lobanov, Sh.A. Nasybullin, I.N. Faizullin and G.P. Sharnin, *Polymer Sci.*, **34**, 411 (1992).
24. N.N. Bakhman and I.N. Lobanov, *Fizika Goreniya i Vzryva*, **27**, 63 (1991).
25. L. Zhou and A.C. Fernandez-Pallo, *Combust. Flame*, **92**, 45 (1993).
26. Sh.A. Nasybullin, M.A. Zaripov and I.N. Faizullin, *Vysokomolekularnye Soedineniya*, (A) **31**, 332 (1989).
27. N.N. Bakhman, V.I. Kodolov, K.I. Larionov and I.N. Lobanov, *Fizika Goreniya i Vzryva*, **24**, 63 (1988).
28. R.F. McAlevy and R.S. Magee, *12th Symposium (Int.) on Combustion* (The Combustion Institute), p. 215 (1969).
29. V.A. Merkulov, R.M. Asseva and K.P. Kuz'menko, *Fizika Goreniya i Vzryva*, **27**, 29 (1991).
30. H.-T. Loh and A.C. Fernandez-Pello, *20th Symposium (Int.) on Combustion* (The Combustion Institute), p. 1575 (1984).
31. R. McIlhagger, B.J. Hill and A.J. Brown, *Fire & Materials*, **12**, 19 (1988).
32. A.S. Campbell, *J. Fire Flammability*, **5**, 167 (1974).
33. A.F. Belyaev and G.V. Lukashenya, *Doklady Akademii Nauk SSSR*, **148**, 1327 (1963).

34. A.M. Kanury, in: *Combustion Experiments in a Zero-gravity Laboratory* (New York), p. 195 (1981).
35. A.C. Fernandez-Pello and T. Hirano, *Combust. Sci. Technology*, **32**, 1 (1983).
36. A.F. Zhevlakov and Yu.M. Groshev, *Fizika Goreniya i Vzryva*, **23**, 36 (1987).
37. L. Zhou, A.C. Fernandez-Pello, and R. Cheng, *Combust. Flame*, **81**, 40 (1990).
38. B.F. Shirokov and N.N. Bakhman, *Fizika Goreniya i Vzryva*, **8**, 247 (1972).
39. V.F. Martynyuk, N.N. Bakhman and I.N. Lobanov, *Fizika Goreniya i Vzryva*, **13**, 176 (1977).
40. N.N. Bakhman, B.N. Kondrikov, S.O. Raubel and L.A. Shutova, *Fizika Goreniya i Vzryva*, **19**, 7 (1983).
41. A.S. Steinberg, V.B. Ulybin, E.I. Dolgov and G.B. Manelis, in: *Goreniya i Vzryva* (Nauka, Moscow), p. 124 (1972).
42. R.H. Essenhigh and W.L. Dreier, *Fuel*, **48**, 330 (1969).
43. G.A. Warshawskii, *Gorenie Kapli Zhidkogo Topliva* (Byuro novoi tekniki NKAP, 1945).
44. G.A.E. Godsave, *4th Symposium (Int.) on Combustion* (The Williams and Wilkins Comp., Baltimore), p. 818 (1953).
45. A. Willimas, *Combust. Flame*, **21**, 1 (1973).
46. V.N. Vorobjev, S.L. Barbot'ko, T.V. Popova, A.S. Steinberg and N.A. Khalturinskii, in: *Abstracts of 1st Int. Conference on Polymers of Low Flammability* (Alma-Ata), V. 1, p. 118 (1990).
47. T.J. Houser and M.V. Peck, in: *Heterogeneous Combustion (Academic Press, New York-London)*, p. 559 (1964).
48. Yu.S. Kichin, S.A. Osvetimskii, and N.N. Bakhman, *Fizika Goreniya i Vzryva*, **9**, 384 (1973).
49. D.J. Hove and R.F. Sawyer, *15th Symposium (Int.) on Combustion* (The Combustion Institute), p. 351 (1974).
50. A.G. Gal'chenko, N.A. Khalturinskii and A.A. Berlin, *Vysokomolekularnye Soedineniya*, (A) **22**, 16 (1980).

Nomenclature

b	—	specimen width
C_{O_2}	—	oxygen content in the environment (molar part)
d	—	diameter of cylindrical or spherical specimen
D	—	diffusion coefficient
K	—	constant in the Eq. 11
l	—	specimen length
$\dot{m} = \rho w$	—	mass burning velocity
p	—	pressure
P	—	porosity of specimen
t	—	time
T	—	temperature
u	—	normal burning rate (NBR)
v	—	glass flow velocity
v'/v	—	turbulence intensity in gas stream
w	—	flame spread velocity over polymer surface
w_c	—	velocity of ceiling flame
w_{cc}	—	flame velocity in co-current gas stream
w_f	—	velocity of floor flame
w_{op}	—	flame velocity in opposite gas stream

Greek letters

β	—	temperature coefficient of flame velocity
Δ	—	specimen thickness
λ	—	heat conductivity
ρ	—	density
φ	—	angle between burning direction and horizontal plane

Indices

cr	—	related to critical conditions
o	—	related to initial state or basic system
un	—	related to underlay (support)

Temperature Effects on Thermodestructive Processes

E.F. VAYNSTEYN, G.E. ZAIKOV and O.F. SHLENSKY

INTRODUCTION

Polymer materials are used in a wide range of working conditions. Low temperatures for a long time and high temperatures for a short time represent typical extremes. Changes are induced by both chemical and physical processes. At relatively low temperatures the dependence of the logarithm of the thin film lifetime τ on $1/T$ are close to Arrhenius one.[1] Deviations from this behaviour are observed at high temperatures.

EXPERIMENTAL METHODS

Thin films of thermoplastic polymer melt were applied to a metal plate preheated to a given temperature. A film of 7 μm thick was formed on the plate surface.[2] Inspite of intensive gas evolution, especially at high temperature, the melt film appeared strongly coupled to the plate surface. The film lifetime on the plate surface was measured either visually or by a thermocouple with its junction situated at the point of decomposition.

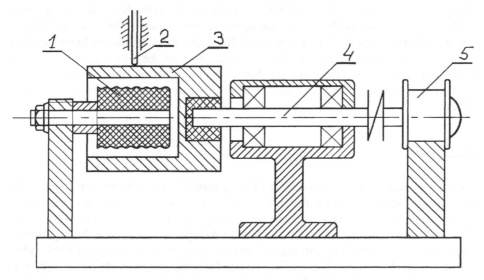

Figure 1 Experimental installation diagram.

219

Figure 2   Photo showing disappearing trace of metal-base polymer film.

The lifetime of the film was measured using IR-light and a video-analysis system based on SEKAM heat vision monitor.[3] Experimental set-ups for visual or thermocouple observation Fig. 1 continues. A moving specimen, the best picture of the melted film was observed in IR light by self-emission of the object, Fig. 2.

A white spot trace is observed which disappears like a comet tail. The length of the tail divided by the rate of movement determines the lifetime with high accuracy. At relatively low temperatures (below 570°C) visual and thermocouple measurements were used. The time required for the film to heat up to the temperature of the plate at relatively high temperatures (below 800°C) was less than 0.01 s (heat rate $5 \cdot 10^4$ grad/s). Experimentally the heat up time can be determined from measurement of the film emission intensity changes in a series of the video pictures. For a film thickness of about 20 μm, the time did not exceed 40 ms. The heating time is proportional to the fourth power of the film thickness.

DISCUSSION

The results of tests, Fig. 3 plotted as lg $(1/\tau)$ against $1/T$ for a number of linear polymers.[3-5] At low temperatures the dependence is close to linear. Linearity is characteristic of a one-staged (Arrhenius equation) or consecutive processes. Below 480°C, the experimental data agree with the literature ones.[1] At higher temperatures the dependence deviates from a straight line curving up sharply towards a limit asymptotical value. Limiting temperatures have characteristic values for every polymer. However, polymers of different chemical structures have relatively similar limiting temperatures. Despite different mechanisms occurring in different polymers all the dependences observed are qualitatively similar.

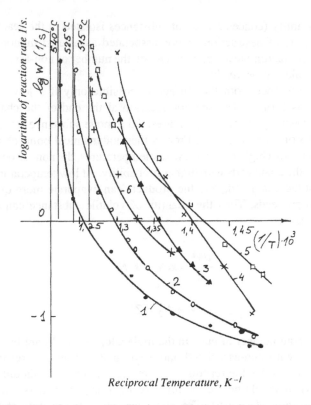

Figure 3   Relationship of lg of polymer thin film inverse life time vs inverse temperature. 1. Impact of resistant polystyrene; 2. Polystyrene (block); 3. Polyethyleneterephthalate; 4. Polyethylene (low density); 5. Polymethylmethacrilate; 6. Polyethylene (high density).

Analysis of the structure and reactivity suggests the pressure of associated and non-associated monomer links leading to parallel processes, Fig. (3).[6] The rate of equilibration, between associated and non-associated monomer links is higher than that of chemical reactions because of differences in the energies of formation of physical (intermolecular) and chemical bonds. As the temperature increases the number of non-associated monomer links increases, the rate constant (K) of the scission of the type

is higher for non-associated monomer links than for associated ones. The reason is that to remove radicals formed in associated monomer links it is necessary to consume energy to tear off intermolecular bonds: otherwise radical recombination ($R^{\cdot} + R^{\cdot} \rightarrow R - R$) should occur. The observed rate constant for associated monomer links can be close to zero. The scission reaction has two rate constants,

$$\frac{dM}{dt} = k_1 n_1 N + k_2 n_2 N \qquad (1)$$

where M is the quantity (concentration) of substances isolated in the reaction; k_1 and k_2 are the rate constants of destruction of non-associated associated monomer links; N is the concentration of macromolecules; n_1 and n_2 are the number of non-associated and associated monomer links, respectively.

Removal of a substance from the surface is possible only if a low-molecular mass volatile substance is formed. The scission reaction at the end of the chain appears the most favorable, especially at low temperatures. To remove substance from the middle of the chain two scissions are necessary. From a thermodynamical point of view the chain end is non-associated to a higher degree. At low temperatures, the non-associated monomer links will locate at the chain ends with higher probability. At low temperatures the reaction should proceed at the chain ends and the products can evaporate more quickly than the chemical reaction proceeds. Then the quantity of volatile substance can be determined by the kinetic equation

$$\frac{dM}{dt} = k_1 N_1 + k_2 N_2 \tag{2}$$

$$N_1 + N_2 = ZN \tag{3}$$

where Z is the average number of ends in the molecule; N_1 and N_2 are the concentrations of macromolecules with non-associated and associated end groups, respectively.

Equation (3) assumes that the reaction rate constants at the ends should be similar and the degree of scission should not be too high. At high degree of transformation the rests of chain can evaporate ($N = $ const). The rate constant of evaporation should obviously increase with the decrease of molecular mass. The rate of evaporation at relatively low temperature is higher than that of the chemical reactions.

The ratio of associated and non-associated monomer links is determined by the energy distribution

$$K = \frac{N_1}{N_2} \tag{4}$$

where K is the thermodynamical equilibrium constant which depends just on temperature.

Equation (4) implies that both the equilibrium and rate constants should not depend on the location of associated and non-associated monomer links and chain length and, consequently, on molecular-mass distribution. As the temperature increases, the difference between stationary and equilibrium states produced should decrease.

Temperature dependences of k_1 and k_2 are described by a Arrhenius equation

$$k_1 = k_1^0 e^{-E_1/RT}; \quad k_2 = k_2^0 e^{-E_2/RT}$$

and that for K the Vant-Hoff equation

$$K = K_0 e^{-H/RT}$$

Solution of the equations yields:

$$M = ZN \cdot \frac{k_1^0 e^{-E_1/RT} \cdot K_o e^{-\Delta H/RT} + k_2^0 e^{-E_2/RT}}{1 + K_o e^{-\Delta H/RT}} \cdot t$$

and the inverse temperature $(1/T)$ dependences of $\lg \tau$ (lifetime) is

$$\lg \tau = \lg M - \lg ZN + \lg \left(1 + K_o e^{-\Delta H/RT}\right) - \lg \left(k_1^0 e^{-E_1/RT} \cdot K_o e^{-\Delta H/RT} + k_2^0 e^{-E_2/RT}\right) \quad (5)$$

To determine the mass losses one should take into account both the molecules themselves and the scission products which can evaporate.

The fact that $K \ll 1$ (low temperatures) leads to a linear dependence of $\lg \tau$ (lifetime) on inverse temperature $(1/T)$, which is observed experimentally.[2] At higher temperatures, K becomes close to 1, and the observed dependence deviates from a linear one.

At higher temperature there will exist a larger number of longer blocks of non-associated monomer links. Suppose, that the probability of intermolecular bond formation should be $p = n_a/n$, where n_a is the average number of associated bonds. Then the probability that at a given bond is not formed, equals $(1 - p)$ (the probabilities of intermolecular bond formation at different parts of chain are supposed equal and independent on its length). If equilibrium in the system is established at a given temperature, the probability of the existence of the block from m non-associated monomer links equals:

$$p_m = p^2 (1 - p)^m \quad (6)$$

The probability of the existence of blocks of t associated monomer links equals

$$p_t = p^t (1 - p)^2 \quad (7)$$

The size of the block with associated monomer links is specified as the number of the latter located in succession in the chain and bordered by non-associated monomer links. The concentration of block of different size depends on their length and are described by a diminishing geometrical progression. The maximum block length can equal the number of monomer links if one monomer link forms one intermolecular bond. The free energy gain on the breaking of non-associated blocks is caused by an increase in the bending entropy of the chain ends formed. The magnitude of the entropy change depending on the length of ends formed and initial size of the blocks. If molecules of low-molecular mass are formed through scission of ends one should take into account just molecule number changes. The bending entropy per link in the final state is not greater than in the initial chain. The maximally attained entropy gain per bond (monomer link) equals R. In reality it is always lower. So a definite length of a non-associated block is necessary to make the chain break thermodynamically possible. The more the block length, the higher the scission rate constant. The probability of intermolecular bond formation p is connected with the equilibrium constant as follows

$$p = 1/(1 + K) \quad (8)$$

The ratio of associated and non-associated monomer links is determined by the thermodynamics of intermolecular bond formation. Assume all the components of the free energy are additive, then the free energy of chain formation without intermolecular bonds ζ_H equals

$$\zeta_H = \zeta_x + \zeta_{\text{bend}} \tag{9}$$

where ζ_x is the free energies of formation of stretched molecules and ζ_{bend} is the free energy of molecule bending.

Monomer links can exist in various conformations, so one should take into account the free energy of various conformers and the possibility of their transpositions in the chain. At high temperatures, the transition energy between conformers is not high and to simplify we need not consider this component. The free energy of chain formation ζ_a with n intermolecular bonds equals

$$\zeta_a = \zeta_x + n\zeta_b - RT \cdot \ln W + \sum_i \zeta_{\text{bend}\,i} \tag{10}$$

where ζ_b is the free energy of intermolecular bond formation; W is the number of methods of formation of n intermolecular bonds along the chain, containing n_0 bonds assuming all bonds are equivalent, ζ_{bend} is the free energy of bending of a block, containing i non-associated monomer links. The block size changes from 1 up to the maximal value n_0.

The bending free energy leads to inequivalency of intermolecular bonds, so it is more valid to evaluate ζ_a according to a Gibbs' distribution.

The minimal number of intermolecular bonds, can be obtained from the condition $\Delta\zeta = \zeta_a - \zeta_n$, equals

$$n_{\min} = \left(\zeta_{\text{bend}} + RT \cdot \ln W + \sum_i \zeta_{\text{bend}\,i} \right) / \zeta_b$$

The limiting quantities of intermolecular bonds necessary to keep the molecule in the associate[7] or adsorbed[8] are known from literature.

Temperature being increased, ζ_b/RT decreases tending to zero, and ζ_{bend}/RT increases tending to the limit with $n_{\min}$ increasing and the number of intermolecular bonds decreasing.

The system free energy ζ_{form}, possessing various numbers of intermolecular bonds in chains, is determined by the equation

$$1 = \sum_i p_i = p_0 + p_1 + \cdots + p_{n_0}$$

where

$$p_i = Ae^{-\zeta_i / RT}, \quad A = e^{-\zeta_{\text{form}} / RT}$$

The change of the free energy in the process, which defines the equilibrium constant, equals

$$\Delta\zeta = \zeta_a - \zeta_n = -RT \cdot \ln K$$

At high temperatures the dependence of $\lg K$ on $1/T$ is not described by a linear Vant-Hoff equation,[9] being a diminishing function, tending to infinity at temperatures above which molecules do not associate.

When "free rotation" of bonds in the chain appears, that we have

$$\sum_i \zeta_{i\,\text{bend}} = (n_0 - n)RT$$

simultaneously the local equilibrium constants become equal for different places of chains.

Pyrolysis of polystrene at low temperatures gives more than 90% of styrene in the reaction products,[10] the degradation of polystyrene, $\overline{M} = 600000$ at temperatures above 530°C gives chain fragments of various molecular masses ranging from dimers to $\overline{M} = 3000$ as reaction products.

The kinetic equations describing the scission process should take into account changes with temperature of both reaction abilities of blocks of various lengths, block distribution of non-associated monomer links by sizes and block concentrations. Unfortunately, such a scheme does not exist.

In experiments we considered the mass losses by the whole film and the kinetics scheme should account for the formation of molecules which can evaporate. The ability to react at different places being equal and independent on the chain length, the description of the process should consider not only two rate constants of scission but rate constants of evaporation of the molecules formed. The following kinds of molecules exist in the system:

1. The molecules which decompose with time. The rate of their concentration decrease

$$\frac{dN_n}{dt} = K_1(1-p)nN_n + K_2 pnN_n = \tilde{K}nN_n \tag{11}$$

where N_n is the concentration of molecules with n monomer links.

As a rule they are molecules of maximal size

$$\tilde{K} = k_1(1-p) + k_2 p$$

Time dependence of concentration for such molecules looks as follows

$$N_n = N_n^0 e^{-n_0 \tilde{K} t} \tag{12}$$

where N_n^0 is the initial concentration of molecules n_0 monomer links in length.

2. Molecules, possessing i monomer links ($1 < i < n_0$), can not only decompose but be formed from larger molecules; (l is the number of monomer links in the largest volatile

molecule). The decomposition rate is proportional to i. The formation from larger chains is proportional to doubling the quantity of the latter. The equation for the reaction rate is as follows

$$\frac{dN_i}{dt} = \tilde{K} \cdot i \cdot N_i - 2\tilde{K} \cdot \sum_{j=i+1}^{i=n} N_j \qquad (13)$$

For the simplest case (all the molecules at the beginning of the reaction have equal molecular masses) the solution of the equation is

$$N_i = N_n^0 \left[(n-i+1)e^{-\tilde{K}it} - 2(n-i)e^{-\tilde{K}(i+1)t} + (n-i-1)e^{-\tilde{K}(i+2)t} \right] \qquad (14)$$

The Eq. (14) shows that the process is Markovian as the number of molecules N_i at a given time is affected by their decomposition and formation of new molecules with number of monomer links $(i+1)$ and $(i+2)$.

Concentrations of the largest molecules decrease with time, those of intermediates change from zero ($t=0$), pass a maximum and goes back to zero ($t \to \infty$). The time to reach the maxima N_i value, from the condition $dN_i dt = 0$, equals

$$t_{max} = -\frac{1}{\tilde{K}} \cdot \ln \frac{(i+1)(n-i) - i\sqrt{(n-i)^2 + i(i+2)}}{(i+2)(n-i-1)}$$

The maxima concentration value N_i is defined by the substitution $t = t_{max}$ in Eq. (14). On the basis of the law of conservation of matter it was shown,[11] that

$$\sum_{j=i+1}^{j=n} N_j = \left[(n-i)e^{-\tilde{K}(i+1)t} - (n-i-1)e^{-\tilde{K}(i+2)t} \right] N_n^0 \qquad (15)$$

The time to reach the maxima value of $\sum_{j=1+1}^{j=n} N_j$ is defined as

$$t_{max} = 1 / \tilde{K} \cdot \ln \frac{(i+2)(n-i-1)}{(i+1)(n-i)}$$

Taking into account the principle of independence of chemical reactions, the destruction of macromolecules of a given molecular-mass distribution can be considered as the sum of processes in molecules with definite molecular masses.

3. If molecules consisting of 1 and less monomer links can evaporate, the rate of decrease of their concentration can be written as follows

$$\frac{dN_i}{dt} = (\tilde{K} + k_i) \cdot N_i - 2\tilde{K} \cdot \sum_{j=i+1}^{n} N_j \qquad (16)$$

where k_i is the rate constant of evaporation of molecules with i monomer links ($i \leqslant 1$). The concentration of evaporating molecules, changes from zero ($t=0$) to zero ($t \to \infty$) passing through a maximum. This constant depends on temperature by Knutsen-Langmuir law[12]

$$k_i = \frac{k_n \exp(-\Delta H / RT)}{\sqrt{2\pi R / i}}$$

and on the molecular mass of the evaporating molecules

$$k_i = A\sqrt{i+1} \cdot e^{-b(i+1)}$$

where A and b are parameters for a given homologous series depending on temperature and enthalpy (ΔH).

4. There also exist molecules of very small sizes which evaporate and are formed from larger molecules.

$$\frac{dN_{l_0}}{dt} = k_{l_0} \cdot N_{l_0} - 2\tilde{K} \cdot \sum_{j=l+1}^{j=n} N_j \tag{17}$$

If molecules can't leave the reaction volume (evaporate), then Eqs. (16, 17) does not contain terms with k_i.

To describe the rate process one should consider changes in concentrations of the molecules of all types, i.e. connect Eqs. (11, 13, 16, 17) in a system. A set of linear differential equations with constant coefficients should be produced, the number of the equations corresponding to the number of types of molecules. The solution of the system of equations (11, 13, 16, 17) includes the concentrations of molecules of various lengths as a sum of terms, containing exponentials with indexes $\tilde{K}$ and k_i.

The shorter chains, the larger number of terms are included into the solution. Taking into account the accuracy of experimental data only the initial terms in the expansion are considered.

At low degradation times ($t \leqslant 1/k$) the concentration of non-evaporating molecules is

$$N_n = N_n^0 \cdot (1 - n\tilde{K}t), \quad N_i = 2\tilde{K}N_n^0 t \tag{18}$$

The mass of the molecule

$$M = \sum_{j=l+1}^{n} N_j \cdot j \cdot M_0 \tag{19}$$

where M_0 is the molecular mass of a monomer link.

Mutual solution of Eqs. (18) and (19) and simplifications gives

$$M = M_0' \left[1 - \frac{(l+2)(l+1)}{n} \tilde{K}t \right]$$

where $M_0' = n_0 N_n^0 M_0$ is the initial mass of the substance.

If $t > 1/k$, the system contains a small number of molecules with high molecular masses. As the process is exponential in character, it is advisable to consider a single term with an exponential index. In this case we have

$$N_i = N_n^0 (n - i + 1) \exp (-\tilde{K}it) \tag{20}$$

where $l + 1 < i < n$.
Substituting (20) and (12) into (19) we have

$$M = M_0' \frac{(n - l)(l + 2)}{n + 1} \cdot e^{-\tilde{K}(l+1)t} \tag{21}$$

One should note that the solution of Eq. (21) for volatile components is a linear approximation and coincides with (18). As a rule, for practical problems one can consider just the terms up to t^2

$$N_i = N_n^0 \tilde{K}t [2 + (n - 2 - 3i) \tilde{K}t - \tilde{K}it] \tag{22}$$

The mass of volatile molecules

$$M_{n_i} = M_0' - M_i, \quad M_{n_i} = M_0 \cdot \frac{(l + 2)(l + 1)}{n} \cdot \tilde{K} \cdot t$$

The concentration changes of volatile fragments are defined by the equation

$$\frac{d\tilde{N}_i}{dt} \cong k_i N_i$$

where $i \leq l$.
  Hence

$$\tilde{N}_i = \int_0^t k_i N_i \, dt$$

At low t from Eq. (22) we obtain:

$$\tilde{N}_i = N_n^0 \tilde{K}k_i t^2$$

The mass of evaporated fragments equals

$$M_n = \sum_{i=0}^{i=l} N_i \cdot i \cdot M_0 = N_n^0 \tilde{K}M_0 t^2 \cdot \sum_{i=0}^{i=l} iA\sqrt{(i+1)} \cdot e^{-b(i+1)} = M_0' / n \cdot K \cdot t^2 \cdot B(l) \tag{23}$$

where

$$B(l) = A \cdot \sum_{i=0}^{i=l} i^{3/2} e^{-b(i+1)}$$

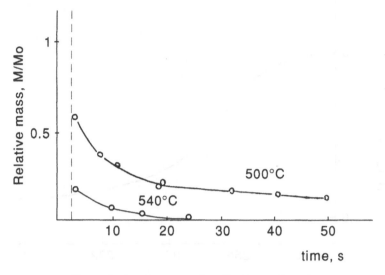

Figure 4   Sample mass vs time (for linear polymers).

From the above it follows that the mass of evaporating fragments at the beginning of the process increases slowly because of the small amount of low molar mass molecules formed. At the end of the process the number of molecules is small, so the evaporation is also low. The total time dependence of the mass losses is S-shaped consistent with experimental data, Fig. 4.

When the reaction takes place just at the chain ends a polymerization process occurs and the largest molecules just decompose

$$\frac{dN_n}{dt} = Z \cdot \tilde{K} \cdot N_n$$

The decomposition is supposed to occur at the chain ends with the same rate constants.

$$N_n = N_n^0 e^{-Z\tilde{K}t}$$

Non-evaporating chains, containing i monomer links are formed from the large ones and are removed during the reaction at the chain ends. The equation for their removal is

$$\frac{dN_i}{dt} = Z\tilde{K}N_i - Z\tilde{K}N_{i+1}$$

where $l < i < m$.

It is supposed that the total number of ends does not change. If at the initial moment there existed just the chains of one size, N_n, the change of their concentration with time is described by the equation

$$N_{n-j} = (ZK)^{j-1} \cdot N_n^0 \cdot t^j \cdot e^{-Z\tilde{K}t}, \quad N_{n-1} = Z\tilde{K}N_n^0 \cdot e^{-Z\tilde{K}t}$$

where $2 \leqslant j < m - l$.

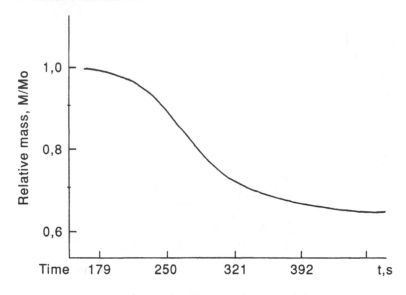

Figure 5   Degradation degree vs time.

The time dependence of the concentration of molecules with various lengths (except N_n) has a maximum value and its concentration at $t = j / Z\tilde{K}$, equals

$$N_{m-j}^{\max} = \frac{N_n^0}{(j-1)!} \cdot j^{j-1} \cdot e^{-j}$$

Maximum concentrations of chains of various lengths decrease as the chain length decreases.
    The equation of their removal is

$$\frac{dN_l}{dt} = (Z\tilde{K} + K_l)N_l - Z\tilde{K}_{l+1}$$

where K_l is the rate constant of evaporation.
    It is supposed, that molecules formed at the end of chains, evaporate practically immediately. The solution of the system of equation, can be represented as a sum of terms e^{-ZKt} and $e^{-(Z\tilde{K}+K_l)t}$. The number of terms increases with decrease of the degree of polymerization.
    If at the moment of time t the average number n' of monomer links is broken, n_1' being associated and n_2' being non-associated ($n' = n_1' + n_2'$), we can introduce respective probabilities of scission $p_1 = n_1' / n'$ and $p_2 = n_2' / n'$, taking into consideration differences in reaction rates of associated and non-associated monomer links.
    Solution of a system of equations with $n_1 + n_2 = n$ ($n \gg 1$), we have $p_1 = n'/n$. The total probability of the scission of any bond equals[13]

$$p = p_1 p[H_1] + p_2 p[H_2] \tag{24}$$

where H_1 is the event, characterizing the break of an associated monomer link; H_2 is that for non-associated.

Conditional probabilities of location of associated $p[H_1]$ and non-associated $p[H_2]$ monomer links in the chain are determined as

$$p[H_1] = k/1+k, \quad p[H_2] = 1/1+k$$

The total probability in this case equals

$$p = \frac{k}{1+k} \cdot \frac{n'}{n} \cdot \frac{(1+k)\beta}{(1+k\beta)} + \frac{1}{1+k} \cdot \frac{n'}{n} \cdot \frac{(1+k)}{(1+k\beta)} = \frac{n'}{n}$$

The probability of scission is the same as in a homogeneous infinitely long chain with non-associated monomer links.[14]

Equation (24) allows evaluation of the probability of bond break in inhomogeneous macromolecules, if H_1 and H_2 — events, characterizing the scission of bonds of various kinds.

If the products remain in the reaction volume,

$$\frac{dn}{dt} = k_1 n_1 + k_2 n_2, \quad n_1 + n_2 = n, \quad k_1 \neq k_2$$

$$\frac{n_1}{n_2} = k = k_0 \cdot \exp(-\Delta H / RT)$$

one can get the scission probability for any monomer link depends on time

$$p = \frac{n'}{n} = 1 - \exp\left(\frac{k_1 K + k_2}{1+K} \cdot t\right)$$

The probability of destruction of products of a given length located in the system at a given moment of time at a fixed degree of transformation. In accordance with Bernoulli by the law of geometric distribution we have

$$p(l) = p(1-p)^{l-1}$$

where $p(l)$ is the probability of occurrence of the fragment with l monomer links, containing $(l-1)$ chemical bonds.

The mathematical expectation (M) and dispersion (D) of such a distribution are[13]

$$M = \frac{1-p}{p}, \quad D = \frac{1-p}{p^2}$$

At the same time to evaluate the kinetics of changes of the molecular-mass distribution it is better to use the Kuhn distribution[15]

$$p'(l) = p^2 (1-p)^{l-1}$$

where $p'(l)$ is the probability of release of i monomer links in the form of fragments $p = n/N$. The sum of probabilities for the geometric distribution equals 1, and for the Kuhn distribution is p.

Concentrations of monomer links, contained in the fragments of length l in both distributions are

$$Cp = Np^2(1-p)^{l-1}$$

The probability of occurrence of fragments of length l is maximal at $lp = 1$ and $nl = N$ and equals

$$p_l^{max} = p(1-p)^{\frac{1-p}{p}} = \frac{n'}{n} \cdot \left(\frac{N-n}{N}\right)^{\frac{n-n'}{n}}$$

The time to achieve maximal probability equals

$$\tau_l^{max} = \frac{1+K}{k_1 K + k_2} \cdot n \cdot l(1-1/l)$$

The number of broken links being described by a binomial distribution, the probability for n' bonds in the molecule to be broken is determined by the formula

$$P_n = C_n^{n'} \cdot p^{n'} \cdot (1-p)^{(n-n')}$$

Coefficient $C_n^{n'}$ characterizes the possibility of breaking bonds in different places of a chain. Hence the relative mass of fragments containing l monomer links is

$$\frac{M_l}{M_0} = \frac{1}{n+1} \cdot \sum_{i=0}^{i=n} C_n^i p^{i+1} (1-p)^{(n-i+l+1)} (i+1)$$

$$= \frac{1}{n+1}[p(1-p)^{l-1} + p^2(1-p)^{l-2}(n-n'+1)] \qquad (25)$$

As $n \gg 1$, the expression (25) is simplified

$$\frac{M_l}{M_0} = \frac{1}{n} \cdot [p(1-p)^{l-1} + p^2(1-p)^{l-2}(n-n')]$$

Mass losses induced by evaporation of fragments of any length from a single monomer link to l_{max} are

$$\frac{M_n}{M_0} = \sum_{i=1}^{i=l_{max}} 1/n \, [p(1-p)^{i-1} + p^2(1-p)^{i-2}(n-n')]$$

$$= \frac{1-(1-p)^l(1+pl)}{p^2} \cdot \left[1/n + \frac{n-n'}{n} \cdot \frac{p^2}{1-p}\right] \qquad (26)$$

For low number of breaks $(n' \ll n)$ the equation of mass losses looks as follows

$$\frac{M_n}{M_0} = 1 - (1-p)^l (1+pl)$$

Equation (26) allows determination of the number of scission by mass losses and, consequently, the molecular-mass distribution of reaction products, if the energy necessary for chain break is supplied so fast that evaporation of the destruction products can be neglected.

At relatively high temperatures during short time periods when destruction does not occur there exist both long blocks from non-associated monomer links and short blocks from associated ones. Evidently, the reactions at the chain ends can be neglected. In large blocks simultaneous breaks of several chain parts are thermodynamically possible. According to the theory of the activated complex the constant increase is due presumably to increase of the free energy for formation of initial state per link. The rate constant of scission can not exceed the value, determined by bond oscillations. As the length of non-associated blocks increase, different parts become statistically independent. For such long blocks simultaneous reactions at several places of chain appears possible and avalanche-like polymer decomposition occurs. Mass loss in this case should be defined by the rate of evaporation or carrying away the chain fragments formed.

In the literature, polyethylene and polytetrafluoroethylene should run though an explosion due to high negative free energy of the process.[17] At low temperatures such process did not occur. Explosions occur at temperatures, where intermolecular interactions can be neglected and confirms the correctness of interpretation of the role of intermolecular interaction in the degradation.

The kinetic equations of the process at temperatures close to limit, one should consider the distribution by the length of non-associated blocks. The equation of the destruction rate looks as follows

$$-\frac{dW}{dt} = k_1(1-p)ZN + k_2 ZpN + \sum_{i=m}^{i=n_0-2} k_i i p^i (1-p)^2 N + k_{n-1}(n_0-1)^{n_0-1} pN + k_n n_o p^{n_0} N \qquad (27)$$

The last equation considers the principle of independence of chemical reactions. Two last terms in it describe destruction, if the chain contains one non-associated monomer link or associated monomer links are absent. The low limit of summation m characterizes the possibility of non-associated block break only beginning from a given length. If the reaction at the chain ends cannot be neglected, the rate equation of chain end formations looks like

$$\frac{dN}{dt} = \sum_{i=n}^{i=n_0-2} k_i N(1-p)^2 p^i$$

At constant temperature and scission constants being independent on block lengths, we have

$$N_t = N_n^0 \exp \left[k(1-p)^2 \sum_{i=m}^{i=n-2} p^i t \right]$$

At temperatures equal to the limiting ones, all the terms in the Eq. (27) except the last can be neglected. At these temperatures, p is close to unity and the rate constant depends weakly on temperature. The time of decomposition for a given degree of transformation equals

$$t = \frac{nKp^{nW}}{1-W}$$

The film exists at the surface only at W values, characterized by the accuracy of the experiment. The time of decomposition becomes equal

$$t = \frac{nK}{1-W}$$

At the limiting temperatures the free energies of formation of associated and non-associated monomer links are equal to

$$D = RT_{\lim} \tag{28}$$

At $T_{\lim}$ has a simple physical sense; at this temperature the free energy of formation of intermolecular bonds becomes equal to heat energy. Energies of intermolecular interaction,[19] coincide satisfactorily.

At the stages where chain participation occurs changes of system entropy, due to changes of mobility should be considered. Rate constants at this such stages will depend on temperature and lead to "the chain effect", i.e. the rate constant will increase with increase of mobility of the chain part.[20]

The reaction rates at various temperatures are lower than the rates of equilibration, it is impossible to describe the process by the dependence of reaction rate on temperature and therefore on the equilibration rate in the sample without calculation of the different rate constants of block reactions, changes in ratios between blocks and their lengths, change of the reaction direction. Attempts to describe the whole kinetics of the processes disregarding the role of intermolecular interactions appeared unsuccessful.[21]

Small non-associated blocks are capable of reacting with neighboring chains. Radicals, formed in the process, will react with neighboring chains with higher probability. In associated blocks the reactions with removal of side groups ($-H$, $-CH_3$, etc.) are also possible. For flexible chains scission is probable, with a gain in entropy and a free energy decrease.

Mechanical load can affect the degradation in various ways.[24] As the lifetime of non-associated fragments is small, if the mechanical load is fed during a very short time or is stored beforehand, the temperature of explosive (avalanche-like) destruction should decrease significantly. If the energy is consumed just to break intermolecular bonds, the temperature of explosive destruction is defined by the equation

$$A + RT = RT_{\lim}$$

N.S. Enikolopov et al.[25] have observed the sharp increase of destruction rate constant at mechanical load application. It should be noted that temperature affects both k_0 and the chain flexibility.[26]

If the mechanical load is constant, the energy is used to change the flexibility of the system and to break intermolecular bonds. According to the above-said it can induce the increase of the observed rate of the process.[27] The increase of the load may at first induce deformation of valent angles and then the bond lengths, either. It increases the observed process rate.[28] All above-said concerns both organic and non-organic polymers.

REFERENCES

1. O.F. Shlensky, A.A. Matyukhin and E.F. Vainshtein, *J. Thermal. Anal.*, V. 31, p. 107 (1986).
2. O.F. Shlensky, E.F. Vainshtein and A.A. Matyukhin, *J. Thermal. Anal.*, V. 34, p. 645 (1988).
3. D.N. Yunoev, A.A. Lash, E.F. Vainshtein, *et al.*, *Teplofizika vysokihk temperatur*, V. 27, N 2, p. 369 (1988).
4. O.F. Shlensky and E.F. Vainshtein, *Dokl. Akad. nauk SSSR*, V. 281, N 3, p. 164 (1984).
5. E.F. Vainshtein and O.F. Shlensky, *Plastmassy*, N 4, p. 45 (1986).
6. N.M. Emanuel, G.E. Zaikov and Z.K. Maizus, *Oxidation of Organic Compounds. Medium Effect in Radical Reactions.* Oxford, Pergamon Press, p. 612 (1984).
7. V.A. Kabanov and I.M. Papisov, *Vysokomol. Soed.*, V. A21, N2, p. 243 (1979).
8. V.M. Komarov, Thesis, Moscow, Inst. of Chem. Phys., p. 28 (1982).
9. T.B. Varshavskaya and E.F. Vainshtein, *Termodinamika organicheskikh soedinenii* (Thermodynamics of Organic Compounds), Gor'kii, Gor'kii University Publ., p. 189, Russian (1989).
10. R.M. Aseeva and G.E. Zaikov, *Combustion of Polymer Materials*, Münich, Karl Hanser Verlag, p. 390 (1985).
11. E.A. Baranova, E.F. Vainshtein and A.A. Matyukhin in: *Modulation and Automatization of Technological Processes in Agriculture* (Modelirovanie i avtomatizatsiya technologicheskikh protsessov v cel'skom khozyaistve), ed. E.F. Vainshtein, Moscow, Gosagroprom, p. 10 Russian (1987).
12. Yu.V. Polezhaev and F.B. Yurevich, *Teplovaya zashchita* (Thermal Protection) Moscow, Energiya, p. 392, Russian (1976).
13. I.N. Bronshtein and K.A. Senendyaev, *Spravochnik po matematike* (Handbook in Mathematiks), 4th ed., Moscow, Nauka, p. 544 (1986).
14. W. Kuhn, *Ber. Bd.*, **63**, S. 1503 (1930).
15. N. Grassie, *Khimiya protsessov destruktsii polimerov* (Chemistry of Polymers Degradation) Moscow, IL, p. 251, Russian (1959).
16. V.N. Skripov, B.N. Sinitsyn, P.A. Pavlov, *et al.*, *Teplofizicheskie svoistva zhidkostei v metastabil'nom sostoyanii* (Thermophysical Properties of Liquid in Metha-Stable State), Moscow, Khimiya, p. 287, Russian (1980).
17. H. Savada, *Termodinamika polimerizatsii* (Thermodynamics of Polymerization), Moscow, Khimiya, p. 312 (1979).
18. E.F. Vainshtein, O.F. Shlensky and N.N. Lyashikova, *Inzhenerno-Fiz. Zh.*, V. 53, N 5, p. 774 (1987).
19. A.A. Ascadsky, *Structure, svoistva teplostoikikh polimerov* (Structure and Properties of Thermostable Polymers), Moscow, Khimiya, p. 320, Russian (1981).
20. E.F. Vainshtein, Thesis, Moscow, Ins. of Chem. Phys., p. 317 (1981).
21. O.F. Shlensky, L.N. Aksenov and A.G. Shashkov, *Teplophizika razlogayushchikhsya materialov* (Thermophysics of Degraded Materials), Moscow, Khimiya, p. 187, Russian (1980).
22. L.D. Landay and E.M. Livshits, *Statisticheskaya pizika* (Statistical Physics), Moscow, Nauka, p. 560, Russian (1964).
23. M.G. Sklyar, *Intensifikatsiya koksovaniya i kachestvo koksov*, (Intensification of Coke Formation and Quality of Cokes), Moscow, Metallurgiya, p. 997, Russian (1976).

24. A.A. Popov, N.Ya. Rapoport and G.E. Zaikov, *Okislenie orientirovannykh n napryazhennykh polimerov* (Oxidation of Stressed Polymers), Moscow, Khimiya, p. 232 (1987).
25. N.S. Enikolopov and N.A. Fridman, *Dokl. Akad. nauk SSSR*, V. 290, N 1, p. 99 (1986).
26. A.A. Popov, N.Ya. Rapoport and G.E. Zaikov, *Effect of Stress on Oxidation of Polymers*, New York, Gordon and Breach, p. 360 (1990).
27. N.M. Emanuel, G.E. Zaikov and V.A. Kritsman, *Tsepnye reaktsii, Istoricheskii aspect* (Chain Reactions, Historical Aspect), Moscow Nauka, p. 336, Russian (1989).
28. A.A. Popov, Thesis Dr. of Sci., Moscow, Inst. of Chem. Phys., p. 320 (1987).
29. G.E. Zaikov, *International Journal of Polymeric Materials*, V. 24, N1-4, , p. 1–38 (1994).

Reviews of Conferences

Reviews of Conferences in Russia

G.E. ZAIKOV

A conference on the "Quality and Resource-saving Technology in Rubber Industry" was held in Yaroslavl from 16–20 September 1992. Despite the very large volume of these goods produced their very low quality makes them uncompetitive with those abroad. The properties of thermoplastic rubbers based on various rubber-thermoplast compositions were described by A.A. Dontsov (Research Institute of Elastomeric Materials and Goods, Moscow). The role of molecular and technological characteristics of polymers, types and content of vulcanizing agents, plasticisers, special additives was also discussed. The effect of rubber-thermoplastic interface on the structure and properties of thermoplastic rubbers, reproducibility of parameters during recycling were particularly stressed.

E.E. Potapov dealt with problems of chemical modification of elasto-plastics, materials and included the development of multicomponent systems development of new promoters-modifiers for elastoplastics adhesion to various materials as well as diffusion-activated modifying systems capable of adding new properties to the interface of elastoplastics, and the modification of synthetic polyisoprene with the use of biologically active compounds.

Z.V. Onishchenko (Institute of Chemical Technology, Dnepropetrovsk) considered the possibilities of improving the rubber quality by modifying their structure and chemistry using various additives with epoxy, hydroxyl, methylolic and amino-groups.

J.A. Tutorsky (Lomonosov Institute of Fine Chemical Technology, Moscow) described aspects of dispersion hardening of polymeric composites. Ultradispersed mineral, particles (100–300 Å) can affect the structure and properties of rubber. The effect of surface-active substances increasing the adhesion of the mineral filler to the polymeric matrix was also described.

V.A. Shershnev (Lomonosov Institute of Fine Chemical Technology, Moscow) reported problems of improving rubber quality by controlling the structures of the vulcanizing agents of the elastoplastics mixtures. Covulcanizing agents consisting of polyisoprene and chlorobutyl-rubber and of polychloropropene and ternary ethylene-propylene copolymer were considered.

V.V. Markov (Lomonosov Institute of Fine Chemical Technology, Moscow) reported modification of an elastic filler (finely ground rubber powder) with the use of solution of a sulfonamide accelerator in a plasticizer. This kind of modification improves dispersion and adhesive interaction of a filler with the matrix, increases the strength and resistance to expansion of rubber mixtures.

THE FIRST WORKSHOP ON STABILITY OF MATERIALS UNDER EXPLOITATION

The first Workshop on Stability of Materials under Exploitation was held on November 18–22, 1991 in Moszhenka (Moscow district, town of Zvenigorod).

O. Karpukhin emphasized the high level of understanding reflected in the present theory of aging of polymers, a huge amount of experimental data, and good facilities for modern study of this problem.

O. Startsev (AUIAE), dealt with the natural tests which were carried out in different climatic stations. Dr. Startsev represented comparative parameters of stations located in different regions of the (former) USSR and abroad. He emphasized the wide range of the variation of conditions used in these tests.

A. Mirzeov and N. Konkin were concerned with climatic conditions in four test stations in the Republic of Tadjikistan. These stations are located in different climatic regions which are as follows: (i) a region with a very hot desert climate (Aivadgetown, 200 m over sea-level). (ii) dry and hot industrial region with high levels of pollution (Deshanbe-City, 800 m over sea-level), (iii) a region with a moderate moisture level (Khorag-town, 2300 m over sea-level), and (iv) a region of dry and cold climate in a mountain region. It was shown that the process of aging in (ii) was progressing 1.5–2 times faster in comparison with other climatic regions (Batumi, Yakutsk, Zvenigorod, Murmansk).

G. Makarov (ICP) presented the results of the tests of polymeric materials, which were carried out in climatic stations of the Republic of Cuba. The differences in the mechanisms of aging of low-density polyethylene was noted for the regions of Sea-Coast and an area in the central part of the island. The contribution of the process of biodegradation was significant for the last case.

The process of atmospherical aging of pigment coatings were considered N. Mayorova and T. Veremeeva. All kinds of aging were divided into three types.

1. Aging of polymer material itself follows by destruction of adhesive compound. This type is usual for the affects of light temperature and moisture, which cause thermal, thermooxidative, hydrolytic, and photo degradation of polymers.
2. Destruction of adhesive compounds follows by degradation of the coating is usually the affects of temperature and moisture when sun light is restricted.
3. Combination of previous two types due to the multifactoral affects of atmosphere. It was emphasized that determination of the dominated type of aging is necessary in order to solve the problem of durability of vanish dye coatings.

Dr. Gorshenin described the Centre of Climatic Tests of CIFE which has at its disposal a complex of hot chambers, cold chamber, moisture chambers, sun radiation, low pressure, and dynamic dust.

THE WORKSHOP ON PREDICTION AND INSURANCE OF THE CAPACITY FOR WORK ON POLYMERIC MATERIALS

The Workshop was held on December 7, 1992 in Moscow and organized by the "Polysert" Centre of Institute of Chemical Physics (ICP) of the Russian Academy of Sciences (this is now Russian Academy of Sciences) together with Russian House of Knowledge, Moscow Institute of Aviation Technology (MIAT) and "Plastic" State Company. About 150 participants represented 20 Research Centres of the former USSR.

Opening remarks of the Chairman Professor V. Vinogradov (MIAT emphasized the high level of knowledge of the theory of aging of polymers, the huge amount of experimental data, and good facilities for modern study of this problem. The lecture of Professor Karpukhin was concerned with the scientific development of the certification of quality of polymeric materials. The importance of such certification is caused by the wide variety of application of polymers and corresponding problems of ecology.

To ensure a solution to the problem a database of materials and compounds was developed in the Centre of Synthetic Materials and reported by Doctor Papirov. Dr. A. Pomerantsev (ICP) described the use of personal computers and their applications for study of the physico-chemical processes of aging. Dr. Pomerantsev described the software "Kinetic Trunk" which enables interpretation of kinetic measurements, selection of typical kinetic, mechanical, and diffusive models, estimation of unknown parameters, analyze statistical data, etc.

Dr. Goncharenko (MIAT) described a new approach to estimation of quality of polymeric materials, while new physical methods for the evaluation of quality and certification of non-metallic materials were reported in the lecture of Dr. Kvacheva (Institute of Aviation Industry).

The effect of the technological stress on the capacity for work of polymer articles was considered in the lecture of Professor Vinogradov.

Dr. Sudarushkin discussed new technological equipment, which provides a higher quality of plastic goods.

New methods of estimation and prediction of the light stability of plastics were reported in details in lectures of Dr. Belikov and Dr. Almaeva from "Plastic" State Company.

The cooperative effect of stress and aggressive liquid media on high density polyethylene was described by Professor Valiotti ("Plastic").

The results of climatic tests on a wide variety of polymeric and textile materials were considered in the lecture of Dr. Anisimova (Moscow Textile Academy).

Participants of the Workshop reached some conclusions which were based on the discussion of problems of prediction of the capacity for work of polymeric materials.

THE FIRST ALL UNION SCIENTIFIC AND TECHNOLOGICAL CONFERENCE

The First All-Union Scientific and Technological Conference on polymeric materials and technological processing them into goods was held at the All-Union Research Institute of Inter-sectoral Information (Moscow) from 26 to 28 November 1991. The first presentation was made by Professor G.E. Zaikov (Institute of Chemical Physics, Moscow) on the subject "Destruction and Stabilization of Polymeric Materials: Researches, implementation, industrial manufacturing".

Yu. I. Lipatov, Academy of Sciences of Ukrain, (Institute of High Molecular Compounds, Kiev) discussed research on a new class of polymeric material based on interpenetrating networks.

Improvement of the thermal and thermooxydizing stability of rubbers were discussed by Professor E.F. Vainstein (Institute of Chemical Physics, Moscow). Application of kinetic data on thermal and thermooxydizing destruction of polymers helped increase the thermal and temperature stability of rubbers by 10–50°C.

V.V. Protasov described the technological processes for production of goods from polymeric materials and their compounding. E.F. Veinstein, dealt with the ecological aspects of polymer materials-based goods production, the problems of reprocessing of plastic and rubber wastes, as well as the utilization of polymeric materials, sound technologies for reprocessing the polymers, polymeric mixtures and composites, the problems of reprocessing the wastes of polymeric materials based on multitonnage polymers were also considered.

IV MEETING ON FIRE RETARDANT POLYMERS

G.E. Zaikov

Fire Retardant Polymers were held from September 9–11, 1992 in Albert-Ludwigs-University of Freiburg. I. McNeil (University of Glasgow, Scotland) discussed the relationship between thermal degradation and flammability of polymers. E. Zaikov (Institute of Chemical Physics of Russian Academy of Sciences, Moscow) dealt with some aspects of polymer pyrolysis and included problems of kinetics and mechanisms of pyrolysis, formation of charcoal and mechanism of retardation of polymer combustion.

A.B. Boscoletto dealt with surface and bulk characterization of fire retardant poly(phenyelen ether)- high impact polystyrene in the UL-94 test. Flame retardant Chemical Mechanisms of FR-Viscose fibres and blends with polyester were discussed by P. Nousiainen. Invariant values of kinetic parameters-evaluation of fire retardancy application to the PP-APP/PER System was discussed by Dr. J.M. Leroy.

G. Camino (Dipartamen to di Chimica, Torino University, Italy) spoke about mechanistic aspects of in timescent fire retardant systems. The potential application and the prospect of X-Ray Photoelectron Spectroscopy (XPS, or ESCA) to the fire retardance mechanism of polymers was discussed by Professor J.Q. Wang from Beijing Institute of Technology (Beijing, People's Republic of China). Dr. G.M. Anthony (FMC, Manchester, England) reported on the chemistry of smoke suppression in polyurethane foams using TGA-IR Residue Analysis Chemical and Physical properties of foamed cokes and their effect on inflammability were discussed by Dr. E.V. Gnedin from Karpov Research Institute, Moscow, Russia. Flame retardant polymers — current status and future trends were analyzed in the lecture of Professor J. Troitzsch from the Fire Protection Service, Wiesbaden, Germany and fire retardancy in polyamide 6,6 plastics: additives and their mechanisms of action was discussed by Dr. S. Bodrero Rhone-Poulenc Recherches, Saint-Fons, France. The synthesis and fire retardant properties of vinyl polymers bearing phosphorous Groups was given by Dr. J.M. Catala (Institute Charles Sadron/CRM, Strasbourg, France).

Dr. D. Munz (Bakelite GmbH, Duisburg, Germany — reviewed research carried out on Bakelite flame retarding materials for advanced composites. Dr. H. Schreiber, Gurit-Essex AG., Freienbach, Switzerland) discussed new fire retardant halogen free polymers. A way to halogen free, flame retardant laminates for electronic application was discussed by Dr. W. Rogler, Siemens AG, Erlagen, Germany. Dr. R.N. Rothon from ICI Chemicals

& Polymers, Runcorn, Cheshire, UK discussed the effects of particle size on the performance of PMMA filled with aluminium hydroxide in a variety of fire tests.

How are standards reflecting reality was discussed by Professor W. Becker, BASF AG, AWETA Brandschutztechnik, Ludwigshafen, Germany). Fire retardant polymers in the future; safety by tests or engineering was presented by D. Woolley, Building Research Establishment, Borehamwood, UK). Temperature-resolved pyrolysis mass spectrometry of fire retardant polymer blends was discussed by R. Luijk, J.J. Boon, University of Amsterdam, Department of Environmental and Toxicological Chemistry, Amsterdam, The Netherlands). PUR-sandwich panels in the Building industry-development of fire regulations in Europe (R. Walter, Bayer AG, Leverkusen, Germany) and laboratory systems to evaluate the toxicity of polymers combustion products and Hazard Assessment were presented by (J.M. Jouany, UFR Médicine & Pharmacie de Rouen, St. Etienne du Rouvray, France) accompanied by extensive discussion.

THE INTERNATIONAL CONFERENCE ON REGULATION OF POLYMERIC MATERIALS STABILITY WAS HELD ON OCTOBER 12–15, 1992 IN MOSCOW IN THE INSTITUTE OF CHEMICAL PHYSICS OF RUSSIAN ACADEMY OF SCIENCES

The topics of the Conference included the problems of thermal, thermo-oxidative, photo-oxidative and irradiative degradation, hydrolysis of polymers, biodegradation, mechano-degradation, degradation during the process of treatment, combustion of polymers and synthesis of antipirenes and stabilizers. There were discussed the problems of the prediction of polymer stability, their secondary treatment, development of self-degradable polymers, environmental study, application of polymers in different fields of industry and agriculture.

E. Burlakova, the Head of Division of Chemical and Biological Kinetics, Institute of Chemical Physics, Russian Academy of Sciences, Russia opened the Conference.

The problem of photodegradation and light stabilization of polymers was discussed by G. Geuskens (University of Brussels, Belgium) who considered the mechanism of action of hindered amine light stabilizers.

L. Rychla (Polymer Institute, Slovak Academy of Sciences, Czech and Slovak Federative Republic), discussed chemiluminescence from oxidative degradation of polymers and application of this effect for the study of kinetic control and the mechanism of polymer oxidation.

W. Starnes, Jr. (College of William and Mary, Maryland, USA), considered the problems of improvement of thermostability of PVC and the implications of "1,2-Dichloroethyl branches in poly(vinyl chloride) for thermal stability and the mechanism of polymerization. Among other reports, contributed in this session we should mention the lectures of Professor K. Minsker ("Characteristics effects in processes of halogen-containing polymers degradation and stabilization"), Doctor S. Kolesov ("PVC thermal stability in blends with poly(olefins)") and Dr. R. Biglova ("Synthesis of new stabilizers through polymer-analogous conversions"), all from the Bashkirian State University) as well as the lecture of Professor B. Troitskii (Institute of Organometallic Chemistry, Russian Academy of Sciences, Nizhnii Novgorod) in which the mathematical models of the initial stage of the thermal degradation of poly(vinyl chloride) in the presence of HCl was considered.

P. Stott (Uniroyal Chemical Company, Connecticut, USA) discussed the theory and application of synergistic interactions between different classes of radical scavenging antioxidants. The disorder in polymers and its role in oxidation processes was considered by Yu. Shlyapnikov (Institute of Chemical Physics). The influence of the mesomorphic state on the durability of polymer coatings was reported by E. Rosantsev mesomorphic state on the durability of polymer coatings was reported by E. Rosantsev from the Moscow Institute for Applied Biotechnology.

Some aspects of thermal degradation of methyl methacrylate were reported by I. McNeill, University of Glasgow. Transfer stabilization of thermostable polymers was discussed in the lecture of Yu. Sazanov from the Institute of Macromolecules of the Russian Academy of Sciences. A. Rusanov (Institute of Element-Organic Compounds, Russian Academy of Sciences) considered m-Carborane-containing polyamides as "stabilizers for poly-heteroarylenes".

A. Askadskii (Institute of Element-Organic Compounds, Russian Academy of Sciences) discussed the influence of chemical structure on the thermal characteristics of polymers. Computer simulation of thermostable polymers caused great interest. Thermal and thermo-oxidative stability of polyether-based polymer electrolytes was discussed by G. Cameron (University of Aberdeen, UK). Polymer stabilization was discussed by S. Al-Malaika (Aston University, UK) on the role of complexes of metals in polymer stabilization and initiation of the thermal decomposition. New metal chelates as stabilizers for polymers and low molecular substances were proposed by V. Vinogradova (Institute of Chemical Physics). The lecture of N. Mukmeneva (Institute of Chemical Technology, Kazan, Tatarstan) concerned with heteroatom stabilizers for polymers: synthesis and properties. The synthesis and application of novel degradable polymers were reported by Z. Jedlinski and M. Kowalczuk from the Institute of Polymer Chemistry of the Polish Academy of Sciences (Zabrze, Poland).

Special session considered the problems of transport phenomena in polymers. A new method for the measurement of the thermodynamical diffusion coefficients of electrolyte solutions in polymers was proposed by V. Lobo (University of Coimbra, Portugal). The general model of multicomponent transport of low-molecular compounds in glassy polymers and its application for the modelling of simultaneous transport of water and drugs in hydrophilic polymers was discussed by J. Petropoulos ("Demokritos" National Research Centre for Physical Sciences, Greece) and A. Polishchuk (Institute of Chemical Physics). Some aspects of diffusion of water in polymers were reported by A. Iordanskii (Institute of Chemical Physics). Fire retardant mechanisms of additives for polymeric materials based on metal halides was discussed by L. Costa (University of Turin, Italy). The lecture of Donskoi (Institute of Aviation Materials, Russia) and R. Aseeva (Institute of Chemical Physics) concerned with the thermal stability and flammability of sulphochlorinated polyethylene compositions.

The XVI All-European Colloqium "Chemical Technology of Polymers" was held from March 11–13, 1992 in Aachen (Germany). (Institute of Aviation Materials, Russia) and R. Aseeva (Institute of Chemical Physics) concerned with the thermal stability and flammability of sulphochlorinated polyethylene compositions.

The XVI All-European Colloqium "Chemical Technology of Polymers" was held from March 11–13, 1992 in Aachen (Germany).

The conference was divided into a series of blocks.

Block 1. "Recycling of Plastics" and was devoted to material properties of recycled polymers (mechanical properties, polymer contamination), processing of mixed thermoplastic waste (extrusion, die casting process) and synthetic gas production from mixed plastic waste by degrading extrusion (synthetic gas, decomposition).

Block 2. was devoted to processing and properties of polymer blend.

The problems "Application of modern strategies and measuring technologies to improve product quality" was discussed in *Block 3*. *Block 4* "Quality-oriented manufacturing of rubber mouldings: material-moulding-process". *Block 5* was "RIM technology-ways towards safe process control". *Block 6* was dedicated to BMC injection moulding. *Block 7* was devoted to injection moulding made transparent new developments in process simulation material testing-preparation for efficient material application were represented in *Block 8.*

A special *Block 9* was dedicated to Fibre composites for medium and large-scale series. Design of components for extrusion lines was discussed in *Block 10*. One of the blocks (*Block 11*) was devoted to moulding design by means of modern software.

The problems of forming, joining, refining-processing for product quality were discussed in *Block 12.*

Block 13 was devoted to special processes step up attractivity of injection moulding.

The last block (no. 14) was devoted to dimensioning, manufacturing and examination of continuous fibre reinforced plastics.

14TH ANNUAL INTERNATIONAL CONFERENCE ON ADVANCES IN THE STABILIZATION AND DEGRADATION OF POLYMERS

The 14th conference on aging and stabilization of polymers was held from May 25 to 27, 1992 in Luzern (Switzerland).

The conference consisted of 15 plenary lectures and a few short reports on extension of storage and reliable usage of polymer products, prediction of the polymer "life-span", creation of express methods for evaluation of polymers stability, the use of degradation as a polymer modification method, giving properties to polymer products, the secondary use of polymers, their secondary processing and regeneration of monomers, reduction of polymer combustibility, incineration of polymer wastes and environmental control, creation of new polymers with improved stability to different kinds of degradation, creation of polymeric materials with given lifetimes, synthesis and studying the mechanism of action of different stabilizers, and mixed synergistic and antagonistic effects of additives in polymers.

A lecture dealt with mechanisms of energy dissipation by ultraviolet absorbers (J. Catalan and R. Sastre, University of Madrid, Spain). The inter-relationship of structure, physical properties and stabilization performance of hindered amine light stabilizers was covered by S.B. Samuels (Union Carbide Chemicals and Plastics Co., USA).

Great interest was aroused by the paper on hindered amine stabilizers (W. Drake, Ciba-Geigy AG, Switzerland). Ultra-acceleration of the photoaging of polypropylene for the automative industry was reported by M. Barrois from Renault Direction Generale, France.

One lecture dealt with chemiluminescence studies of the early stage of polypropylene photooxidation and stabilization (G.A. George, M. Celina, M. Ghaemy, The University of Queensland, Australia). Hindered amine light stabilizers (influence of substituents and molecular weight on efficiency) was covered by P. Hrdlovic from Institute of Polymers, Slovak Academy of Sciences, Czechoslovakia. A lecture dealt with determination of peracids in photooxidation of polypropylene (J. Sedlar and A. Zahradnickova, Research Institute of Macromolecular Chemistry, Czechoslovakia). Photochemical behaviour in the solid phase of polymers bearing pendant acyloxyimino groups and their application to photoresists was reported by M. Tsunooka from University of Osaka, Japan. Two lectures were devoted to degradable polymers: the report of A. Andrady from Research Triangle Institute (USA) dealt with performance of environmentally degradable polymers under outdoor marine exposure and J. de Bruin from Delft Technical University (The Netherlands) dealt with application at fracture mechanism to degraded polymers.

The influence of temperature, thickness and stabilizer system on the long term heat stability of polypropylene were covered by P. Gijsman, J.M.A. Jansen and J.A.J.M. Vincent (DSM Research, The Netherlands), the mechano-chemical degradation of polymers and its after-effects was discussed in report of J. Sohma (Kanagawa University, Japan). L. Rosik (coauthor Z. Horak) from research institute of synthetic rubber, Czechoslovakia, described selected problems, connected with the aging, degradation and stabilization of flame-retardant styremic plastics.

PVC degradation: R.J. Meier and B.J. Kip from DSM Research (The Netherlands) spoke about a quantum chemical study of degradation and maximum polyene lengths in PVC, and H. Votter (Ciba-Geigy, GmbH, Germany) told about new PVC heat stabilizers as alternative to Cd- and Pb-containing systems.

INTERNATIONAL CONFERENCE ON COMBUSTION

The International Conference on Combustion was held in Russia on June 21–26, 1993. The first part of the Conference was conducted in Moscow (June 21–23), and the second was in St. Petersburg (June 24–26). The Conference was dedicated to the memory of the greatest Soviet physicists. Professor Yakov B. Zel'dovich, who made an essential contribution in the development of the theory of combustion and explosion, chemical and physical kinetics, physical chemistry and chemical physics. The purposes of this Conference was to celebrate the 50th anniversary of Zel'dovich's fundamental paper "on Theory of Combustion of Powders and Explosives" (1942). The conference highlighted theoretical and experimental research on combustion and intrachamber processes.

(i)   mathematical and physical problems is simulation of combustion and internal ballistic processes.
(ii) nonsteady combustion and pressure waves in practical systems.
(iii) mechanisms and kinetics of ignition, flame propagation and extinction.
(iv) combustion in extreme conditions (high and small pressure, gravity, irradiation, critical diameter, etc.).
(v)   new experimental methods for diagnostics and investigations of combustion and intrachamber processes.
(vi) physical and biological analogs of combustion.
(vii) transition from combustion to detonation.
(viii) wave phenomena in internal ballistics.

The Conference has shown the progressive development of the theory of Combustion and the study of its practical aspects. It is planned to conduct such Conference every 3–4 years.

"THE RESULTS AND PERSPECTIVES OF APPLICATION OF NATURAL AND SYNTHETIC HIGH-MOLECULAR COMPOUNDS IN FOOD PRODUCTION" WAS HELD IN SUSDAL FROM 2–4 DECEMBER 1991

The Conference dealt with problems of production of foodstuffs, chemical aspects of reprocessing of biological raw materials, human ecology and biotechnology, biotechnological aspects of food production, touched upon the chemical aspects of the problems of reprocessing of biological raw materials into foodstuffs and the possibilities of the application of natural and synthetic high molecular compounds in food production.

Professor G.E. Zaikov (Institute of Chemical Physics, Moscow) devoted his presentation to the current state of and perspectives for further development of research in the filed of polymers aging, used in the production of foodstuffs. Food biopolymers as a basis for ecologically pure foodstuffs were discussed. Professor A.P. Nechaev (Moscow Institute of Food Industry) and A.L. Peshekhonova.

Professor V.E. Goul (MIAB) discussed the application of polymeric materials for packaging films. Ways of improving the protection of polymeric coatings in biologically active media were discussed by Professor L.A. Sukhareva (MIAB).

V.V. Abramov and N.G. Zharkova, dealt with multilayer packing films, equipment for their manufacturing and combined packaging films used for food products.

The problems of utilization of membrane technology for purification of substances in the food production processes, development of installations for obtaining fresh water with the use of membrane technology, the technology of manufacturing packages from polymers were also considered.

THE IXTH ALL-UNION SCIENTIFIC SYMPOSIUM ON "SYNTHETIC POLYMERS FOR MEDICAL PURPOSES"

T.E. Rudakova and G.E. Zaikov

The symposium was held in the city of Zvenigorod (Moscow region) from 5 to 7 December 1991. The IXth All-Union Symposium on "Synthetic Polymers for Medical Purposes" was organized by the USSR Academy of Sciences, the Topchiev Institute of Petrochemical Synthesis, the Scientific Board on High-molecular Compounds within the Division of General and Industrial Chemistry of the Presidium of the Academy of Sciences, and by the GKNT Scientific Board "Synthetic Polymers for Medical Purposes". The symposium was sponsored by Research Institute "Medpolymer", Research Industrial Association "Biotekhnologiya", Scientific Industrial Center of Biotechnology and Engineering.

Professor E.F. Panarin, (St. Petersburg) opened the first morning session as a vice-chairman of the Organizing Committee.

A plenary presentation by N.A. Platae and A.E. Vasiliev (Moscow) discussed transdermal and therapeutic systems which are preparations for a new generation of continuous and steady delivery of remedial substances through undamaged skin as a treatment of chronical diseases.

Professor U.N. Musaev (Tashkent) considered the synthesis and preparation of polymeric forms of pharmacologically active monomers. The dependence of pharmacological activity on macromolecular characteristics of polymers and co-polymers demonstrated the use of pectin compounds obtained from cotton serum, apple husks, sugar beet pulp as polymeric carriers was described. These synthetic preparations were found to be highly efficient in tests on experimental animals. Professor R.D. Katsarava (Tbilisi) reported on new biocompatible water-soluble carriers of α-amino-acid-based biological substances.

Professor I.K. Grigoryants with G.A. Trikhanova, dealt with the physico-chemical aspects of the controlled formation of biologically and chemically active substances. New approaches to the development of systems of directed transport of medicines based on the use of self-accumulating supermolecular complexes were reported by Professor E.S. Severin in collaboration with Professor V.Yu. Alakhov, A.V. Kabanov. Professor R. Barbucci (Italy) presented a paper on the synthesis, physico-chemical and biological characteristics of materials capable of binding large amounts of heparin.

Professor E.F. Panarin, presented interesting data on the correlation of biologically activity of glucocorticoid hormones by synthetic polymers. L.L. Stoitskaya, reported on macromolecular therapeutic systems of antiviral and anti-tumorigenic action. A very interesting paper on interpolyelectrolyte reactions as a basis for a new method of diagnosis was presented by Professor A.B. Zuzin prepared in collaboration with V.I. Izumrudov. Professor G.P. Vlasov (St. Petersburg) reported on synthetic immunoreagents. New methods for testing the carboplastics for their application in medicine were reported by T.E. Rudakova and G.E. Zaikov. About 132 exposition papers were presented at the sessions. In addition to presentations made at the symposium, 26 additional papers were published in "Abstracts" of the symposium.

Progress in Polymer Science in Japan

Preface

The reviews in this section cover papers presented at the 43rd Annual Meeting (Spring) and Symposium on Macromolecules (Fall) of the Society of Polymer Science, Japan (SPSJ) which were held, respectively, in Negoya on May 25–27, 1994 and in Fukuoka on October 12–14, 1994. There were 1,543 presentations at the Spring Meeting and 1,195 presentations at the Fall Symposium.

The Spring Meeting covered only general topics, functional polymers, bio-related polymers, high performance polymers, polymer engineering and technology, polymer structure, polymer dynamic and rheology, physical properties of polymer solid and polymer solution, surface and interfacial properties, and polymer synthesis and polymer reaction. The Fall Symposium covered 643 general topics, 552 special topics and 7 invited lectures. The papers presented in the special topics covered metallocene catalysts (20), hybrid polymer materials for medical use (52), polymer gels (77), optoelectronic materials (75), polymer alloy (47), polymer liquid crystals (57), engineering plastics (32), dispersion phenomena for polymers (35), supramolecular design and molecular recognition (74), ecomaterial (51), and surface characterization of organic materials (32).

Every review article concentrates on presentations made at both meetings. The support of the Board of Directors of the SPSJ and, in particular, Mr. Takeshi Takahiko, Secretary General of the SPSJ, is acknowledged by the Regional Editors.

REFERENCES

References which appear in brackets in the text are pages numbers in *Polymer preprints, Japan* (Japanese edition) volume 43 (1994). Following the page numbers, E signifies the page number for the English abstract in *Polymer Preprints, Japan* (English Edition) volume 43 (1994).

Any inquiry about the Meeting on References, please contact with SPSJ office following the **FAX: 81–3–3545–8560**

Ionic Polymerizations

TAMOTSU HASHIMOTO

Department of Materials Science and Engineering, Faculty of Engineering, Fukui University, Bunkyo, Fukui 910, Japan

KEYWORDS: Cationic polymerization/Anionic polymerization/Coordination polymerization/Living polymerization/Vinyl polymerization

This review concerns cationic, anionic and coordination polymerizations of vinyl and related monomers and new polymer syntheses thereby. Some of the highlights were selected from the papers presented at 43rd Annual Meeting and Symposium of the Society of Polymer Science Japan (SPSJ) in 1994.

1 CATIONIC POLYMERIZATION

Living Cationic Polymerization

Although the sorts of monomers and initiators of living vinyl cationic polymerizations have rapidly increased recently, the nature of the living propagating species is still not fully understood in many cases. T. Higashimura and M. Sawamoto analyzed the model growing species of the actual living polymerizations of isobutyl vinyl ether (IBVE), ethyl vinyl ether (EVE), and 2-chloroethyl vinyl ether (CEVE) initiated by their HCl-adducts (1) coupled with $SnCl_4$ in the presence of nBu_4NCl by *in situ* multinucler (^{1}H, ^{13}C, and ^{119}Sn) NMR spectroscopy (210, E57; 1900, E1140). As model reactions, mixtures of HCl-vinyl ether (IBVE, EVE, and CEVE) adducts (1a, 1b, and 1c, respectively) and $SnCl_4$/nBu_4NCl ([1]0 = 200 mM; [$SnCl_4$]0 = 100 mM; [nBu_4NCl]0 = 140 mM) were employed [Eq. (1)]. In ^{1}H NMR measurements in CD_2Cl_2 at −78°C, for example, the α-methine protons of 1a, 1b and 1c absorb at ca. 5.7 ppm. On mixing $SnCl_4$, the methine signals of 1a and 1b considerably shifted downfield to 9.0–9.5 ppm. According to the researchers, these downfield shifts are due to the Lewis acid-assisted formation of carbocations from the adducts 1a and 1b, and the carbocationic species, 2a and 2b, are in a rapid exchange equilibrium with 1a and 1b, respectively. In contrast, the methine absorption of 1c only slightly (less than 0.5 ppm) shifted downfield by the addition of $SnCl_4$. Thus, the concentration of the carbocations in 1c/$SnCl_4$ is much lower than in 1a/$SnCl_4$ and 1b/$SnCl_4$, in other words, the carbocation concentration under the salt-free conditions is dependent on the monomer structure.

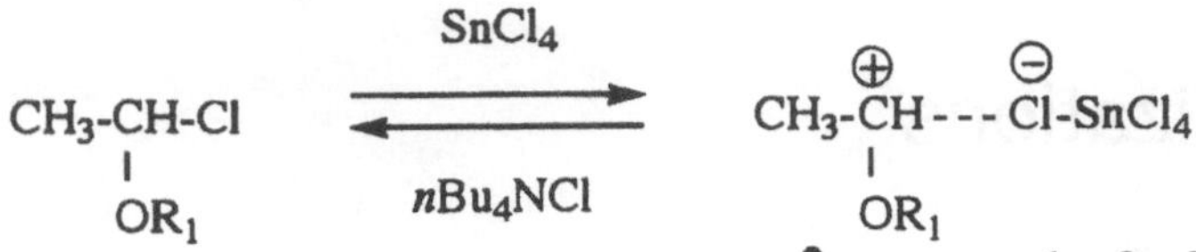

$$\text{CH}_3\text{-CH-Cl} \quad \underset{n\text{Bu}_4\text{NCl}}{\overset{\text{SnCl}_4}{\rightleftarrows}} \quad \overset{\oplus}{\text{CH}_3\text{-CH}} \text{---} \overset{\ominus}{\text{Cl-SnCl}_4} \qquad (1)$$

$$\underset{\text{OR}_1}{|} \qquad\qquad\qquad\qquad \underset{\text{OR}_1}{|}$$

$$\mathbf{1} \qquad\qquad\qquad\qquad\qquad \mathbf{2}$$

1a, 2a: $R_1 = CH_2CH(CH_3)_2$

1b, 2b: $R_1 = CH_2CH_3$

1c, 2c: $R_1 = CH_2CH_2Cl$

Under these salt-free conditions, the actual polymerizations of IBVE and CEVE are not living, whereas the EVE polymerization is living. However, the former two non-living polymerizations can be converted into living processes by addition of $n\text{Bu}_4\text{NCl}$. On addition of $n\text{Bu}_4\text{NCl}$ to the mixtures of **1a** or **1c** and SnCl_4, the downfield resonances return to their original positions for the covalent precursor **1a** or **1c**, respectively. This shows that the critical carbocation concentration for the living polymerization to occur also depends on the sort of monomers; lowering the carbocation concentration is inevitable for IBVE and CEVE but not for EVE to form a living polymer.

Effects of bulkiness of counter anion in the living polymerizations of vinyl ethers were reported (205, E55, 1908, E1142). T. Hashimoto and T. Kodaira employed a series of acetic acid derivatives [$R_2\text{COOH}$: $R_2 = CH_3$, CH_3CH_2, $(CH_3)_2CH$, $(CH_3)_3C$] in conjunction with SnBr_4 as an initiator for the polymerizations of IBVE in toluene at 22°C. The propagating end was schematically illustrated as species **3**. For all cases, livingness of the polymerizations was confirmed by linear increases of the molecular weight of the obtained polymers against monomer conversion. The molecular weight distributions (MWDs) of the polymers were very narrow ($\overline{M}_w / \overline{M}_n \cong 1.1$) for $R_2 = CH_3$, CH_3CH_2, and $(CH_3)_2CH$, but they slightly broadened ($\overline{M}_w / \overline{M}_n \cong 1.3$) for $R_2 = (CH_3)_3C$. Also, the polymerization rates with $(CH_3)_3CCOOH$ were much greater than those with the other three acids [R_2 in $R_2\text{COOH}$ (relative rate): CH_3 (1) < CH_3CH_2 (1.1) < $(CH_3)_3CH$ (1.6) < $(CH_3)_3C$ (6.4)]. The steric hindrance of a highly crowded counter anion like $(CH_3)_3COO^-$ weakened the interaction with the carbocation, thus, accelerated the propagation and possibly caused side reactions to render the polymer MWDs broader. On the other hand, the bulkiness of counter anion did not affect obviously the stereoregularity of the obtained polymers (meso: 66–67%) under the present reaction conditions.

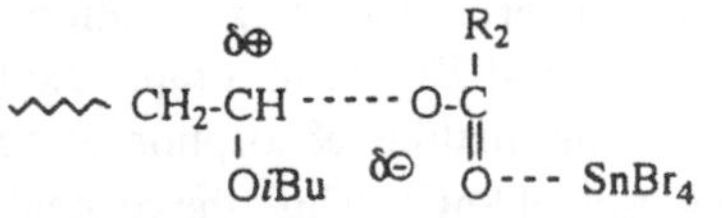

$$\underset{\text{O}i\text{Bu}}{\overset{\delta\oplus}{\sim\!\sim\!\sim \text{CH}_2\text{-CH}}} \text{-----} \underset{\overset{\delta\ominus}{\text{O}}\text{--- SnBr}_4}{\overset{R_2}{\underset{\parallel}{\text{O-C}}}}$$

$$\mathbf{3}$$

$R_2 = CH_3, CH_3CH_2, (CH_3)_2CH, (CH_3)_3C$

$$\underset{\text{O}i\text{Bu}}{\overset{\delta\oplus}{\sim\!\sim\!\sim \text{CH}_2\text{-CH}}} \text{-----} \underset{\text{O-CH}_3}{\overset{R_2}{\text{O=C}}} \overset{\delta\ominus}{\text{----}} \underset{\text{O --- AlEtCl}_2}{\overset{\text{O-C-CH}_3}{\parallel}}$$

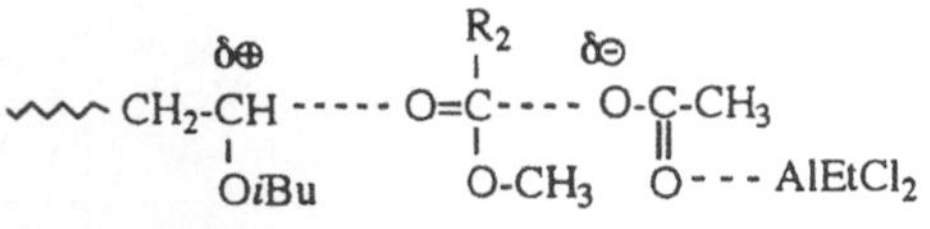

$$\mathbf{4}$$

Scheme 1

Effects of the bulkiness of added base in the living IBVE polymerizations initiated by $CH_3CH(iBu)OCOCH_3/EtAlCl_2$ in hexane at +40°C were investigated by S. Aoshima and E. Kobayashi (1906, E1142). A series of carboxylates [R_2COOCH_3: $R_2 = CH_3$, CH_3CH_2, $(CH_3)_2CH$, $(CH_3)_3C$] were employed and considered to stabilize the growing carbocation (species 4). Although the polymerization was invariably living, its rate increased with increasing steric hindrance of the substituent R_2 of the esters [R_2 in R_2COOCH_3 (relative rate): CH_3 (1) < CH_3CH_2 (1.6) < $(CH_3)_3CH$ (2.3) < $(CH_3)_3C$ (10)]. The role of the added base in stabilizing the carbocation was discussed on the basis of the thermodynamic parameters obtained by kinetic study.

Synthesis of Functional Polymers

T. Kodaira et al. synthesized a series of new liquid crystalline poly(vinyl ethers) by cationic polymerizations of vinyl ethers with two ester groups in the pendant mesogenic units (5a–5d) (3021, E1043). The $HCl/ZnCl_2$ initiating system afforded the corresponding polymers [poly(5)] of narrow MWDs ($\overline{M}_w / \overline{M}_n = 1.1–1.2$). Their liquid crystalline properties strongly depended on alkyl length of the terminal groups of the pendant mesogens; poly(5a) and poly(5b) form nematic phases whereas poly(5c) and poly(5d) form smectic phases, after one heating-cooling cycle.

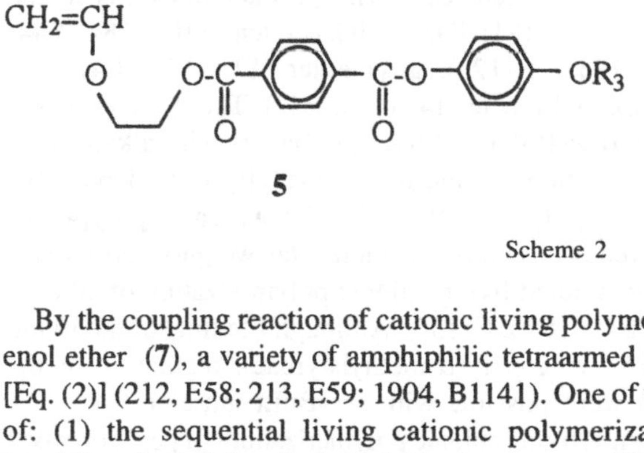

	5a	5b	5c	5d
R_3	CH_3, C_2H_5, C_3H_7, C_4H_9			

Scheme 2

By the coupling reaction of cationic living polymers (e.g., 6) with a tetrafunctional silyl enol ether (7), a variety of amphiphilic tetraarmed star block polymers were synthesized [Eq. (2)] (212, E58; 213, E59; 1904, B1141). One of these by H. Fukui et al. was composed of: (1) the sequential living cationic polymerization of α-methylstyrene and 2-[(t-butyldimethyl)silyloxy]ethyl vinyl ether into an AB-type living block copolymer (6); (2) the coupling of four chains of 6 with the terminator 7 into a tetraarmed polymer (8); and (3) the deprotection of the t-$BuMe_2SiO$- group of 8 into an alcohol. The obtained amphiphilic tetraarmed star polymer with α-methylstyrene-vinyl ether block arm chains [$\overline{DP}_n$ of block segments; α-methylstyrene/vinyl ether = 27/27] is soluble not only in toluene and chloroform but also in ethanol. Such amphiphilicity was absent for the corresponding linear block copolymer with a similar chain length and composition.

$$CH_3\text{-}CH \overbrace{\left(CH_2\text{-}\underset{\underset{\bigcirc}{|}}{\overset{\overset{CH_3}{|}}{C}} \right)_n}^{} \left(CH_2\text{-}CH \right)_m \overset{\delta\oplus}{CH_2}\text{-}\overset{\delta\ominus}{CH} \cdots Cl \cdots SnBr_4$$

6

$$+ \quad \left(C\text{—}CH_2O\text{—}\underset{}{\bigcirc}\text{—}\overset{\overset{OSi(CH_3)_3}{|}}{C}\text{=}CH_2 \right)_4 \quad \longrightarrow \quad \qquad (2)$$

7

8

2 ANIONIC POLYMERIZATION

Living Polymerization of Functionalized Monomers

S. Nakahama *et al.* are continuing their study for living anionic polymerizations of pendant-functionalized monomers (see Figure 1 and references). The pendant functions in the styrene derivatives include methylsilane (9) (181: E43), silylacetylene (10) (184, E44; 1832, E1123), phenylacetylene (11) (1832, E1123), allyl ether (12) (178, E41), N-(benzylidene)aniline (13) (1828, E1122), and nitrile (14) (180, E42). These styrenes and α-methylstyrene can be polymerized into well-defined living polymers with alkali metal countercations in THF at –78°C. Deprotections of the trimethylsilyl group of poly(10) and hydrolysis of the imino group of poly(13) [R_4 = CH(CH$_3$)$_2$] gave poly[4-(3′-butynyl)styrene] and poly(4-formylstyrene) of controlled molecular weights and narrow MWDs, respectively. Y. Nagasaki *et al.* studied living anionic polymerization of silicon-containing α-methylstyrene (15) (185, E45; 1880, E1135). Although life time of the living species of 15 was shorter than that of living poly(α-methylstyrene), regulation of the polymer molecular weight and MWD were possible with sec-BuLi initiator.

Nakahama's research group also presented living anionic polymerizations of perfluoroalkyl methacrylates (16) (176, E40), N,N-diallylacrylamide (17) (1830, E1123), 2-isopropenylbenzoxazole (18) (179, E42), and 1,3-butadiene with a protected group for two hydroxy functions (19) (1826, E1122). For the trifunctional monomer 17, the polymerizations were all quantitative in THF at –78°C with Li$^+$, K$^+$, and Cs$^+$ countercations. However, addition of ZnEt$_2$ was necessary to obtain controlled molecular weight and narrow MWD. The coordination of ZnEt$_2$ with the propagating end may stabilize the active enolate anion, and hence suppress side reactions such as a nucleophilic attack on the carbonyl group in the monomer and polymer.

Synthesis of End-functionalized Polymers

Alkyl halides bearing an epoxy group (20) (183, E44) and a 1,3-butadiene unit (21) (1836,

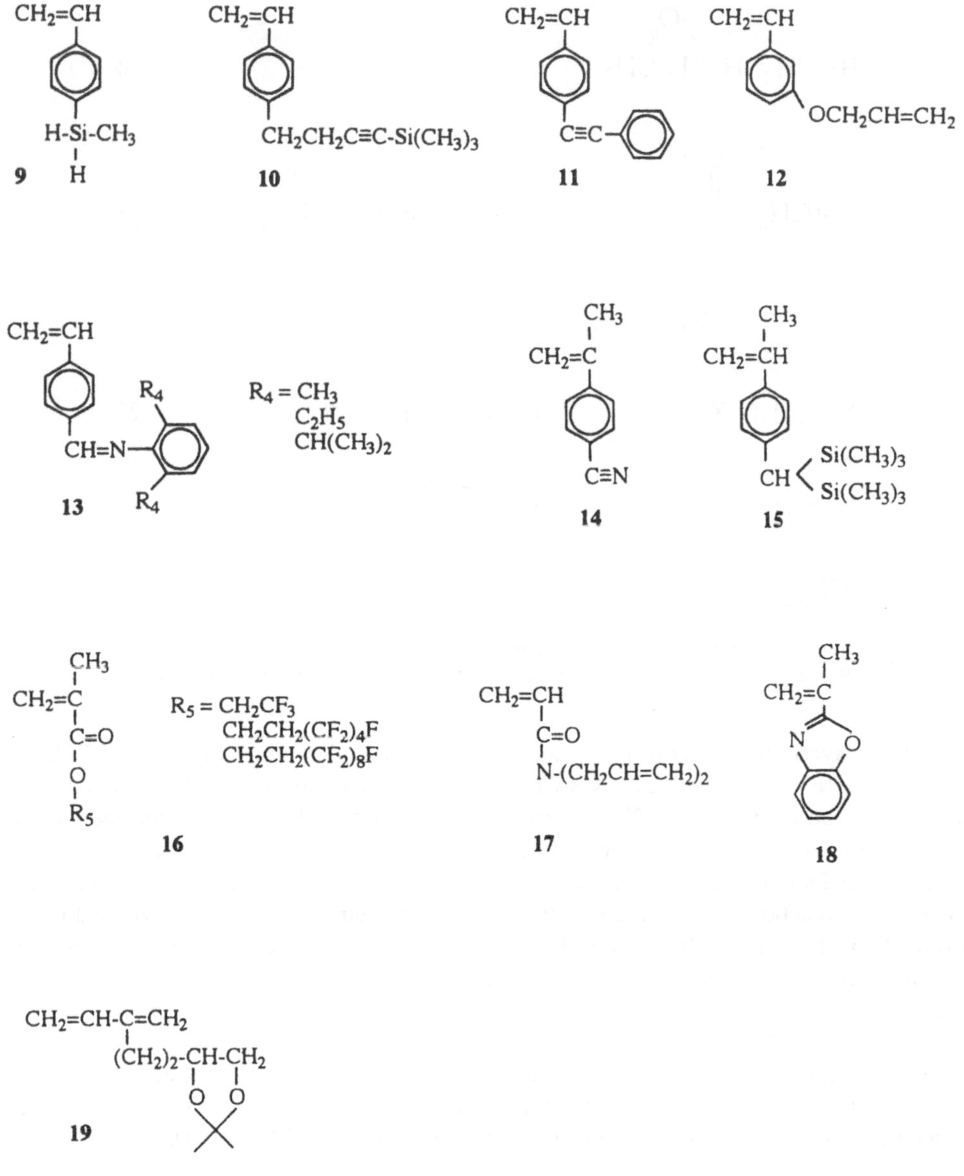

$CH_2=CH$

$H-Si-CH_3$

9 H

$CH_2=CH$

$CH_2CH_2C\equiv C-Si(CH_3)_3$

10

$CH_2=CH$

$C\equiv C-$

11

$CH_2=CH$

$OCH_2CH=CH_2$

12

$CH_2=CH$

$CH=N-$

R_4

R_4

13

$R_4 = CH_3$
$\quad C_2H_5$
$\quad CH(CH_3)_2$

CH_3

$CH_2=C$

$C\equiv N$

14

CH_3

$CH_2=CH$

$CH \big\langle \begin{array}{l} Si(CH_3)_3 \\ Si(CH_3)_3 \end{array}$

15

CH_3

$CH_2=C$

$C=O$

O

R_5

16

$R_5 = CH_2CF_3$
$\quad CH_2CH_2(CF_2)_4F$
$\quad CH_2CH_2(CF_2)_8F$

$CH_2=CH$

$C=O$

$N-(CH_2CH=CH_2)_2$

17

CH_3

$CH_2=C$

$N \quad O$

18

$CH_2=CH-C=CH_2$

$(CH_2)_2-CH-CH_2$

$O \quad O$

19

Figure 1   Functionalized monomers for anionic polymerization.
References: **9** (181, E43); **10** (184, E44; 1832, E1123); **11** (1832, E1123); **12** (178, E41); **13** (1828, E1122); **14** (180, E42); **15** (185, E45; 1880, E1135); **16** (176, E40); **17** (1830, E1123); **18** (179, E42); **19** (1826, E1122).

E1124) were used as end-capping agents for anionic living polymers to prepare end-functionalized polymers. Despite the existence of the anionically polymerizable functions, these quenchers cleanly terminated living polymers of styrene and isoprene (countercation: Li$^+$) to achieve nearly quantitative attachment of the ω-end polymerizable functional groups.

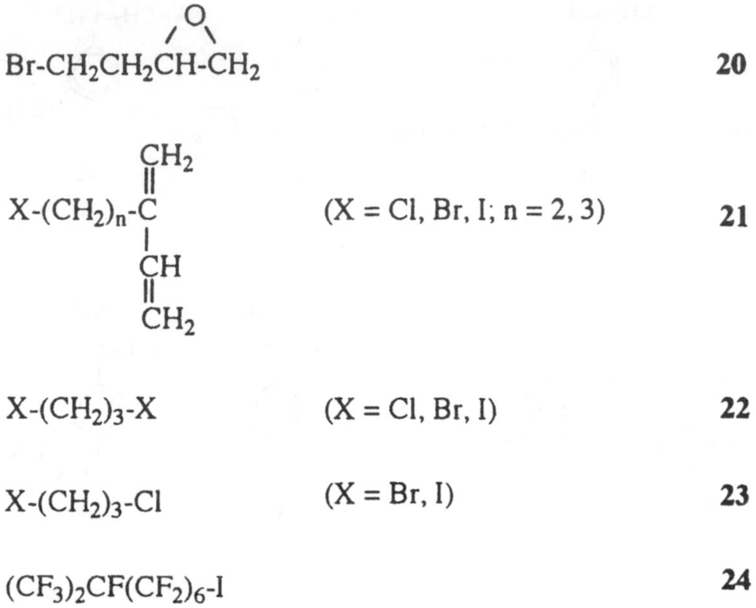

$$Br\text{-}CH_2CH_2CH\text{-}CH_2 \quad \overset{O}{\diagdown}$$

20

$$X\text{-}(CH_2)_n\text{-}\overset{\displaystyle CH_2}{\underset{\displaystyle CH}{\overset{\displaystyle \|}{C}}}\quad\quad (X = Cl, Br, I; n = 2, 3)$$
$$\overset{\|}{CH_2}$$

21

$$X\text{-}(CH_2)_3\text{-}X \quad\quad (X = Cl, Br, I)$$

22

$$X\text{-}(CH_2)_3\text{-}Cl \quad\quad (X = Br, I)$$

23

$$(CF_3)_2CF(CF_2)_6\text{-}I$$

24

Figure 2   End-capping agents for anionic living polymers.
References: **20** (183, E44); **21** (1836, E1124); **22** and **23** (1838, E1125); **24** (1834, E1124).

For the synthesis of halogen-terminated polymers, end-capping of polystyryllithium with 1,3-dichloropropane (**22**: X = Cl), 1-chloro-3-bromopropane (**23**: X = Br), and 1-chloro-3-iodopropane (**23**: X = I) were studied and found to afford chlorine-capped polystyrenes in a THF-hexane mixed solvent at $-78°C$ by S. Nakahama et al. (1838, E1125). 1,3-Dibromopropane (**22**: X = Br) and 1,3-diiodopropane (**22**: X = I), in contrast, caused side reactions with living polystyrene under the similar conditions, where dimeric products of the living polymer were formed together with the desired polymers with one halide terminal (1838, E1125). Bromide and iodide groups are apparently too reactive to undergo a selective substitution of one of the two identical halide groups in 1,3-dihalopropanes (**22**).

According to A. Hirao and S. Nakahama, the perfluoroalkyl iodide **24** can also quantitatively terminate living poly(t-butyl methacrylate) anion (countercation: Li^+) to produce a well-defined polymer with a perfluoroalkyl end group (1834, E1124).

New Initiating Systems

Divalent lantanide complexes are known as mild one-electron reducing agents. R. Nomura and T. Endo polymerized alkyl methacrylates by samarium (II) iodide (SmI_2 (in the presence of hexamethylphosphoramide (HMPA) in THF (1922, E1146). The polymers of molecular weight of 1600–20300 and polydispersity of 1.13–1.19 were obtained in nearly quantitative yield. The researchers claimed that the polymerizations proceeded via the bifunctional propagating anions that was generated in an initiation process by SmI_2-mediated reduction and coupling of the methacrylates [Eq. (3)].

$$CH_2=\overset{\overset{\displaystyle CH_3}{|}}{\underset{\underset{\displaystyle CO_2R}{|}}{C}} \xrightarrow[\text{HMPA}]{SmI_2} \ominus\overset{\overset{\displaystyle CH_3}{|}}{\underset{\underset{\displaystyle CO_2R}{|}}{C}}-CH_2-CH_2-\overset{\overset{\displaystyle CH_3}{|}}{\underset{\underset{\displaystyle CO_2R}{|}}{C}}\ominus \quad \overset{\overset{\displaystyle CH_2=\overset{\overset{\displaystyle CH_3}{|}}{CH}}{\underset{\displaystyle CO_2R}{|}}}{\Longrightarrow} \quad \overset{\overset{\displaystyle CH_3}{|}}{\underset{\underset{\displaystyle CO_2R}{|}}{C}}-CH_2-\overset{\overset{\displaystyle CH_3}{|}}{\underset{\underset{\displaystyle CO_2R}{|}}{C}}-CH_2-CH_2-\overset{\overset{\displaystyle CH_3}{|}}{\underset{\underset{\displaystyle CO_2R}{|}}{C}}-(CH_2-\overset{\overset{\displaystyle CH_3}{|}}{\underset{\underset{\displaystyle CO_2R}{|}}{C}})$$

(HMPA: hexamethylphosphoramide; R = Me, *t*Bu, benzyl)   (3)

Among stereoregular polymers, the formations of heterotactic polymers require a higher order of stereoregulation than those for isotactic and syndiotactic polymers, because meso and racemic additions should occur in an alternate manner. T. Kitayama and K. Hatada achieved heterotactic triad content of 87% for poly(ethyl methacrylate) and poly(butyl methacrylate) and this value is the highest ever reported for homopolymers (170, E37; 1840, E1125). The initiating system was composed of t-C_4H_9Li and bis(2,6-di-t-butylphenoxy)methylaluminum and the polymerizations were carried out in toluene at −78°C. The use of excess of the phenoxide over t-C_4H_9Li and the low temperature were critical factors for the formation of heterotactic polymers in high yield. The polymerizations proceeded in living manner.

3   COORDINATION POLYMERIZATIONS

New Living Polymerizations of Phenylacetylenes

Y. Kishimoto and R. Noyori reported stereo specific living polymerizations of phenylacetylene and its derivatives using $Rh(C\equiv CC_6H_5)(2,5$-norbornadiene)$[P(C_6H_5)_3]_2$ (**25**) as a catalyst [Eq. (4)] (1890, E1138). For example, the complex **25** in conjunction with 4-dimethylaminopyridine induced living polymerization of phenylacetylene in ether or THF at room temperature to give the polymers with a regular cis-transoidal structure and narrow MWD in almost quantitative yield. On the basis of structural analysis of the product polymers using $C_6H_5{}^{13}C\equiv{}^{13}CH$ as a monomer, the researchers claimed the insertion

25

(4)

(X = H, OCH₃, COOCH₃)

polymerization mechanism instead of the metathesis pathway. Another interesting feature of this living system is that the living polymer with an active end can be isolated in a solid state under argon atmosphere. The isolated poly(phenylacetylene) initiated the second polymerization of p-methoxyphenylacetylene in ether in the presence of 1 equivalent triphenylphosphine to produce a block copolymer.

T. Masuda *et al.* developed MoOCl$_4$-based new catalyst for the metathesis living polymerization of substituted acetylenes (1894, E1139). Polymerization of [o-(trifluoromethyl)phenyl]acetylene with the MoOCl$_4$-EtAl$_3$ catalytic system in anisole at 30°C gave the polymers with broad MWD ($\overline{M}_w / \overline{M}_n = 2.90$). Addition of a four fold excess of EtOH over MoOCl$_4$ narrowed MWD of the polymer up to $\overline{M}_w / \overline{M}_n = 1.02$. Livingness of the polymerization with the MoOCl$_4$-Et$_3$Al-EtOH (molar ratio 1:1:4) thre component catalyst was confirmed by the proportionality of the polymer molecular weight to monomer conversion and further increase of the polymer molecular weight upon the monomer re-addition to the completely polymerized reaction mixture.

Polymer Dynamics and Rheology

NORIO NEMOTO

Department of Applied Physics, Faculty of Engineering, Kyushu University, Hakozaki,
Fukuoka 812 Japan

This review mainly deals with 37 papers presented in one session of selected theme, "Polymer and Diffusion", in the 43rd Fall Symposium and picks up a few important topics presented in the session "Polymer Dynamics and Rheology".

POLYMER AND DIFFUSION

T. Nose, an organizer of this session, gave comprehensive remarks on diffusion of materials, which is a fundamental dynamical property related to their thermal motion in various environments (3613). He classified contributed papers to three parts; (1) diffusion of polymer as well as low molecular weight substance in polymeric systems, (2) diffusion in specific environment, and (3) mass transport accompanied with nonlinear pattern formation. This section summarizes significant achievements in respective areas. Suitable books for readers to learn more about DLS are listed in the Reference section.[1,2]

1 Diffusion of Polymer and Low Molecular Weight Substance In Polymeric System

Diffusion of macromolecules in dilute solution is the most basic and most thoroughly studied field both theoretically and experimentally utilizing the dynamic light scattering (DLS) technique. Nevertheless, there remains disagreement between theory and experiment. Y. Tsunashima (3616, E1274) showed that DLS measurement under shear flow was useful for examination of the effect of hydrodynamic interaction on translational diffusion of polymer in dilute solution. N. Nemoto, A. Koike *et al.* (3614, E1273) studied diffusion and sedimentation properties of poly(macromonomer), a new type of branched polymer which became available due to advance in polymerization techniques. The shape of the poly(macromonomer) as a whole was found to be quantitatively described by the prolate ellipsoid model with the planar zigzag backbone chain and Gaussian branched chains.

DLS under direct electric field called electrophoretic dynamic light scattering permits simultaneous measurements on electrophoretic mobility μ and translational diffusion coefficient D_t of macroions in polyelectrolyte solutions. Y. Maruyama, R. Hayakawa *et al.* (3624, E1276) proposed that DLS with sinusoidal electric field could give more detailed information on the dynamical behavior of macroions. They developed a new method for measuring the time correlation function of scattered light intensity with a new type of correlator. Data analysis for a polystyrene latex solution indicates that amplitude μ and phase δ of the complex mobility $\mu^* = \mu \exp{(i\delta)}$ and apparent diffusion coefficient D_{app} can be obtained in the frequency domain much more easily than by the conventional method.

With increasing polymer concentration, polymer chains begin to overlap one other. DLS proves local concentration fluctuation in the semidilute and the concentrated region whose growth and decay rate can be characterized by the cooperative diffusion coefficient or the gel diffusion coefficient. The recent topic in the semidilute region is the study of the slow mode which appears at the long delay time end in the time profile of the time correlation function and physics underlying the slow mode is of issue. A. Koike, N. Nemoto *et al.* (3652, E1283) made DLS and dynamic viscoelastic measurements on semidilute aqueous solutions of poly(vinyl alcohol)/borax which behave as psuedo gel due to physical crosslinks of –OH-borax complex between chains, and showed that the characteristic decay time of the slow mode was in good agreement with a mechanical relaxation time obtained by fitting the data to the Maxwell model with a single relaxation time, indicating dynamical coupling between stress of the viscoelastic network and concentration fluctuation. At very high concentration, DLS proves density fluctuation as well as concentration fluctuations occur, whereas the former becomes only responsible for DLS power spectrum of polymer melt and glass. Those characteristic features including DLS of deionized dispersion of silica particles are reviewed by N. Nemoto (1685).

The self diffusion coefficient D_s can not be measured by DLS at finite concentration, but is conveniently measured by either forced Rayleigh scattering (FRS) or pulsed field gradient NMR. K. Adachi, H. Yu *et al.* (3626, E1276) measured D_s of bulk polyisoprene as a function of molecular weight. The result of $D_s \sim M^{-3}$ in the entangled region does not agree with the prediction of the tube model of $D_s \sim M^{-2}$. Similar discrepancy have been found for semidilute and concentrated polystyrene solutions. The tube model translates entanglement dynamics, being a mathematically intractable many body problem, into a one body problem on physically reasonable grounds. The above diffusion data clearly indicate that the theory only qualitatively explains dynamical behavior of highly entangled polymeric system, and also that simple modification of the theory like the constraint release model does not lead to final solution of entanglement dynamics related to topological interaction between long polymer chains.

Diffusion of gas and low molecular weight liquid through polymeric film is important not only from the practical point of view but also scientific point of view. Recently, simulation study with molecular dynamics (MD) method has been frequently done on gas diffusion in polymer. H. Takeuchi, K. Okazaki (3634, E1278) estimated mean-square displacement $<r^2(t)>$ of He and CO_2 gas molecules in bulk polymer using MD. They found that stationary diffusive process of $<r^2(t)> \sim t$ was only attained at much longer sampling times compared with the case of self diffusion in the same medium. The gas molecule in the bulk polymer is confined in a cell formed by polymer segments and must jump out from the cell for diffusion. The hopping mechanism with long relaxation time can explain the initial anomalous time dependent behavior. K. Okamoto, Y. Ito (3632, E1278) studied correlation between gas diffusion coefficients and positron annihilation lifetime properties in rubbery and glassy states. M. Watanabe (3642, E1280) showed that electrochemical methods using microelectrodes was useful for an estimate of the diffusion coefficient of redox molecules. These may offer new techniques for diffusion coefficient measurement.

2 Diffusion in Gel

Gel electrophoresis is known as the standard technique for separation of DNA. The mobility μ under constant electric field is almost independent of molecular weight M for low M DNA (Ogston regime). In the intermediate M region, μ is reciprocally proportional to M, which is explained by reptation of DNA through gel network fixed in space (reptation regime). The μ becomes again independent of M in the very high M region so that separation of DNA is practically impossible by conventional gel electrophoresis. Direct observation using fluorescence microscope shows that DNA molecules make repeated stretching and shrinking motions in the network, and DNA with very high M keeps stretched conformation with long relaxation time, which leads to M-independent μ. In order to overcome this unpleasant situation, a few techniques have been developed. T. Kotaka, T. Shikata (3658, E1285) proposed a method in which biased sinusoidal electric field was superimposed on constant field. They showed their method was much more effective than earlier ones. Conformation change of DNA during injection from solution to gel was studied by direct observation using fluorescence microscopy by K. Morita, H. Oana et al. (3656, E1284). At the gel surface, DNA changes shape from globular to V-shaped conformation followed by stretched one, and then penetrates into the gel with a coil form.

In recent years, disposable plastics are under development as substitute for currently used ones made from synthetic polymers for protection of environment of the earth. A promising raw material is protein. A. Koike, E. Doi et al. (3654, E1284) prepared gel and plastic using ovalbumin in egg white by high temperature heat-denaturation and processing. It turned out from dynamic viscoelastic measurements and DLS that the gel had a critical gel structure at protein concentration less than 60% and behaved as hard plastic at higher concentration.

3 Diffusion Related to Phase Transition

K. Kubota, N. Kuwahara et al. (3662, E1286) studied phase separation process of a dilute polystyrene solution with isochronic solvent in a metastable region with static and dynamic light scattering methods. When the sample is quenched by $\delta T = 0.1 \sim 0.4°C$, droplets of condensed PS molecules grow with time in the main dilute phase. Increases in radius $R(t)$ of droplets and light intensity $I(t)$ scattered from the solution were divided into three regimes from their time dependences. A power law was found applicable to the first and third regimes and the second was the crossover regime. The first process with $R(t) \sim t^1$ and $I(t) \sim t^6$ is interpreted as growing process of droplets in which PS molecules form entanglement network. At later times, those droplets gather and fuse into much larger ones without change of total volume of the minor phase. In this Ostwald-ripening region $R(t) \sim t^{1/3}$ and $I(t) \sim t^1$ are observed.

Critical-fluctuation dynamics of polymer blend solution was studied by N. Miyashita. T. Nose (3660, E1285) to confirm dynamical crossover from the mode-decoupling to the mode-coupling region. The relation between the first cumulant Γ_e and the correlation length ξ_t of critical fluctuation was expressed by the model H. The value of ξ_t at which the dynamical crossover occurs is found to be much larger than the theoretical value predicted from the blob theory.

Another type of phase transition is the volume phase transition first found by T. Tanaka. T. Komai and his group (3648, 3650, E1282, E1283) measured mobility μ and diffusion coefficient D of a few probe molecule in poly-(acrylamide) gel and poly(N-isopropyl-acrylamide) gel. At the temperature of the volume phase transition of the latter gel, μ decreased by nearly one order of magnitude, while D first increased and then decreased. This behavior can not be solely explained by large fluctuation of crosslinking density of the gel accompanying the transition. The authors emphasize that specific interaction between the probe molecule and gel network may play a more important role at the phase transition temperature. H. Ushiki, K. Hamano et al. (3666, E1287) directly observed the swelling process of poly(acrylamide) gel at the volume phase transition point and showed peculiar change of the sample shape with time.

4 Nonlinear Pattern Formation

Nonlinear pattern formation or the dissipation structure is frequently observed in a open system far from thermodynamic equilibrium. Owing to very high viscosity in high polymeric system, sufficient amount of hydrodynamic flow which may be a prerequisite for the pattern formation does not take place. S. Sakurai, K. Tanaka et al. (3672, 3674, E1288, E1289) first succeeded in observation of nonlinear pattern formation on the surface of polymer cast films made by evaporation of solvent from dilute solution of styrene-butadiene random copolymer. Honeycomb-like pattern was formed on the film surface when 5% solution was used. This pattern can be observed on the top surface of liquid like water when it is heated from the bottom and called Marangoni convection pattern. Formation of Marangoni pattern suggests that evaporation of solvent is accompanied with convection flow in solution, which gives rise to a distribution of polymer concentration to be fixed in space with enhancement of solution viscosity at a later stage. The same hexagonal pattern persists at polymer concentration of 10 wt% and changed to roll-like pattern at $C = 35$ wt%. More peculiar patterns were observed for block copolymers which undergoes microphase separation only with lamella structure. However, the relation between the macroscopic pattern formation and microphase separation is ambiguous at present.

Target Pattern formation was found for photo-crosslinked polymer mixture which has lower critical solution temperature by A. Harada. T-C. Qui (3678, E1290). When photo-crosslinking was performed at temperature close to the coexistence curve, spinodal decomposition was observed. At temperature close to the glass transition temperature of the mixture, soluble semi-IPN structure was formed. At intermediate temperature, target pattern formation occurred. Although the mechanism of this pattern formation is not well understood, the authors claim that the pattern originates from competition of reaction dynamics of photo-crosslinking with increase in elasticity as the chemical reaction proceeds. H. Nakazawa, K. Sekimoto (3680, E1290) made a simulation study on polymer blend one component of which make crosslinks and network formation. They showed that the target pattern can be obtained with a technique of cell dynamical system.

When fluid with low viscosity is injected onto viscous liquid, viscous fingering occurs at the interface. K. Makino, M. Kawaguchi (3684, E1291) used aqueous polyelectrolyte solution and compressed air to observe the pattern formation. Fingers grew while making side branching. Change in surface tension by addition of isopropyl alcohol surpressed side branching. The shape of the pattern is quite different from that observed in glycerin, indicating the important role of the elasticity of the system.

POLYMER DYNAMICS AND RHEOLOGY

1 Electro-Rheology

The electro-rheology (ER) effect has been studied so far mainly on suspensions of polarizable particles and dielectric liquid as solvent. However, difference of density between particles and solvent leads to sedimentation of the particles and induces instability of the ER suspensions. Therefore, recent interest in the research area of the ER effect moved to homogeneous systems such as polymer solutions and liquid crystalline polymers. T. Uemura, K. Koyama (3395, E1212) found that polypropylene glycohol solutions of urethane polymers exhibited a 'negative ER effect', i.e., a phenomena that solution viscosity and storage modulus decrease when an electric field is applied. They examined the effects of molecular structures such as main chain structure, molecular species of hard segment, and terminal groups on the ER property of the above system. The negative ER effect is shown to be enhanced if the polymer has either a rigid structure at the urethane residue or a branched structure in the main chain. They also showed that the positive ER effect was observed for the polymer having terminal groups with high dielectric constant like cyanide groups. K. Tanaka, R. Akiyama (3397, E1213) studied the ER effect using solutions of two liquid crystalline polymers, poly(γ-benzyl L-glutamate) and poly(hexyl isocyanate). They find that stress increases about three times under DC electric field for the PBLG sample and is saturated above 2.5 kV/mm, while much larger increase in stress is observed for the PHIC sample and is unsaturated up to 3 kV/mm. Direct observation by polarized optical microscope showed that PBLG molecules formed molecular clusters which were not aligned under no electric field and tended to align perpendicular to the electrode surface under electric field. On the other hand, PHIC molecules aligned perpendicular to the electrode surface under no electric field and flow instability was seen above 1.5 kV/mm. Stress increase under electric field is discussed based on these observation.

2 Dynamics of Threadlike Micelles

Electric birefringence is a powerful technique for investigation of the dynamical behavior of long threadlike micelles which are formed by mixing a surfactant and suitable salt and exhibit pronounced viscoelastic properties at low surfactant concentration C_D. A typical example is cetyltrimethylammonium bromide (CTAB) and sodium salicylate (NaSal). T. Shikata, T. Kotaka (3399, E1213) examined formation and growth of threadlike micelle with increase in salt concentration C_S at fixed concentration of $C_D = 0.01$ M. Small spherical CTAB:NaSal micelles for $C_S < 0.004$ M started to grow into short rods at $C_S = 0.0045$ M, then to longer ones with further increase in C_S. A viscoelastic network was formed above $C_S = 0.006$ M for this sample, and the frequency dependence of components of dynamic Kerr constants, K' and K'', became independent of micelle length and can be compared with that of the Rouse mode in the polymer system.

J. Oizumi, R. Hayakawa *et al.* (3435, E1222) used frequency-domain electric bire-fringence spectroscopy for a study of C_S dependence of the same system. The rotational relaxation time and the relaxation time of the induced dipole moment of linear micelles showed similar C_S dependence. Both of them first increased with C_S and took a maximum at $C_S = C_D$. They estimated the diffusion coefficient of salt⁻ ions bound inside the micelle as 3.7×10^{-11} m² s⁻¹, which was much smaller than that of free salt⁻ ions, 5.3×10^{-10} m² s⁻¹.

3   Dynamic Birefringence of Glassy Polymers

K. Osaki and his group have made dynamic birefringence and viscoelastic measurements on many amorphous polymers, and proposed the modified stress optical rule (MSOR) which was successful in explaining their data consistently. This rule assumes that both the complex young modulus E^* and the complex strain-optical coefficient O^* can be decomposed into the rubbery and glassy components, respectively. Results of polyisoprene was successfully interpreted by the MSOR, but polyisobutylene data could not (3415, E1217). They added a new term to the MSOR for data analysis based on the molecular expression of stress proposed by Gao and Weiner. The theory states that stress is composed of contributions from chain orientation (the orientation term), monomer orientation around the chain axis (the rotation term), and fluctuation of local stress tensor (the fluctuation term). Following the theory, they attempted to classify polymers into three types (3415, E1217). The fluctuation and rotation terms are responsible for data in the glassy zone and the high frequency region of the glass-to-rubber transition zone but the degree of these contributions may vary with polymers. According to their classification, the fluctuation term is negligible for polymers with flat units like polycarbonate and the relaxation spectrum is low in the glassy zone. For polymers with thin axisymmetric units like polyisobutylene or bulky irregular units like poly(2-vinyl naphthalate) the spectrum is enhanced by the fluctuation term. Some experiments are proposed to verify the statements.

REFERENCES

1. B.J. Berne and R. Pecora "Dynamic Light Scattering", John Wiley, NY (1976).
2. "Dynamic Light Scattering, The Method and Some Applications", W. Brown ed., Clarendon Press, Oxford (1993).

Polymer Engineering and Technology

HIDEKI YAMANE

Department of Polymer Science and Engineering, Kyoto Institute of Technology, Matsugasaki, Sakyo-ku, Kyoto, 606 Japan

In the SPSJ 43rd Symposium of Macromolecules, Fukuoka, Japan, a wide variety of topics were presented in the sessions related to the Polymer Engineering and Technology. Three topics were chosen for review from these sessions. Those are "The Influence of Short Chain Branching on Creep Behavior of Gel-Spun Polyethylene Fibers", "Water Absorption Behavior of Silica-filled Epoxy Resin Compounds", and "Crystallization Behavior during the Filling Stage in Injection Molding".

1 THE INFLUENCE OF SHORT CHAIN BRANCHING ON CREEP BEHAVIOR OF GEL-SPUN POLYETHYLENE FIBERS

Ultra high molecular weight polyethylenes (UHMWPE) have been gel-spun into fibers with superior mechanical properties and used as high performance industrial materials. However, their creep property limits their application so far. Ohta *et al.* (3143, E1063) reported the influence of the short chain branching on the UHMWPE on this creep property, investigating the mechanism of the creep process by wide angle X-ray diffraction (WAXD) and measurement of the dynamic mechanical properties of gel-spun fibers.[1]

Ultra high molecular weight ethylene-propylene random copolymers with various methyl branch content were gel-spun and drawn to various extents. It was shown that the tensile modulus of the fibers obtained was decreased with the methyl branch content due to the lower draw ratio attainable of the gel fibers with higher methyl branch content, and the creep rate measured at 50°C decreased with the tensile modulus. However introduction of small number of short branching ($\sim$6 CH_3 per 1000 CH_2) strongly suppressed the creep behavior as shown in Fig. 1. The activation energies evaluated from the temperature dependence of the dynamic mechanical properties plotted in Fig. 2 indicated that the fibers possess at least three relaxation processes termed as $\alpha 1$, $\alpha 2$, and $\alpha 3$ from the lower temperature side, respectively. It is well known that the $\alpha 1$ process is associated with inter-crystallites relaxation (grain boundary phenomena) and the $\alpha 2$ process is associated with intra-crystallites relaxation. The authors showed in Figs. 3 and 4 that the activation energies evaluated from the creep rate has the almost equivalent dependence on the methyl branch content to those of $\alpha 2$ process and the *a*-axis size of orthorhombic system revealed by WAXD increases proportionally with the methyl branch content. These results indicate that some portion of the branch sites are embedded in the crystalline region and effectively hinder the chain slippage in the crystalline region, whose molecular motion is identical

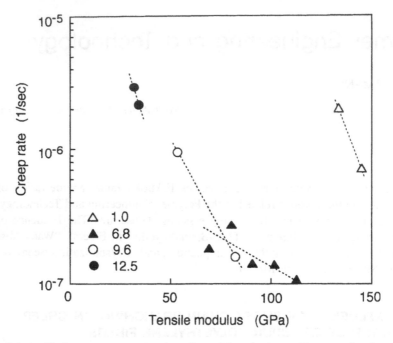

Figure 1   Relationship between creep rate and tensile modulus of gel-spun fibers as a function of average methyl branch content. Numbers in the figure indicate the methyl branch contents.

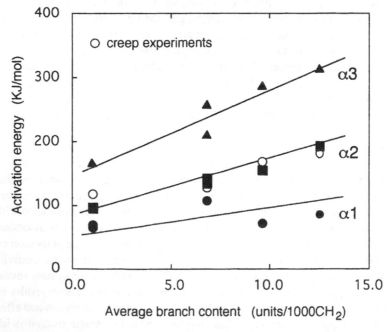

Figure 2   Arrhenius plot for the shift factor from time-temperature superposition of dynamic mechanical properties of the highly branched sample (9.6 CH_3/1000 CH_2).

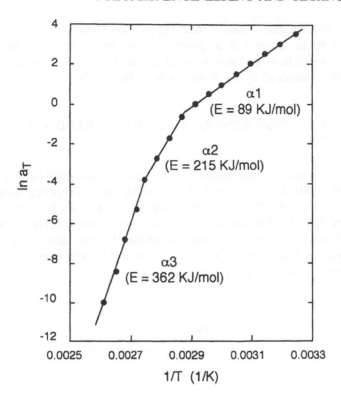

Figure 3   Various activation energies as a function of average branch content.

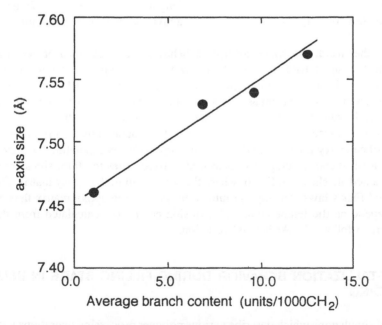

Figure 4   Size of a-axis as a function of average branch content.

to those of the $\alpha2$ relaxation, chain-to-chain slippage or rotation-transition motion of the chain. Although the assignment of $\alpha3$ process has not been established, they suggested that this process would be related to more localized motion in the crystalline region which should be highly influenced by the branches localizing in the crystalline region.

2 WATER ABSORPTION BEHAVIOR OF SILICA-FILLED EPOXY RESIN COMPOUNDS

Silica-filled epoxy resin has been widely used as a sealing compound for semiconductors. However, the reliability of the semiconductors, especially that of very large scale integrated circuit (VLSI) is strongly affected by the water absorbed in the sealing compound.

M. Yamagata [3230, E1079] reported the detailed study on the water absorption behavior of such silica-filled epoxy resin. It has been known that the water absorption in pure epoxy resin obeys Henley's (Eq. 1) and Fick's (Eq. 2) laws and its temperature dependence can be expressed by Arrhenius' equation (Eqs. 3 and 4).

$$c = S \cdot p \tag{1}$$

$$dm/dt = D \cdot A \cdot (dc/dx) \tag{2}$$

$$S(T) = S_0 \cdot \exp(-\Delta H_S/RT) \tag{3}$$

$$D(T) = D_0 \cdot \exp(-E_D/RT) \tag{4}$$

c: concentration	S: solubility	p: pressure
m: amount of matter	t: time	D: diffusivity
A: area	x: length	T: temperature
ΔH_S: molar heat of sorption	ED: activation energy of diffusion	

On the other hand, the water absorption behavior of silica-filled epoxy compounds shows a much complicated behavior although water absorption in pure epoxy resin reaches a certain saturated state after a long period of time, that in silica filled-epoxy compound does not reach to an equilibrium as shown in Fig. 5. Since the water absorption into silica is negligible, the author considered that this non-ideal behavior is attributable to the effect of the interface between silica and epoxy matrix. He separated the contributions of epoxy resin and silica-epoxy interface to the water absorption by extrapolating the water content in the silica-filled epoxy compound measured at fixed times to 100% silica content. The results obtained are shown in Fig. 6, where the water gain in the epoxy matrix obeys ideal Henry's and Fick's laws. The equilibrium water gain in the epoxy matrix thus estimated did not depend on the temperature and diffusion coefficient calculated from the results was shown to follow the Arrhenius' equation.

3 CRYSTALLIZATION BEHAVIOR DURING FILLING STAGE IN INJECTION MOLDING

Numerical simulations which describe various polymer processing operations with a computer aided engineering (CAE) programs have made remarkable advances recently.[2-4]

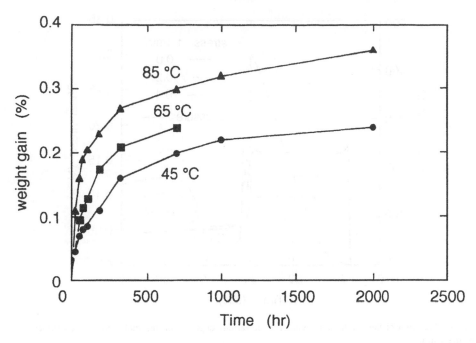

Figure 5   Water sorption behavior of the epoxy resin composite filled with 85 wt% silica at 45, 65, and 85°C at 85% RH.

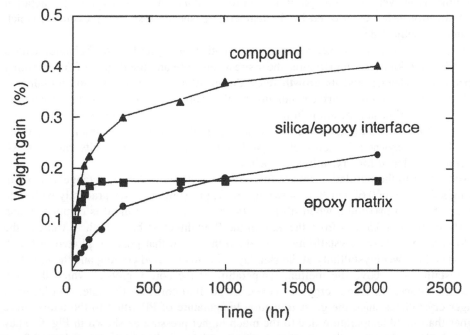

Figure 6   Water sorption behaviors separated into the effect of silica/epoxy resin interface and that of epoxy resin matrix at 85°C at 85% RH.

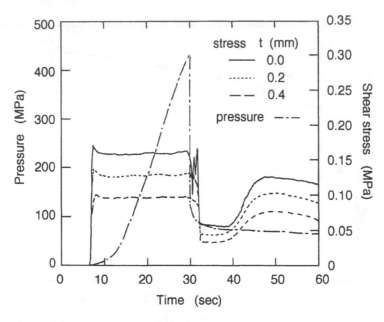

Figure 7 Variation of the crystallinity with time at various depths from the mold surface ($x = 0.5$ cm from the gate).

Especially CAE has almost made it possible to predict the overall processes in injection molding. However most computer programs available do not involve the detailed crystallization behavior affected by the temperature, shear stress, and pressure to which polymer is subjected.[5]

H. Ito *et al.* (3256, E1332) reported a simulation of crystallization behavior with a crystallization kinetics, temperature, shear stress, pressure and the degree of supercooling dependent nucleation and its growth rates, during filling stage in injection molding.

First of all authors obtained the simulation results of pressure, shear stress and temperature variation with time and position in the injection molding part by utilizing a commercial CAE program. Then they simulated the crystallinity profile and the distribution of spherulite size of polypropylene injection molded partly by introducing these results into the CAE program developed for crystallization behavior.

According to the simulation results for 200 mm × 40 mm × 2 mm rectangular mold at the injection rate of 0.69 cm³/sec shown in Fig. 7, pressure increases gradually with time and reaches to a maximum when filling of the polymer into the mold is completed, and the shear stress at 0.5 mm from the gate in the flow direction has a highest value at the mold surface. Relative crystallinity variation with time at that position plotted in Fig. 8 indicates the lower crystallinity at the skin layer due to the rapid cooling and the initiation of crystallization above the melting temperature at the core region at which the high pressure reduces the free energy of nucleation. At 10.0 cm from the gate, skin layer no longer crystallizes since the grass transition temperature of PP sifted to the temperature higher than mold temperature due to the much higher pressure as shown in Fig. 9. They also calculated the spherulite size distribution in thickness direction near and at 10.0 cm from the gate.

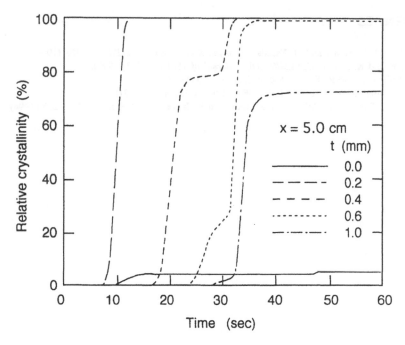

Figure 8   Variation of the crystallinity with time at various depths from the mold surface ($x = 10.0$ cm from the gate).

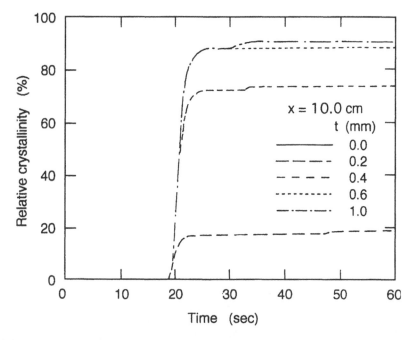

Figure 9   Variation of pressure and shear stress with time at various depths from the mold surface ($x = 0.5$ cm from the gate).

REFERENCES

1.   Y. Ohta, H. Sugiyama and H. Yasuda, *J. Polym. Sci., Polym. Phys. Ed.*, **32**, 261 (1994).
2.   K. Tada and K. Shindo, *SEIKEIKAKOU (Journal of JSPP)*, **2**, 317 (1990).
3.   P.G. Lafleur and M.R. Kamal, *Polym. Eng. Sci.*, **26**(1), 92 (1986).
4.   C.M. Hsiung and M. Cakmak, *Polym. Eng. Sci.*, **31**(19), 1372 (1991).
5.   H. Ito, Y. Tsutsumi, K. Minagawa, K. Tada, and K. Koyama, *SEIKEIKAKOU*, **6**, 265 (1994).

Photosensitive Polymers

MASAHIRO IRIE

Institute of Advanced Material Study, Kyushu University, Kasuga-Koen 6–1, Kasuga, Fukuoka 816, Japan

KEYWORDS: Photoluminescence/Photoresist/Photochromism/Non-linear optics

The present summary covers fundamental studies on primary polymer photochemical processes and their applications. Interests in polymer photochemistry focusses mainly now on photoresist for deep-UV lithography, optical data storage materials, and materials for imaging. Polymers for non-linear optics are also included.

PHOTOPHYSICAL PROCESSES IN POLYMERS

Photostimulated charge separation and its migration along the chain are key processes in polymer photo-conductor. Although extensive studies have been so far carried out to reveal the processes, no one has succeeded directly in measuring the primary processes. H. Miyasaka, T. Moriyama *et al.* (2322, E967) detected the processes by combining the picosecond transient absorption spectra and transient dichroism. The sample they examined was poly(N-vinylcarbazol) solid film doped with a tetracyanobenzene (TCNB) electron acceptor.

The initial charge separation (generation) process was followed by the transient absorption measurement. Upon picosecond laser excitation (532 nm, 15 ps) carbazol cation (780 nm) and TCNB anion (465 nm) were instantaneously produced by the pulse. The charge generation process is a very fast process. Following charge migration process was measured by the transient dichroism.

$$A^- \cdot D^+ \cdot D \cdot D \cdots \xrightarrow{k_{mig}} A^- \cdot D \cdot D^+ \cdot D \cdot D \cdots$$

$$h\nu \uparrow \downarrow k_{CR}$$

$$A^{\delta-} \cdot D^{\delta+} \cdot D \cdot D \cdots$$

When the positive charge migrates along the chain, the dichroism signal is expected to disappear. The initial strong dichroism of the carbazol cation was found to decay in less than 2 ns, while the dichroism of TCNB anion remained almost constant. The migration rate of 4.8×10^8 s^{-1} was evaluated from the decay curve. The result indicates that the hole migration in the polymer film is a rather slow process.

Energy migration in polysilanes has attracted special interest because of σ-conjugated nature of the polymer backbone. The migration process was examined by introducing low energy traps, namely crosslinking points, into the polymer (H. Tachibana, M. Matsumoto *et al.* 2320, E966).

When the energy migration is efficient, the photoexcited energy migrates to the traps, and fluorescence from the traps is considered to dominate the emission. The intrinsic fluorescence from poly(methylphenyl)silane was observed at 3.5 eV at 77 K in the absence of crosslinking agents. In the presence of the agents as much as 10% the fluorescence at 3.5 eV was replaced with a new fluorescence at 2.5 eV, which is the emission from the traps. This result clearly indicates that excited energy in poly(methyl phenyl)silane efficiently migrates in the σ-conjugated polymer backbone and is trapped into the cross-linking point.

PHOTORESIST

Photo-lithography research is currently aiming at development of a new type of photo-resist polymer systems which have sensitivity in the deep-UV region (248 nm of KrF excimer laser line or 193 nm of ArF excimer laser line), and reactive ion etching dura-bility. The sensitivity at shorter wavelengths is indispensable to achieve high lithographic resolution applicable to 256 Mbit or 1 Gbit DRAM. Photoresists so far developed have strong absorption in the deep-UV region and cannot be applied to excimer laser lithography. E. Hasegawa, K. Nakano *et al.* (2418, E997) developed a new photosensitizer, which generates acid by irradiation with 193 nm laser light. The compound has very low absorption even at 193 nm and efficiently decomposes to produce acid. The sensitizer was dispersed in the following polymer.

R=H, CH$_3$

The polymer has dry-etching durability comparable to poly(vinyl phenol), and its absorption coefficient at 193 nm is much less than that of polymers having aromatic side groups. The transmittance of 1 μm polymer film at 193 nm was as high as 65%. The sensitivity of the photoresist polymer system was around 10 mJ/cm^2 and a 0.16 μm line and space negative pattern was generated by irradiation with 193 nm light and following Ar ion etching.

When dialkoxypyrimidine groups are introduced into polymer backbone or sidegroups, the polymers can be used as an excimer laser resist (T. Sugiki, E. Mochizuki, 2422, E998). The alkoxy groups are broken or eliminated from the polymer by photo irradiation in the presence of acid generator, such as triphenylsulfonium triflate.

The polymer having the alkoxypyrimidine groups in the main chain had the sensitivity as high as 4 mJ/cm^2 and γ value of 1.7, while the sensitivity of the polymer having them in the side groups was as low as 1 J/cm^2. The low sensitivity is ascribable to the slow diffusion rate of photo generated acid in the polymer because of its high molecular weight (MW > 20,000).

PHOTOCHROMIC POLYMERS

Although much attention has been paid to use the photochromic polymers as rewritable memory media, the practical use is still limited. One of the reasons of the limitation is lack of non-destructive readout capability. Photochromic reactions, in general, have no reaction threshold, and the memory is destroyed during multiple readout operations. Useful photochromic compounds must have non-destructive readout ability. One way to solve this problem is to let the compound have a thermal reaction threshold.

Diarylethenes having oligothiophene aryl groups were found to show such thermal reaction threshold (M. Irie, T. Eriguchi, 2340, E971).

The quantum yield of the ring-opening reaction decreased remarkably by increasing the number of thiophene unit. The value of diarylethene having three thiophene units was about 1/600 in comparison with that of only one thiophene unit. At room temperature the ring-opening reaction by photoirradiation was scarcely observed. On the other hand, the quantum yield decreased when the reaction temperature was raised. At 150°C the ring-opening quantum yield was as high as 33 times the value at 27°C. The stored information remain stable provided that the memory is read at room temperature, while the memory can be erased by high intensity lasers, which cause a temperature increase of the medium.

Azobenzene chromophores placed on a glass plate was found to control the crystal growth of dichroic dyes (M. Momose, K. Kudo et al., 2352, E974). Anisotropic crystallization of the dye molecules from an aqueous solution was observed on spin-coated poly(4-methacryloylazobenzene), which was exposed in advance to linearly polarized visible light.

The photogenerated anisotropy of the azobenzene chromophores was transferred to the alignment of dye molecules. The orientational order parameter of the dye molecules depends on several conditions. Relative humidity in an operating room influenced the result. Highly oriented dye films were made when the humidity was around 60%. Another factor was the molecular weight of the polymers, and solvents used for the spin-coating procedure. The order parameter of the dichroic dye molecules as high as 0.26 was obtained when the molecular weight of the polymer was 33,000 and p-xylene was used as the solvent. This finding provides a new preparation method of film polarizers and images.

NON-LINEAR OPTICAL PROPERTIES

Polymers having extended π-conjugation in the main chain exhibit a third order non-linear property. Among them polybenzothiazol is an attracting polymer because of its thermal and mechanical durability. T. Yuba, M. Kakimoto *et al.* (2384, E989) tried to prepare a LB film made of polybenzothiazol by the method shown below.

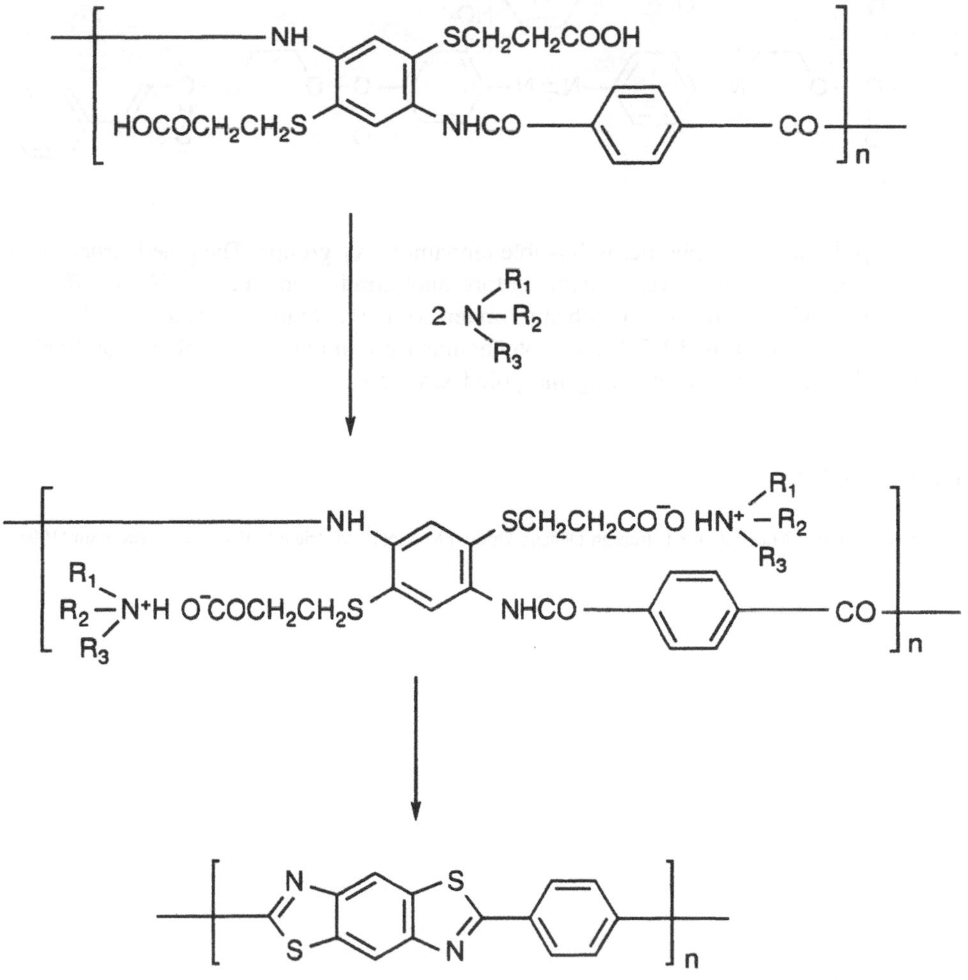

The precursor polyamide was spread on water surface, and transferred onto a glass plate by using LB technique. The LB film was heated at 280°C for 8 h and converted to polybenzothiazol thin films. The film thickness thus obtained was 0.34 nm per monolayer.

Although the precursor LB film did not show any third order susceptibility, the heat treated film gave $\chi^{(3)}$ of 10^{-11}. The susceptibility showed in-plane anisotropy as high as 1.8, depending on the dipping direction. This suggests that the dipping procedure orients the polymer chains on the glass surface.

One of the key technologies for the second-order non-linear optical polymers is how to stabilize the poled structure. K. Kanda, T. Hirayama *et al.* (2382, E988) stabilized the polymer structure by photocrosslinking.

The copolymers have photocrosslinkable cinnamoyloxy groups. The poled structure of the copolymer ($X/Y = 2.3$) was unstable before photoirradiation and the SHG coefficient d_{33} decreased with a half life of 614 h at room temperature. On the other hand, the lifetime increased to as much as 1099 h by photochemical crosslinking. The photocrosslinking was found to be effective in fixing the poled structure.

REFERENCES

1. "Photo-reactive Materials for Ultrahigh Density Optical Memory" M. Irie ed., Elsevier, Amsterdam (1994).

Degradable Polymers

M. TSUNOOKA

College of Engineering, University of Osaka Prefecture

1 SYNTHESIS OF BIODEGRADABLE POLYMERS

Polypropyllactone (PPL) is very interesting polymers from a view point of biodegradable polymers. However, so far high molecular weight PPL ($>10^5$) could not be obtained in the ring opening polymerization of β-propiolactone. H. Nishida and M. Yamashita (3975,E1351) reported that when tetramethyl ammonium acetate was used as a catalyst, a high molecular weight PPL (Mn = 197000, Mw = 414000) could be obtained in a yield of 95.4%. M. Yamashita and N. Hattori (3977, E1352) also reported that the polymerization was a living anionic polymerization and that Mw/Mn was 1.05.

A. Yamaguchi and Y. Hori (4002, E1374) reported that (R)-β-butyrolactone ((R)-β-BL) with enantioselective of 92% was easily polymerized in the presence of catalytic amount of distannoxane complexes to afford high molecular weight poly[(R)-3-hydroxy-butylate] (P[(R)-3HB]). When 1-ethoxy-3-chlorotetrabutyl-distannoxane was used as a catalyst (catalyst/(R)-β-BL = 1/8000 at 100°C for 4 h), P[(R)-3HB] was obtained from (R)-β-BL in a yield of 90%. In a similar manner, a new series of biodegradable copolyesters of (R)-β-BL with ε-caprolatone (ε-CL), δ-valerolactone (δ-VL), β-methyl-δ-valerolactone (β-Me-δ-VL), or L-lactide (L-LA) were easily obtained in exellent yields (Scheme 1 and Table 1). All polyesters have high molecular weight (Mn > 10^5) and the ratios of each units in the copolymer were almost same as the feed ratios of the monomers. The biodegradation of copolyesters composed of (R)-β-BL with ε-CL, δ-VL, β-Me-δ-VL, or L-LA was assessed using films at 25°C in aqueous solutions of a standard activated sludge. Copolymers which have over 80 mol% of the (R)-3HB unit showed almost the same degree of biodegradability in comparison with P[(R)-3HB-co-(R)-3-HV(11%)].

T. Fujimaki and E.Takiyama (3993, E1356) reported that new aliphatic polyester such as poly(ethylene succinate), poly(butylene succinate), and copolymers of poly(butylene succinate/adipate) were produced through polycondensation of glycols with aliphatic dicarboxylic acids, and others. They have melting points in a range of about 90 to 120°C, and resistant to water and common organic solvents. They also can be processed on a conventional equipment for polyethylene at 160–220°C into a variety of goods, which are currently made of polyethylene. Especially, they are processed into thin tubular film having high tensile strength and have heat sealable properties. They are stable under air, but are biodegradable in hot compost, in most soil, in the sea, and in water with activated sludge.

Table 1
Polymerization of (R)-β-BL with various lactones

entry	monomer (feed ratio)	polymer (obsd ratio)	T_m	T_g	M_n	yield, %
1	(R)-β-BU/ε-CL (90/10)	(90/10)	126	−7.4	200 000	94
2	(80/20)	(80/20)	97	−16.5	154 000	97
3	(10/90)	(9/91)	47	−61.1	255 000	99
4	(R)-β-BL/δ-VL (90/10)	(90/10)	113	−4.3	191 000	96
5	(R)-β-BL/β-Me-δ-VL (90/10)	(91/9)	115	−2.8	176 000	93
6	(R)-β-BL/L-LA (90/10)	(83/17)	120	12.3	124 000	83

2 STRUCTURE AND DEGRADATION BEHAVIOR OF BIODEGRADABLE POLYMERS

Y.Tokiwa and S. Komatsu (4042, El385) reported the relation between aliphatic polyesters and biodegradability. The aliphatic polyesters were prepared from diols(1,4-butanediol and ethylene glycol) and dicarboxylic acids (succinic acid and adipic acid) by using Ti(OR)$_4$ as a catalyst. The enzymatic hydrolysis of resulting polyesters was carried out in a phosphate buffer solution at 30°C. The degradability of polyesters using adipic acid was higher than poly(ε-caprolactone), though they had low melting points (50–60°C). Enzymatic degradability of poly(butylene succinate) (mp 112°C) decreased with an increase in their molecular weights and that with Mn of more than 50000 was not degraded at all. On the other hand, enzymatic degradability of copolymers composed of 1,4-butanediol, succinic acid and adipic acid increased with an increase in adipic acid content.

So far, practical biodegradable polymers are aliphatic polyesters such as poly(ε-caprolactone)(PCL), poly-(butylene succinate), poly(β-hydroxybutyrate), and polylactide. From the view point of expanding this kind of biodegradable plastics, A. Iwamoto and Y. Tokiwa (4044, E1385) studied biodegradaton of aliphatic polyurethanes containing ester bonds composed of PCL-diol (Mn = 530), 1,4-butanediol (BD),and 1,6-hexamethylene diisocyante. Melting points of these polyurethanes rosed in proportion to an increase in BD content. The melting points of polyurethane PU-2 (BD/PCL = 25/75) and PU-3 (50/50) were about 100 and 140°C, respectively. When microbial degradability of these polyurethanes were investigated by using 11 strains of PCL-diol degrading bacteria, microbial degradability of these polyurethanes decreased in proportion to an increase of BD content. Among the 11 strains, strain WHR6-10 degraded 90% of PU-2 and converted it into water soluble low molecular weight material in 400 h.

H. Abe and I. Matubra (4004, E1375) reported the relation between poly(3-hydroxybutylate) [P(HB)] stereoisomers and biodegradability. P(3-HB) steroisomers of high molecular weights (Mn > 10^5) were synthesized by ring-opening stereocopolymerization of (R)-β-butyrolactone with (S)-β-butyrolactone at various feed ratios (R/S = 96/4–50/50) in the presence of 1-ethoxy-3-chlorotetrabutyl-distannoxane as a catalyst. The isotactic diad fraction [i] of P(3HB) stereoisomer samples were varied from 0.92 to 0.30. A syndiotactic P(3HB) sample ([i] = 0.30) was obtained by polymerizing a racemic mixture of β-butyrolactone. The glass transition temperatures were 5 ± 2°C, irrespective of the stereoregularity. The melting temperature of P(3HB) samples decreased from 177 to 92°C as the [i] value was decreased from 1.00 to 0.68. The predominant syndiotactic P(3HB) sample ([l] = 0.30) had two endothermic peaks at around 52 and 62°C. The degree of X-ray crystallinity of solvent-cast P(3HB) films decreased from 62 to 40% as the [i] value was decreased from 1.00 to 0.68. The X-ray crystallinities of atactic ([i] = 0.46) and syndiotactic ([i] = 0·30) P(3HB) films were 0 and 26%, respectively. Enzymatic degradation of P(3HB) were carried out by the use of PHB depolymerases. The rates of enzymatic degradation of P(3HB) films ranging in the [i] value from 0.68 to 0.92 were higher than that of bacterial P[(R)-3HB] film ([i] = 1.00). The highest rate of enzymatic hydrolysis was observed at the [i] value of 0.76. The water soluble products liberated during the enzymatic degradation of P(3HB) films were characterized by HPLC analysis. The bacterial P[(R)-3HB] film gave a mixture of monomer and timer of 3-hydroxybutylic acid, but the stereoregular P(3HB) samples of [i] = 0.92–0.68 produced a mixture of monomer, dimer, trimer, and tetramer.

N. Koyama and Y. Doi (4060, E1389) reported the miscibility, morphology, and biodegradability of binary blends of poly[(R)-3-hydroxybutyric acid] (P[(R)3HB]) with poly[(S)-lactic acid] (P[(S)LA]) or poly[(R,S)-lactic acid] (P[(R,S)LA). The miscibility of P[(R)3HB]/P[(S)LA] was found to be strongly dependent on the molecular weight of P[(S)LA] by means of differential scanning calorimetry and optical microscopy. The enzymatic degradation of P[(R)3HB]/P[(S)LA] blend films was carried out by a depolymerase. The rates of enzymatic erosion decreased with an increase in P[(S)LA] content in the blend films. The simple hydrolysis of P[(R)3HB] based blend films was also investigated. Blending P[(S)LA] or P[(R,S)LA] of low molecular rights with P[(R)3HB] accelerated the rate of main-chain scission of P[(R)3HB].

3   EVALUATION OF BIODEGRADABILITY

H. Nishida and S. Suzuki (4070, E1392) reported distribution and identification of polypropyllactone (PPL) degrading microorganisms, and field tests of films. The number of PPL (Mv 153,000, Mn 911,000, and Mw 1,663,000) degrading microorganisms in environments were estimated by counting colonies that formed clear zones on plates containing emulsified PPL. It was found that PPL degrading microorganisms are widespread over a wide variety of environments. Of the total counts, the percentages of PPL degrading microorganisms were in the range of 0.0004–33.3%. A total of 77 strains was isolated from colonies formed with clear-zones on the plates. The majority of the strain was bacteria. A large number of gram positive rods and sporeformers were identified as *Bacillus* spp. e.g. *Bacillus subtilis*. Soil burial tests of PPL films were carried out at 14 place in Japan. It was found that most of undrawing films and uni-axial drawing films disintegrated in soil burial for 1 year.

H. Nishida and S. Suzuki (4072, E1392) reported complete degradation of PPL by microorganisms. Utilization of fibrous PPL were examined with 16 strains isolated from 11 places. The fourteen strains decomposed completely the PPL after 100 days' incubation at 30°C. From measurement of the water soluble total organic carbon in the culture medium, it was indicated that 14 strains converted PPL into CO_2 (and H_2O). Tests over a period of time of the fibrous PPL degradation by strain SP-1 and Y-3 were performed over a 8 day period. Strain Y-3 consumed PPL completely in 7 days. The TOC accumulated in the medium from 3 to 5 days due to faster degradation than assimilation. The pH dropped from 6.85 to 6.50 with PPL decomposition and then reverted to 6.71. These results indicate the complete degradation of PPL. The undrawing films and uni-axial drawing films of PPL were also degraded completely by the some isolated strains. However, each strain did not degrade the films in a similar manner.

4   APPLICATIONS OF BIODEGRADABLE POLYMERS

M. Yamashita and N. Hattori (3979, E1352) investigated processing of polypropyllactone (PPL: Mn = 11700–18300, Mw = 357000–412000) films and their properties. From IR and X-ray diffraction analyses, it was found that the uni-axial drawing film possessed high crystallinity and that the crystal form was the planar zigzag structure. Melting points of the films were in the range of 80–90°C. The bi-axial film exhibited a highest value (89.1°C). The dynamic mechanical measurement of uni-axial film revealed two transition zones at −30 and 2°C. The major zone at 2°C is related to the glass transition. The T-die extruded film showed excellent ductile properties, and it could be extended about 4 to 14.4 times their original length. The tensile strength of T-die extruded film was about 30 MPa, which was equal to that of HDPE or PP undrawing film. The values of tensile strength of the drawing films were increased in the range of 120-482 MPa with an increase in drawing ratio. The heat seal measurements of the uni-axial drawing films gave high seal strength values (3000–5000gf/15mm). All films exhibited high moisture permeability.

T. Imai (3981, E1353) reported an application of a biodegradable polymer, poly(3-hydroxybutylate-co-3-hydroxyvalerate) to telephone and railway cards. The cards were made of paper board coated with biodegradable polymers. The coating method employed a T-die type multi-layer extruder, which was used for coating LDPE. The operating temperature range was narrower than for the conventional coating. There was no problem in printing and card punching. The physical characteristics of the cards were superior to the conventional paper cards and the cards were substantially biodegradable in a few months when they were in an activated sludge tank.

H. Umeyama and M. Taniguchi (4052, E1387) investigated an application of BIONOLLE to bottles. Several physical properties (processability, mechanical strength, and gas barrier properties) and biodegradability of strech-blown BIONOLLE bottle were studied. It was proved that strech-blown BIONOLLE bottles had superior processability, mechanical properties (impact strength), optical properties (haze) and gas (O_2 and H_2O) barrier properties.

Hybrid Polymer Materials for Medical Use

KOHEI KUGO

Department of Applied Chemistry, Konan University, Okamoto 8-9-1, Higashinada-ku, Kobe 658, Japan

The 43rd Polymer Symposium of the Society of Polymer Science, Japan, was held in Kyushu on October 12–14 in 1994. The session for hybrid biomedical materials, which was primarily focused on "Biocompatibility", "Cell Culture", "Hybrid Artificial Organs", and "Drug Delivery Systems", consisted of two invited lectures and 57 papers. S.C. Kim, Korea Advanced Institute of Science & Technology, presented the "Hydrophilic/Hydrophobic IPN Materials" research (1672, E790; 3 references). Hydrophilic-hydrophobic interpenetrating polymer network (IPN) materials were prepared by the simultaneous polymerization method. Hydrophilic polyurethane (PU) network was prepared by forming the NCO terminated prepolymer first by reacting the polyethylene glycol with molecular weight of 1,000 and hexamethylene diisocyanate. The NCO terminated prepolymer was then reacted with trimethylolpropane (TMP) and 1,4 butanediol (BD) mixture to form the hydrophilic polyurethane network. Hydrophobic polystyrene (PS) network was prepared by reacting the styrene monomer, divinyl benzene and benzoyl peroxide mixture in bulk. The IPN sheets and tubes were prepared for the blood compatibility experiment and their surface morphology was found to form a continuous PU matrix with dispersed PS domains in the size range 0.1 to 0.7 μm. Platelet adhesion and deformation on the IPN surface decreased depending on the PS composition. *Ex vivo* A-A shunt test with rabbit showed prolonged blood occlusion times of 104 minutes compared to the 45 minutes occlusion time with Biomer using 1.5 mm I.D. tubes. W.J. Cho, Kosin University, presented the "Synthesis and Biological Activity of Polymers Containing Epoxy and Carbonyl Group" research (1681, E797; 9 references). In this lecture a new biologically active polymer composed of exo-3,6-epoxy-1,2,3,6-tetrahydrophthalic anhydride (ETA) and exo-3,6-epoxy-1,2,3,6-tetrahydrophthalic glycinylimide (ETGI) was expected to show high biological activity and low toxicity because of the structural feature and polyanionic character after hydrolysis. The antitumor activities of the synthesized monomers and polymers against Sarcoma 180 were evaluated using tumor bearing BALB/C mice.

This review summarizes some topics for fundamentals and applications of hybrid biomedical polymers in the session.

CELL-POLYMER INTERACTION AND BLOOD COMPATIBILITY

M. Akashi *et al.* (2093, E816), Kagoshima University, evaluated the cell-polymer interaction by cytokines' mRNA expression using reverse transcription-polymerase chain reaction

(RT-PCR) method. Human myeloid (HL-60) cell was differentiated from macrophage, and then cultured on tissue culture polystyrene (TCPS), polyethylene (PE), silicone (Silastic), acrylamide (AAm) grafted PE, N,N-dimethylaminopropylacrylamide (DMAPAA) grafted PE, and p-styrenesulfonic acid sodium salt (NaSS) grafted PE. After incubation for the prescribed time, the RNA was isolated from HL-60 cells and reverse-transcribed to cDNA. cDNA fragments were amplified by sequence specific primers. The products of RT-PCR were fractionated by electrophoresis in agarose gel. The expression of IL-1β was observed for TCPS and PE stimulated by lipopolysaccharide included in 10% serum, or TCPS and sllicone without serum. The cells were spread on the surface of NaSS grafted PE, but IL-1β expression was not observed on its surface. On the other hand, IL-1β expression was observed for AAm grafted PE, on which attached cells were few. The results suggested that the expression of IL-1β mRNA was not influenced by the cell attachment and spreading. Comparing to experimental results on ELISA, they concluded that RT-PCR method is one of the novel research tools for evaluation of interaction between cells and polymeric materials. They (2099, E817) also studied the synthesis of functional microspheres using macromonomer and the subsequent immobilization of protein onto the microsphere surfaces for developing medical diagnostics. The aim of this work is focused on the decrease of random adsorption of proteins and the inhibition of desorption of proteins. Poly(ethylene glycol) (PEG) and tertiary butyl methacrylate (t-BMA) macromonomers were used as water-soluble macromonomer, and styrene was used as hydrophobic comonomer. Microspheres were prepared via free radical copolymerization of those macromonomers with styrene. The microspheres obtained were monodispersed in nano meter order size. The immobilization of bovine serum albumin (BSA) onto microspheres was carried out by the condensation reaction using water-soluble carbodiimide (WSC) at 4°C for 24 h. The amount of BSA immobilized onto microspheres was about 0.05 μg/cm^2, and the random adsorption of BSA was thoroughly inhibited on PEG microspheres. The agglutination induced by the addition of anti-BSA antibody were determined by the measurement of turbidity.

H. Ihara and C. Hirayama *et al.* (2103, E818), Kumamoto University, prepared spherical polymer adsorbents for endotoxin adsorption by suspension copolymerization of N,N-dimethylaminopropylacrylamide (DMAPAA) with N-allylacrylamide (AAA). The amino group contents and the pore size of the adsorbents were easily adjusted by changing the monomer ratio and the diluent ratio. The more amino groups are introduced, the larger is the endotoxin-adsorbing capacity of the adsorbent, and the smaller the pore size (molecular mass exclusion, M_{lim}) of the adsorbent, the less acidic protein such as bovine serum albumin are adsorbed. When M_{lim} was smaller than 300 (as the molecular mass of polysaccharide) and the amino group content as 4.5 mequiv./g, the DMAPAA-AAA adsorbent showed a high endotoxin-removing activity at an ionic strength of $\mu = 0.05$–0.2 and pH6–8. The adsorbent was also able to remove endotoxin from a protein solution, naturally contaminated with endotoxin, at $\mu = 0.05$ and pH7 without affecting the recovery of the protein.

Surface modification of segmented polyurethane (SPU)s was carried out by K. Ishihara and N. Nakabayashi *et al.* (2488, E875), Tokyo Medical and Dental University, using new blood compatible polymers having both phospholipid polar group and urethane bond in the side chains. The polymers were composed of 2-methacryloyloxyethyl phosphorylcholine

(MPC), n-butyl methacrylate (BMA) and methacrylate with urethane bond (MU). The MPC copolymers were soluble in ethanol. The SPU membranes were immersed in the ethanol solution containing the MPC copolymer and dried *in vacuo* for coating. The surface was completely covered with the MPC copolymer which was confirmed by an X-ray photo-electron spectroscopic analysis. The polymer coated on the membrane surface was hardly detached in water, ethanol and 40% aqueous solution of ethanol compared to poly(MPC-co-BMA) which did not have the MU moiety. Therefore, the MU moieties had the affinity to the SPU. The surface modification of the SPUs effectively suppressed platelet adhesion after contact with platelet-rich plasma for 180 min. They (2490, E875) also immobilized MPC onto a cellulose hemodialysis membrane surface through the coating of cellulose-graft-MPC (MGC), which was synthesized by graft copolymerization of MPC on a water-soluble cellulose using cerium ion as an initiator. The MGC on the surface was not significantly detached even after immersion in water. The permeability and mechanical strength of the membrane coated with the MGC did not decrease compared to those of the original membrane. The MGC coated cellulose membrane displayed excellent blood compatibility such as prevention of platelet adhesion and aggregation after contact with platelet-rich plasma. From these results, they concluded that the MGC was a useful material for improving the blood compatibility on the cellulose hemodialysis membrane.

Y. Okahata *et al.* (2508, E880), Tokyo Institute of Technology, investigated the role of serum for adhesion and spreading of Hela cells to a synthetic glycolipid film. Hela cells selectively adhered and spread on a lactose-bearing glycolipid film in serum medium, but not in serum free medium. The cell spreading also occurred on the film pre-coated with serum, but did not occur on that pre-coated with bovine serum albumin. The cell spreading in serum medium was inhibited in the presence of lactose, but not in the presence of maltose. Serum adsorption was quantitatively detected by use of a quartz-crystal micro balance (QCM). Serum adsorption to a lactose-bearing glycolipid was suppressed compared with that to a maltose-bearing glycolipid and a phosphatidylcholine. In the presence of lactose, the amount of serum adsorbed to lactose-bearing glycolipid film increased. It was found that Hela cells specifically interact with the lactose-bearing glycolipid on which serum hardly adsorbed, but not the maltose-bearing glycolipid or the phosphatidylcholine on which serum largely adsorbed.

N. Ogata and T. Okano *et al.* (2510, E880), Sophia University and Tokyo Women's Medical College, investigated the effects of contents of phenylboronic acid groups on cell culture. Five copolymers having various contents of phenylboronic acid were synthesized by radical copolymerization of isopropylacrylamide (I), *N,N'*-dimethylamino-propy-lacrylamide (A), and acrylamide phenylboronic acid (B) (IAB-2, IAB-8, IAB-9, IAB-12, and AB-14; each number indicates the molar ratio of phenylboronic acid groups in the copolymers). With increasing content of phenylboronic acid groups in the copolymers, the growth of bovine aortic endothelial cells (BAECs) was suppressed. However, it was observed that the endothelial cells formed networks on each of the IAB copolymers with various contents of phenylboronic acid groups. Moreover, capillary structure was observed on IAB-2. This morphological differentiation of BAEC is caused by the interaction between the cells and polymer surfaces, that is, complexation between boronic acid groups in IAB copolymer and sugar moiety on cell membranes. They suggested that IAB copolymers having boronic acid may control the morphology of BAECs.

DRUG DELIVERY SYSTEMS AND HYDROGEL

K. Sakai and T. Okano *et al.* (2520, E883), Waseda University and Tokyo Women's Medical College, prepared a novel thermosensitive hydrogel with a comb structure in which poly(*N*-isopropylacrylamide) (PIPAAm) chains are grafted onto crosslinked networks of PIPAAm. Swelling-deswelling kinetics of comb type PIPAAm gel was investigated and compared to that of conventional PIPAAm homopolymer gel. Furthermore, deswelling kinetics was discussed in terms of changes in water structure determined by DSC measurements. The shrinking rate of conventional PIPAAm homopolymer gel is limited by water permeation from the gel interior through the collapsed polymer skin, and the gel keeps water more than one month. On the other hand, the comb-type PIPAAm grafted gel shrunk much more rapidly to its equilibrium state. Entrapped water was rapidly squeezed out from the gel interior. This is supported by the changes in the amount of free water within two types of gel matrices determined by DSC. The amount of free water in the comb-type PIPAAm gel decreased dramatically after temperature was increased to 40°C, although bound water content was maintained almost constant and low. Therefore, drastic swelling change observed for comb-type PIPAAm gel was attributed to the significant change in free water content in the gel. Rapid shrinking of comb-type PIPAAm gel is due to the immediate dehydration of PIPAAm graft chain in the gel matrix followed by the subsequent hydrophobic interaction between dehydrated graft chains preceding the shrinking of the PIPAAm network. Rapid shrinking kinetics of the comb-type PIPAAm gel resulted from the lack of collapsed polymer skin on the gel surface after temperature increase was also observed.

H. Iwata and Y. Ikada *et al.* (2536, E887), Kyoto University, proposed a novel microbead which can maintain the function of a bioartificial pancreas (BAP) in nonobese diabetic (NOD) mice for a long term. The novel microbead gel was composed of the agarose and poly(styrene sulfonic acid) (PSSa), of which surface was covered with carboxymethyl cellulose to provide anionic charges. The islets of Langerhans in the microbeads were intraperitoneally implanted. Graft functioning periods depended on the concentration of PSSa and the kind of polyanions. All 5 recipients of islets encapsulated in the microbeads, which were composed of 5% agarose and 5% PSSa and of which outermost surfaces were covered with carboxymethyl cellulose, demonstrated more than 60 days normoglycemic periods. On the other hand, control mice receiving transplants of unenclosed islets or islets in agarose microbeads showed less than 12 days normoglycemic periods. They concluded that the agarose/PSSa microbeads is one of the most promising semipermeable membranes which enable xenotransplantation of islets in spontaneous diabetes mellitus.

T. Akaike and K. Kobayashi *et al.* (2540, E888), Kanagawa Academy of Science and Technology, used lactose-carrying polystyrene (PVLA) and LDL from rabbit fed with semisynthetic diet containing cholesterol as ligands of endocytic receptor. Hepatocytes preferentially adhered to plastic dishes modified with PVLA and LDL, as well as those with collagen. The dishes with PVLA and LDL promoted the formation of three-dimensional multicellular aggregates, and supported stable primary culture of the cells during a period of over 8 days. An attempt has been made to develop a new phenotypic assay system using ligands of the endocytic receptors. Flow cytometry of rat hepatocytes, Hep G2 cells, Chang Liver cells, and LEC rat cells was carried out by using PVLA and LDL as probes which were derived with fluorescein isothiocyanate. These ligands were effective as markers for phenotypic assay using neither antibodies nor radiolabeled substances, and the flow cytometry with fluorescent ligands distinguished normal rat cells from hepatomas and LEC rat cells involving functional disorders. Tokyo Institute of Technology (2542, E888), also attempted to design a new polymer containing double ligand to endow a new function to the material, and used MIN6 cells derived from transgenic mouse which was superior to the other insulinoma cell line. They found that MIN6 cells recognized the copolymers which contained the sulfonylurea better than polystyrene dishes. This result suggested that MIN6 cells recognized the sulfonylurea and then insulin was released by effect of hypoglycemic sulfonylurea. This is not only effective on insulin release but also on morphological change of the MIN6 cells.

An anticancer drug, *cis*-dichlorodiamineplatinum(II) (CDDP), was conjugated to poly(ethylene glycol-aspartate) block copolymer by K. Kataoka and M. Yokoyama *et al.* (2554, E891), Science University of Tokyo, for a novel vehicle design that may be useful for targeting to tumors. The conjugates of PEG-P(Asp(CDDP)) were formed by mixing CDDP with the block copolymer in water at various molar ratios of CDDP to Asp units of PEG-P(Asp). The reaction mixture were kept at 37°C for 48 hours. The conjugates were mixed with the *o*-phenylenediamine reagent dissolved in DMF and the absorbance was measured at 703 nm for determination of the free platinum(II) compounds. Considerable decrease in the peaks corresponding to carboxyl proton of aspartic acid was observed by ^{1}H NMR when the molar ratios of CDDP to Asp units were in the range of 0.8 to 1.0. The CDDP content of the conjugate was approximately 55% when the molar ratio of CDDP to Asp units was 1.0. The conjugates obtained after ultrafiltration (molecular weight cut off 100,000) were observed to form a micelle structure possessing diameter of approximately 30 nm with narrow distribution in 0.1 M phosphate-buffer saline at 37°C by dynamic light scattering. The micelle obtained was not disrupted even in the presence of SDS, indicating its high stability. They (2560, E893) also prepared a novel system of DNA vector, soluble complexes of DNA with poly(ethylene glycol)-poly(lysine) AB type block copolymer, and studied the thermal denaturation behavior. They obtained stable and soluble PEG-poly(lysine)/DNA complexes even at the electrostatic equivalence at low DNA concentration as 25 µg/ml. Addition of the PEG-poly(lysine) block copolymers to DNA resulted in a stabilization of the helix against thermal denaturation as in the case of poly(L-lysine). Thermal denaturation monitored at 260 nm revealed a biphasic transition profile in 50% (v/v) methanol-1 mM PBS buffer (pH 7.40). DNA fraction showing lower transition temperature based on melting of DNA decreased with addition of the PEG-poly(lysine) while the fraction having a higher temperature transition based on denaturation of complex increased. At the electrostatic equivalence the lower transition disappeared and only the transition due to the complex was observed. These results indicate that stoichiometric complex was formed between the DNA and PEG-poly(lysine) to associate into supramolecular assembly. Furthermore, DNA in the PEG-poly(lysine)/DNA complex was completely dissociated by addition of 20eq. excess of poly-L-aspartic acid. Poly-L-aspartic acid replaced DNA in the complex with PEG-poly(lysine) and resulted in the formation of free DNA.

K. Akiyoshi and J. Sunamoto *et al.* (2558, E892), Kyoto University, investigated the complexation between Adriamycin (ADR) and cholesterol-bearing pullulans (CHP) by UV, fluorescence, and CD measurements. The nanoparticles of the CHP-ADR complex were isolated by ultra filtration. One CHP self-aggregate incorporated 90-480 ADR molecules depending on the substitution degree of cholesterol moieties of CHP. The complex showed an excellent colloidal stability and no leakage of ADR from the complex was observed at all even after a week.

Y. Tabata and Y. Ikada *et al.* (2562, E893), Kyoto University, reported the sustained release of basic fibroblast growth factor from gelatin hydrogels based on polyion complexation. Acidic gelatin was crosslinked with a water-soluble carbodiimide to prepare biodegradable hydrogels. The prepared gelatin hydrogels were dried and then swollen again in the aqueous solution containing basic fibroblast growth factor (bFGF) to prepare bFGF-incorporated gelatin hydrogels. The bFGF-incorporating hydrogel was implanted subcutaneously into the mouse back to assess the extent of vascularization by determining the amount of tissue hemoglobin around the implantation site. When the released bFGF was analyzed on a heparin affinity HPLC, bFGF was found to be released *in vitro* from the bFGF-incorporated gelatin hydrogels within the first day up to about 40%, but thereafter no release was observed. Gelatin hydrogels incorporating bFGF induced neovascularization and the amount of tissue hemoglobin increased within 1 day after hydrogel implantation to reach a maximum and returned to the initial level of hemoglobin at day 14. On the contrary, phosphate-buffered saline solution (PBS) with and without bFGF could not increase the amount of hemoglobin at the injection site. The bFGF-incorporated gelatin hydrogels were degraded with time, completely being resolved within 14 days. These findings indicate that bFGF can be released from the gelatin hydrogel as a result of biodegradation of gelatin, leading to induction of vascularization. In addition, no increase in the hemoglobin amount was observed by implantation of bFGF-free gelatin hydrogels. They concluded that the biodegradable hydrogel prepared from acidic gelatin is a promising formulation enabling the release of biologically active bFGF.

Biopolymers

KENICHI HATANAKA

Department of Biomolecular Engineering, Faculty of Bioscience and Biotechnology, Tokyo Institute of Technology, Nagatsuta-cho, Midori-ku, Yokohama 226, Japan

SYNTHESIS OF QUATERNARY CHITOSAN HAVING GALACTOSE RESIDUES AND ITS POLYELECTROLYTE COMPLEX FORMATION WITH DNA

It is well-known that saccharides play important roles in the biological recognition of cell surfaces and may be expected to be applied to cellular recognition devices. Recently, gene therapy by delivering genes to target cells has become a topic of interest in medicinal chemistry. There were many reports about gene derivery systems using polyelectrolyte complex of DNA and polycation such as poly(L-lysine). Chitosan is a cationic natural polysaccharide and can form polyelectrolyte complex with DNA. It is expected to be used as a carrier of DNA in gene derivery system. To achieve an efficient gene derivery system by receptor-mediated endocytosis, J. Murata *et al.* (2835, E961) synthesized novel cationic polysaccharide derivative having recognizable saccharide residues, *N,N,N*-trimethyl-6-*O*-carboxymethyl-chitosan/galactose conjugate, and investigated its polyelectrolyte complex formation behavior with DNA and cellular recognition ability.

As shown in the following scheme, amino group of chitosan was methylated to prepare quarternary amine with methyl iodide. Then, 6-OH of *N,N,N*-trimethyl-chitosan (TM-chitosan) was carboxymethylated to give *N,N,N*-trimethyl-6-O-carboxymethyl-chitosan (TMCM-chitosan). On the other hand, the lactone which was prepared by the oxidation of lactose was reacted with 1,6-hexanediamine to give *N*-lactosyl-1,6-hexanediamine. TMCM-chitosan/galactose conjugate was synthesized by coupling reaction of TMCM-chitosan and *N*-lactosyl-1,6-hexanediamine.

UV spectra of TM-chitosan and other chitosan derivatives with DNA indicated that the complexation of polycation and polyanion occurred. The complex did not precipitate even after many hours. The formation of the ion complex of TMCM-chitosan/galactose conjugate with DNA, and cell recognition by the conjugate were investigated.

GRAFT COPOLYMERS HAVING AMYLOSE SIDE CHAINS: SYNTHESIS VIA ENZYME-CATALYZED POLYMERIZATION AND FUNCTIONS

Efficient preparation of polysaccharide, of which structure is well-defined, by enzyme-catalyzed synthetic reactions is one of the most attractive subjects in the field of fine polymer synthesis. K. Kobayashi *et al.* (2868, E984) synthesized graft copolymers having amylose as side chains by the combination of radical polymerization and enzyme-catalyzed polymerization by the use of phosphorylase. There are many kinds of graft copolymers having polysaccharides as the main chains, while graft copolymers having polysaccharides as side chains are rarely prepared.

Amylose-substituted styrene macromonomers ($DP_n = 24$, 40, and 150) (Vinylbenzyl-Amylose-Amide, 2) were synthesized via enzyme catalyzed polymerization of glucose 1-phosphate onto maltopentaose-substituted styrene primer (1) using phosphorylase as catalyst. Homopolymerization and copolymerization of the macromonomer using AIBN as initiator gave an uncommon type of graft copolymers consisting of synthetic polymer backbones and polysaccharide grafts. These graft-copolymers have uniform length of the amylose side chains.

The interaction of amylose chains in these graft copolymers with iodine was investigated by UV spectroscopy. The maximum wavelength (λ_{max}) of the complexes was shifted to longer wavelength with increases of DP_n and concentrations of amylose moieties of the copolymers.

A graft copolymer (DP_n = 50) was also synthesized by an alternative route via the radical polymerization of 1 followed by enzyme catalyzed polymerization. The copolymer showed the strong interaction with iodine: λ_{max} = 613 nm was longer than the expected value (λ_{max} = 574 nm) for amylose of DP_n = 50. It was reasonably assumed that the distribution of DP in the amylose chains became broad.

CHEMICAL STRUCTURAL CONTRIBUTION TO MANIFESTATION OF POTENT ANTI-HIV ACTIVITY ON SULFATED ALKYL OLIGOSACCHARIDE

The synthesis of sulfated alkyl oligosaccharides with potent inhibitory effects on HIV infection have been investigated. K. Katsuraya and T. Uryu et al. (2855, E982) examined the anti-HIV activity and anticoagulant activity of several sulfated alkyl laminari-oligosaccharides which were individually prepared by use of five pure oligosaccharides of laminari-pentaose through laminari-nonaose, and the relationship between the chemical structure and biological activities. Influences of the length of the alkyl chain and the number of glucose residues on the biological activities were examined to optimize the chemical structure for a potential AIDS drug. On the other hand, in order to probe the contribution of alkyl portion, branched alkyl chain and perfluoroalkyl chain which has high hydrophobicity were also examined.

Synthesis of peracetyl alkyl laminari-oligosaccharide as carried out with β-peracetyl laminari-oligosaccharides and corresponding alcohols by using stannic tetrachloride as a Lewis acid catalyst. All of the sulfated alkyl laminari-oligosaccharides except for those having short alkyl chains such as n-butyl group, exhibited potent inhibitory effects on HIV infection. The anti-HIV activity of sulfated dodecyl and sulfated octadecyl laminaripentaoside had a considerably high cytotoxicity. It was revealed that sulfated perfluoroalkyl laminari-oligosaccharides had high anti-HIV activities and negligible cytotoxicities.

It is speculated that the anti-HIV effect of sulfated alkyl oligosaccharides is attributable to conformational change of gp120, which is an envelope glycoprotein of HIV and can bind to CD4 of helper T cell, owing to an interaction between sulfated alkyl oligosaccharide and gp120. It can be assumed that sulfated alkyl oligosaccharide binding site in gp120 is cationic region including lysine and arginine residues, that is, 506–518 amino acid residues in the sequence of gp120. The cationic oligopeptide region has almost the same size as pentasaccharide chain. This speculation is consistent with the experimental results of steady effect of sulfated alkyl oligosaccharide which is greater than pentasaccharide. Since the anti-HIV activity of sulfated alkyl oligosaccharide is independent of the ratio of hydrophobic region to hydrophilic region, it might be indicated that the sulfated alkyl oligosaccharide did not act as a surface active agent, the alkyl part interacting with the lipid membrane of HIV.

PHOSPHOLIPID MEMBRANE

It is well-known that biomembrane is composed of phospholipid and protein. And self organization and other functions of phospholipid are of great interest. Thus, it is quite interesting to prepare and characterize phospholipid analog and the membrane composed of phospholipid analog.

T. Sakaguchi et al. (2765, E944) reported the synthesis of acrylamide monomer and polymer containing dioleyl or stearyl moieties and a phosphatidylcholine analogous moieties. The monomer and the polymer were confirmed by IR and ^{1}H-NMR spectroscopies. The structure of the polymer was investigated by X-ray diffraction method. The build-up membrane with the polymer was prepared by LB method. And the bilayer of phospholipid analog containing dioleylglyceride was made by patch clamp method. The capacitance and electrical resistance of bilayer were also measured. M. Uekawa et al. (2767, E944) synthesized vinilic phospholipid analog containing diglyceride originated from egg yolk. They detected channels or current fluctuation in bilayer membranes composed of polymeric phospholipid analog.

T. Chen et al. (2769, E945) synthesized a series of acrylamide monomers containing phosphatidylcholine analogues. Some monomers holding long chain alkyl group as hydrophobic group were homopolymerized and copolymerized with vinyl laurate, acrylamide, and so on by using AIBN as initiator. The properties of monomers and polymers were investigated by DSC and X-ray diffraction method. In addition, the stable monolayers of those monomers and polymers at the air/water interface were made with the mixture of chlorobenzene and methanol. The π-A curves for those monolayers were determined at different temperature and different pH. The LB multilayers of those monomers and polymers are assembled onto a silicon wafer plate in the Y-mode. Depending on the difference of deposition ratio, the balance of hydrophobic and hydrophilic regions could be adjusted by copolymerization, and the LB multilayers with high quality were obtained. On the other hand, water soluble phospholipid acrylamide monomers holding short alkyl group were used for making hydrogel. The gelatin was carried out with MBAA (methylene-bis-acrylamide) as crosslinking agent in pure water. The dependence of temperature, solvent, or ion concentration of these hydrogels were measured by variations of swelling ratio.

HYDROPHOBIZED POLYAMINO ACID: SYNTHESIS AND CHARACTERIZATION OF CHOLESTEROL-BEARING POLY-L-LYSINE

Self-aggregate of hydrophobized water soluble polymer is of great interest with respect to understanding the behavior of biological macromolecules in living system. Recently, it has been reported that hydrophobized polysaccharides form nanosize hydrogels upon self-aggregation. A. Ueminami *et al.* (2783, E948) described the synthesis of cholesterol-bearing poly-L-lysine (CHPLL) and the solution properties in water.

CHPLL

PLL(hydrobromide) ($M_w = 49000$) and cholesteryl *N*-(6-isocyanatohexyl)-carbamate was reacted in DMSO/triethylamine for 24 h at 25°C. The reaction mixture was dialyzed against 0.1 N HCl and pure water. Hydrophobized PLL that contains 2.2 cholesterol groups per 100 amino acid residues (CHPLL-49-2.2(hydrochloride)) was obtained. The solution of CHPLL(hydrochloride) was obtained by suspending CHPLL(hydrochloride) in water and by sonication at 40 W for 30 min. Spherical nanosize particles were observed by negatively stained electron microscopy. The molecular weight and R_G of the aggregate were determined by SLS to be 9.6×10^5 and approximately 69 nm, respectively. The aggregation number was calculated to be approximately 14. Several hydrophobic substances were incorporated to the hydrophobic domain in the aggregates. CD measurement indicates that CHPLL self-aggregates show a coil — α-helix transition at lower pH than does PLL.

Physical Properties of Polymeric Solids

TISATO KAJIYAMA and ATSUSHI TAKAHARA

Department of Chemical Science and Technology, Faculty of Engineering, Kyushu University, Hakozaki, Higashi-ku, Fukuoka 812-81, Japan

1 CRYSTALLIZATION

Scanning probe microscopy (SPM) is a novel technique to investigate materials surfaces with high resolution. SPM can make the image of surface morphology and surface properties on the basis of the interaction between sharp cantilever tip and the sample surface. The application of SPM to the structure study of polymer crystals has produced new data on the aggregation structure of polymeric solids. A. Ishitani, H. Miyaji *et al.* (3911, E1357) measured the thickness of an isolated polyethylene (PE) single crystal by atomic force microscopy (AFM). Hexagon PE single crystals were grown from an octane solution. As was expected from Lauritzen-Hoffman theory,[1] the thickness difference between {110} and {100} growth sectors was clearly observed by AFM. The thickness difference was interpreted as an effect of the difference in the fold structure between those sectors (fold directions). From the crystallization temperature dependence of lamellar thickness, it was revealed that the magnitude of free energy of the fold surface in the {110} growth sector was 30% larger than that in the {100} sector.

T. Kajiyama, A. Takahara *et al.* (3913, E1357) studied the frictional force anisotropy of the PE single crystal surface on the basis of friction force microscopy (FFM). Fractionated HDPE with $Mw = 10$ k and the unfractionated HDPE with $Mw = 520$ k were used for the preparation of solution grown single crystals. The magnitude of frictional force was measured as a function of scanning angle, θ against the {110} growth surface. The frictional force for the single crystal prepared from unfractionated HDPE with $Mw = 520$ k was independent of θ, and it was larger than that for fractionated HDPE with $Mw = 10$ k. However, the single crystal prepared from fractionated HDPE with $Mw = 10$ k showed the scanning direction dependence of the magnitude of frictional force. These results apparently indicates that the unfractionated HDPE single crystal with $Mw = 520$ k formed loose loop chain folding with loose loop and random reentry, whereas the fractionated HDPE with $Mw = 10$ k showed the short regular fold.

Polymer blends of amorphous polyolefin and crystalline block copolyolefin have been paid great attention for practical application. However, the phase separated structure of this blend system has not yet been clarified K. Sakurai, W.J. MacKnight *et al.* (3327, E1203) studied the crystallization behavior of binary blends prepared form monodisperse diblock ethylene-propylene copolymer (DEP) with $Mw = 113$ k and two monodisperse atactic polypropylenes (APP) with $Mw = 15$ and 190 k. Mixing of these polyolefins did not affect the melting behavior of polyethylene, but drastically changed the crystallization behavior depending on the molecular weight and composition of the blend. The addition of APP with Mw = 15 k to DEP made the primary crystallization peak shifted to lower temperature. In the composition range of more than 50 wt% of APP, the crystallization peak slightly shifted to higher temperature for all the blends. In the case of APP with Mw = 150 k, the crystallization peak did not shift as much. Isothermal crystallization was carried out for the blends and was analyzed on the basis of Avrami theory. It became apparent from their analysis that the Avrami exponent of the APP (Mw = 15 k) blend system was reduced by blending and each crystallization peak might correspond to the different growth geometries. The phase-separated state of the blend in the molten state was investigated by small-angle X-ray scattering (SAXS). DEP formed lamellar structure above Tm. The state of phase separation was dependent on the molecular weight of APP in the blend system. The morphology of the blend system with APP of Mw = 150 k did not depend on the APP content. However, the morphology of the blend system with APP of Mw = 15 k changed with the APP content. The phase-separated structure was closely related to the growth geometry in the blend system.

2 THERMAL MOLECULAR MOTION

The physical properties of glassy polymers strongly depend on the size and distribution of free volume. The positron annihilation technique is a novel technique which can measure the size and distribution of free volume. M. Ban, M. Kyotoo et al. (3529, E1252) characterized the size and distribution of free volume for monodisperse polystyrenes (PSs) with Mn of 5.7 k and 1524 k by positron annihilation. The lifetime spectra of positron were decomposed into three components. These components were designated as τ_1, τ_2 and τ_3 in the order increasing lifetime. No temperature dependence was observed in the magnitude of τ_1 and τ_2 whereas τ_3 increased with temperature. τ_1 was mainly attributed to the self-annihilation of para-positron (para-Ps) and the τ_2 to the annihilation of positrons. Since the longest lifetime, τ_3 was found to range between 1.7 ns and 2.1 ns, this component was ascribed to the pick-off annihilation of ortho-Ps. The magnitude of τ_3 increased with temperature by thermal expansion. On the other hand, the intensity of the τ_3 component, l_3 and S-parameter which is a measure of Doppler-broadening showed the drastic temperature dependence. The large temperature dependence of l_3 suggested that relative concentration of the number of free-volume in PS was also changed due to the local-mode motion below the glass transition temperature.

The physical properties of random copolymers have been predicted from the properties of the corresponding homopolymers. However, if the copolymers have been prepared from the monomers with completely different physicochemical properties, the properties of the copolymers do not follow the additivity law. T. Ougisawa, T. Inoue et al. (3523, E1251) measured the Pressure-Volume-Temperature (PVT) behavior of poly(styrene- acrylonitrile) (SAN) and estimated the pressure and copolymer composition (ϕ) dependencies of both glass transition temperature (T_g) and thermal expansion coefficient (α). The ϕ dependence of α_l and $\Delta\alpha_{T_g}$ ($=\alpha_l - \alpha_g$, l:liquid, g:glass) was not additive and these values were larger than the additive ones. To discuss this behavior thermodynamically, the characteristics parameters in random copolymers were estimated by fitting PVT data to the equation of state theory by Flory and coworkers.[2] The ϕ dependence of both characteristic volume and pressure was almost additive. However, the ϕ dependence of the characteristic temperature, T^* deviated from the additivity law remarkably and the values of T^* in copolymers were similar to that of polystyrene. The external degree of freedom, C was evaluated from T^*. The external degree of freedom was larger in copolymers and therefore, the copolymer chains were more flexible than homopolymers. They also discussed the effect of chain flexibility on the glass transition behavior on the basis of the free volume theory. The magnitude of ΔC_p at T_g might reflect the chain flexibility.[3] ΔC_p at T_g increased with an introduction of AN. One can expect that larger flexibility in random copolymer chains affects the physical properties such as viscoelasticity and phase behavior in mixtures with other polymers.

Ionic conductive polymers were prepared by doping alkali metal ion into polyether in order to improve the ionic conductivity in solid state electrolyte. Since ionic conductivity depends on the mobility of alkali metal ion, it is necessary to understand the mechanism of thermal molecular motion for polyether doped with alkali metal ions. T. Furukawa et al. (3513, E1248) measured the linear and third nonlinear complex conductivities (σ_1^* and σ_3^*) as functions of frequency and temperature for poly(propylene oxide) (PPO) with molecular weight of 1000 doped with $LiClO_4$. The frequency spectra of σ_1^* consisted of three regions associated with electrode polarization, DC conductivity and dielectric relaxation in the order of increasing frequency. In the region of electrode polarization, σ_1^* showed a Debye-type relaxation while σ_3^* depended upon frequency in a complex manner including sign reversal of its real and imaginary components. Computer simulation has successfully reproduced the observed spectra of both σ_1^* and σ_3^* under the contribution of ionic mobility and ionic diffusion. It was found that analysis of σ_3^* allowed separate estimation of the mobility and density of ionic carriers.

Solid-state nuclear magnetic resonance (NMR) becomes a powerful technique to characterize thermal molecular motion. F. Ishii, I. Seo et al. (3495, E1224) measured the NMR spectra of the cast film of the alternative copolymers of vinylidene cyanide (VDCN) and vinylbutyrate (VBu) in the temperature range of 98–470 K. The second moment, $<\Delta H^2>$ showed four relaxation process. The γ_1 and γ_2 processes were observed at around 123 and 223 K, respectively. At 98 K, the conformation of main chain and the side chain of VBu were deflected by 9° from trans and rigid trans, respectively. Only the methyl group in the side chain rotated freely around C_{3V} axis. The main chain and the $C\alpha$ H_2 group in the side chain was in rigid below 263 K. The $C\beta$ H_2CH_3 group in the side chain oscillated around the $C\alpha$ $C\beta$ axis with amplitude of 64° between 108 to 203 K. Then, it started to rotate freely between 203 K and 283 K. This process corresponded to the γ_2 process. The β-narrowing process was observed between 283 K and 418 K. The motional narrowing of $<\Delta H^2>$ was attributed to the local mode motion of main chain, accompanied with the reorientation of side chain which rotated about 40°. Between 418 and 468 K, $<\Delta H^2>$ narrowed extremely. This process is so-called α-process. The magnitude of $<\Delta H^2>$ at 468 K was one fourth of that at 98 K. Therefore, this indicated that the whole molecular chain rotated freely around their molecular axes. The narrowing temperature of 463 K corresponded to the temperature at which the dielectric constant increased anomalously.

M. Kobayashi, T. Adachi (3331, E1204) studied thermal molecular motion in the three crystal modifications (R: rhombohedral, I: intermediate, M: monoclinic) of linear oligomers of polytetrafluoroethylene (PTFE) such as n-perfluoroeicosane, $C_{20}F_{42}$ and n-perfluorotetracosane, $C_{24}F_{50}$ on the basis of the infrared band profiles. M-I and I-R transforms were observed at around 143 and 200 K, respectively. For the absorptions of the v_3 progression bands, the polarization components parallel and perpendicular to the chain axes were measured and the temperature dependencies of the full width half height (FWHH) of the components were analyzed according to the site hopping theory. The FWHHs of the parallel components, which can be ascribable to the helix reversal motion accompanied with fluctuation of the parallel transition dipoles, were found to decrease continuously with lowering temperature range, approaching the small limiting value (less than 2 cm^{-1}) in the M phase. The activation energy of the helix reversal in the R and I phases of $C_{20}F_{42}$ were evaluated as 5.7 and 5.3 kJ mol^{-1}, respectively. The FWHHs of the perpendicular components exhibited discontinuous change around M-I transition. This was related to a lateral molecular fluctuation that induces large displacements of the molecules from isotropic hexagonal structure to anisotropic monoclinic structure at the I-M phase transition.

3   MECHANICAL PROPERTIES

Recently a new family of polyolefins synthesized by metallocene catalyst have been extensively developed. (Ethylene/α-olefin) copolymer is one of the most important target in polymer industry. Metallocene catalyzed (ethylene/α-olefin copolymer) have a narrow molecular weight distribution and a narrow comonomer distributions compared with Zieglar catalyzed one. T. Ohhama, A. Turuta (3531, 3553, E1253) studied properties and structure of metallocene and Zieglar catalyzed linear low density polyethylene (LLDPE). The LLDPE film prepared by metallocene catalyst (LL-A) showed higher transparency, better sealing performance and higher impact strength compared with LLDPE synthesized by Ziegler catalyst (LL-B). However, the Elemendorf tear strength of LL-A was inferior to LL-B. It was revealed that the Elemendorf tear strength changed with the extent of plastic deformation in the vicinity of fracture plane. Because LL-A showed less plastic deformation than LL-B, the tear strength of LL-A was lower than that of LL-B. The difference in mechanical behavior between LL-A and LL-B may be associated with the strain-hardening induced at large deformation region.

Super olefin polymer (SOP) has been widely used for automobile outer parts because of its excellent mechanical properties along with good injection moldability. Y. Fujita, T. Nomura et al. (3451, E1233) studied the crystalline structure of polypropylene phase in the injection molded SOP and its relationship to the mechanical behavior. Wide angle X-ray diffraction studies and transmission electron microscopic (TEM) observations clarified that b-axis, which corresponded to the growth direction of PP lamella, oriented vertically to the surface even though in a core layer. This can be ascribed to the shear induced orientation crystallization. Stress-strain behavior of this oriented PP in ethylene-propylene rubber (EPR) matrix was analyzed by finite element method (FEM) assuming that PP crystallite is dispersed in an amorphous elastomeric phase. It was revealed that modulus and stiffness along the direction perpendicular to the film surface were higher than those parallel to the film surface. On the basis of the observed crystalline structure and FEM analysis, it can be concluded that the crystalline structure in SOP is preferable for not only surface hardness but elastomeric properties against the stress parallel to the surface.

Fatigue strength of polymeric materials is an important property related to the practical performance. Most of the fatigue study have been done based on fracture mechanical analysis which was initially applied to metals. Little attention has been paid on the relationship between viscoelasticity of polymeric materials and fatigue behaviors. The fatigue criterion on the basis of dynamic viscoelastic energy loss has been proposed.[4,5] The total hysteresis loss, H_T was calculated from the apparent magnitude of loss modulus, E''. The amount of generated heat, H_H was estimated from the temperature rise of the specimen under cyclic fatigue. Since the difference, H_s ($= H_T - H_H$) increased with an increase in strain amplitude and a decrease in fatigue lifetime, it was assumed that the magnitude of H_s corresponded to the energy loss dissipated for structural change leading to fatigue fracture. The relationship between average of H_s, $H_{s,av}$ and fatigue lifetime was described by the following equation in the case of brittle failure.

$$(H_{s,av} - H_0) t_f = C \tag{1}$$

where H_0 and C were the hysteresis for the fatigue limit and the total hysteresis loss up to fatigue failure. This equation means that fatigue failure occurs when the effective energy loss dissipated for the structural change reaches a certain magnitude, C. From this equation, the fatigue lifetime of polymers can be predicted quantitatively on the basis of fatigue test for short period without fatigue fracture of the specimen.[4,5] The application of this equation to the composite material is practically important for the assessment of safety of structural component. T. Kajiyama, A. Takahara et al. (4423, E1187) applied this fatigue criterion to nylon 6/montomorillonite composites (NCH) with 2 wt% montmorillonite content. Transmission electron microscopic (TEM) observation of ultrathin section of NCH revealed that silicate layer is dispersed in a molecular order in matrix nylon 6. Also, the wide angle X-ray diffraction and dynamic viscoelastic measurements revealed the strong interaction between nylon 6 and silicate layer. The fatigue criterion has been successfully applied to the NCH as well as nylon 6. The magnitude of C in Eq. (1) was greater than that of nylon 6. These results indicate that NCH has the excellent fatigue resistance, because the silicate layers of montomorillonite are uniformly dispersed in nylon 6 which acts as a reinforcing component without increasing in number of mechanical defect.

4 FERROELECTRIC POLYMERS

Polymers having a large dipole moment and a polar conformation have a possibility to be ferroelectrics. Copolymers of vinylidene fluoride and trifluoroethylene [P(VDF-TrFE)] exhibit the strongest piezoelectric activities among piezoelectric polymers ever known, which maybe due to the extended-chain lamellar crystal highly developed during annealing in the paraelectric phase. H. Ohigashi, K. Omote et al. (3323, E1202) reported the formation of a large single crystal of P(VDF/TrFE) by annealing the drawn film in the paraelectric phase. The film was studied on the basis of WAXD, TEM, scanning electron microscopy (SEM) and polarized optical microscopy. Under the crossed polarizers, the film was uniform and any lamellar crystals were not observed. WAXD showed that the chain axis (c-axis) is parallel to the drawn axis, the a-axis normal to the film surface and the b-axis parallel to the film surface. The (110) was parallel to the film surface after poling. The single crystalline film exhibited a sharp D-E hysteresis loop, strong piezoelectric activity and high modulus along the chain axis. H. Ohigashi, K. Omote et al. (3487, E1242) also studied the dielectric relaxation behavior parallel and perpendicular to the surface of single crystalline films of P(VDF/TrFE) in a temperatures range around the ferroelectric to paraelectric phase transition (Curie transition). At temperature around the Curie transition, T_c, the parallel piezoelectric constant increased remarkably, and the relaxation strength at T_c was about 63. Since the P(VDF/TrFE) with all trans conformation does not have the polarization parallel to the main chain direction, the TGTG conformation with large polarization appeared at T_c. At T_c, the ε' perpendicular to the film surface measured at 1 MHz showed the maximum, whereas one measured at 100 MHz did the minimum. This corresponded to the critical slowing down phenomena. Also, the Cole-Cole plot of dielectric constant along the thickness direction at 403 K suggested the single-relaxation process. The dielectric relaxation time parallel to the chain direction was about 4 times as large as that perpendicular to the chain axis. The difference in relaxation behavior can be ascribed to the difference in state of thermal molecular motion. At T_c, trans and gauche conformation is present in the same molecular chain. A cooperative rotational molecular motion among neighboring molecules occurs, whereas flip-flop motion of the gauche-rich part occurs along the chain direction. The latter showed a longer relaxation time than that along the thickness direction.

K. Tashiro, M. Kobayashi et al. (3479, E1240) carried out the computer simulation in order to clarify the mechanism of phase transition in P(VDF/TrFE). The energetically stable molecular and crystal structures have been searched for the PVDF forms I, II, and III and for P(VDF/TrFE) on the basis of a lattice energy minimization technique. The deflection angle 2σ from the planar zigzag conformation has been reasonably reproduced for the PVDF form l. For a series of P(VDF/TrFE) with different VDF content, a systematic change in the unit cell parameters and the lattice mode frequencies were reproduced well. The results were consistent with the X-ray diffraction study and IR, Raman spectroscopic data. The all-trans conformation was found to be more stable than the $(T_g)_3$ helical form for any model of isotactic, syndiotactic, and atactic configurations of polytrifluoroetylene (PTrFE). The calculated result for PTrFE corresponded to the crystal structure of the cooled phase.

Recently, the ferroelectricity of odd nylons has been observed by several researchers. They exhibit D-E hysteresis and have spontaneous polarization which can be reversed by an electric field. However, ferroelectric behavior depends strongly on the preparation condition. Y. Takahashi, T. Furukawa et al. (3491, E1243) studied the ferroelectric behavior of nylon 11 which was prepared by quenching the melt and uniaxially drawn at room temperature. When the drawn film was annealed, the polarization decreased with annealing temperature while the coercive field increased. If the sample was annealed at high temperature, no hysteresis loop was observed and the crystal structure changed to triclinic unit cell. The switching curve showed the dynamical progress of the polarization reversal. It increased steeply with time and was saturated at 2Pr, twice the magnitude of polarization. The value of Pr increased with the field applied and was saturated at ca 40 mC m^{-1}. The switching time decreased with an increase in the magnitude of electric field. The ferroelectricity of nylon 11 was only observed for the sample which was quenched and uniaxially drawn. The drawn sample gave a broad X-ray diffraction peak suggesting the lack in regularity in hydrogen-bonding. As the intramolecular hydrogen bonding for quenched nylon 11 might be weak, the polarization reversal of amide group might be the possible origin of ferroelectricity for quenched-drawn nylon 11.

The aliphatic polyureas with odd-number carbon chains showed the ferroelectric behavior similar to odd nylon. Its piezoelectric properties depended on the crystallinity and the degree of the formation of hydrogen bonds. A novel method for the preparation of piezoelectric thin films by vapor deposition polymerization (VDP) has been proposed by Fukada *et al.*[6] M. Iijima, E. Fukada *et al.* (3469, E1237) prepared polyurea 5 and polyurea 9 thin films from 1,5-diaminoheptane and 1,5-diisocyanatenonane and 1,9-diaminoheptane and 1,9-diisocyanatenonane by VDP at the substrate temperature of 233 or 193 K. The WAXD and infrared study revealed that the polyurea prepared at 233 K had a much stronger hydrogen-bonding than that at 193 K. Also, polyurea prepared at 233 K had much more structural regularity than that prepared at 193 K. The dielectric and dynamic viscoelastic measurements revealed the relaxation process of polyurea 5 at 123 K (γ), 223 K (β) and 373–423 K (α). The α relaxation temperature for polyurea 5 was higher than that for polyurea 9 due to the presence of strong hydrogen-bonding. The piezoelectric constant, e_{31} of polyurea 5 increased with the poling temperature above the temperature region of the α-relaxation. The e_{31} of polyurea 5 prepared at 233 K followed by poling at 423 K was much stabler than that of polyurea 9. This is due to the presence of much strong molecular interaction in the polyurea 5 compared with the polyurea 9.

5   SURFACE AND INTERFACE STRUCTURE AND PROPERTIES

Surface and interface of polymers play important roles in various industrial applications. However, the surface structure-surface property relationships of polymeric solids have not been clarified yet. In 1994, the session on "Surface Characterization of Organic Materials" has been arranged and 32 papers have been presented.

AFM has become a widely used instrument for the observation of nm-scale surface topology.[7] AFM profiles the topology of the sample by scanning a probe over the sample surface with a constant force between surface and probe tip. If the sample was deformed sinusoidally, the dynamic viscoelastic behavior at the outermost surface can be evaluated by measuring the cantilever deflection under such a condition. Also, the scanning over the surface in the X-Y direction provides the image of dynamic viscoelastic functions with a nm-scale resolution. T. Kajiyama, A. Takahara *et al.* (3925, E1361) imaged the dynamic viscoelastic properties of polymeric solids by scanning viscoelasticity microscope (SVM).[8] When the tip is in a repulsive force region of the force curve, the surface of the specimen might be deformed by the indentation of the tip. The modulation of the vertical sample position leads to that of the force between tip and sample. This force modulation results in an indentation of the tip against the sample depending on its viscoelastic properties. The phase difference between the modulation signals of the sample position and modulated force response corresponds to the mechanical loss tangent, tan δ. The phase-separated structures of polymer blend films and immobilized mixed organosilane monolayers were clearly imaged by SVM with a lateral resolution of ca. 10 nm. The magnitude of tan δ evaluated for the outermost surface of monodisperse polystyrene (PS) at 293 K (tan $\delta = 0.10$) was much larger than that for the bulk PS (tan $\delta = 0.01$ at 110 Hz). This result apparently indicates that PS chains at the PS-air interface should have larger molecular mobility compared with that in the bulk, due to their larger free volume fraction induced by the localization of chain ends at the PS-air interface.

Chemical force microscopy or molecular force microscopy can carry out by using a standard AFM which has a cantilever tip modified with some functional group. T. Nakagawa (3927, E1361) measured the chemical force acting between a chemically modified cantilever tip and a chemically modified silicone wafer. The chemically modified cantilever tip and the silicone wafer were prepared by the chemisorption of organosilane from a solution. Using the chemically modified tips, the monolayer consisting of hydrocarbon chains could be discriminated from that consisting of perfluorocarbon chains by measuring the adhesive force in ethanol. If the solid surface and cantilever tip has the same chemical structure, the largest adhesive force was observed. The origin of interaction between the chemically modified tips and the sample surface depended on the chemical structure of the monolayers and the environmental liquids where the interactions were measured. If the interaction between monolayer and environmental liquid was strong, the adhesion force became small due to the weak aggregation of hydrocarbon or fluorocarbon chains.

The polymeric blend film with thickness less than twice as large as the radius of gyration of the longest component can be defined as the two-dimensional ultrathin blend film. T. Kajiyama, A. Takahara *et al.* (3937, E1364) investigated surface structure of [polystyrene/ poly(vinyl methyl ether) (PS/PVME)] blend ultrathin film in comparison with that of thick film. The phase separation temperature of the (PS/PVME) blend film on hydrophilic substrate decreased with a decrease in the film thickness The ultrathin blend film with thickness less than 30 nm did not show any distinct cloud point. X-ray photoelectron spectroscopic (XPS) measurement showed the enrichment of the PVME component with lower surface free energy at the air-solid interface. However, the PVME weight fraction at the surface decreased with a decrease in the film thickness thinner than ca. 30 nm. AFM observation revealed that the two-dimensional (PS/PVME) ultrathin blend film with the film thickness thinner than ca. 30 nm was in an apparent phase-separated state. SVM observation revealed that the droplet-like domains were composed of a PVME rich phase. The formation of phase-separated domain can be explained by the factors such as the adsorption of PS onto hydrophilic substrate, the negative spreading coefficient of PVME on the PS matrix and the recovery of conformational entropy loss of PVME chains.

Photophysical and photochemical dynamics of organic solid surfaces and interfaces has received much attention and has been studied by second harmonic generation methods and time-resolved total internal reflection fluorescence spectroscopy. Transient absorption spectroscopy is indispensable for analyzing a variety of surface and interface photoprocess. H. Fukumura, H. Masuhara *et al.* (3959, E1369) developed a new picosecond transient regular reflection spectroscopy (TRRS) system on the basis of specular reflection at a surface and interface. The dynamics on β-cooper phthalocyanine (β-CuPc) solid surfaces (vacuum and/or the atmosphere) and β-CuPc/methyl viologen (MV$^+$)/water system was studied on the basis of transient extinction coefficient change spectra obtained by the Kramer-Kronig transformation. Under vacuum, photoexcitation energy rapidly converted to thermal energy upon exciton-exciton annihilation, and the "hot band" due to the vibrationally excited ground state was observed. At the interface with an aqueous solution, photoelectron ejection to water was induced. The ejected electron recombined quickly with the CuPc cation within 10 ps, leading to the "hot band." When an electron acceptor molecule was added in water, the ejected electron was trapped by the molecule, which was also, confirmed spectroscopically.

Surface structure of multiphase polymers in an aqueous environment is particularly important for biomedical application. K. Senshu, S. Nakahama (3945, E1366) prepared six kinds of hydrophilic-hydrophobic block copolymers containing poly(2-hydroxyethyl methacrylate) (PHEMA) and poly(2,3-dihydroxypropyl methacrylate) (PDIMA) as a hydrophilic segment and polystyrene (PS), poly(4-octylstyrene) (POctSt) and polyisoprene (PIsp) as a hydrophobic one by a living anion polymerization. The stained ultrathin section of these block copolymers films under dry and wet conditions were observed with TEM. TEM observation of the ascast films of all block copolymers showed a clear lamellar or cylindrical structure in bulk, and the outermost surface was covered predominantly with hydrophobic segment. After immersing in water for 60 min, the top surface layers of hydrophobic segment disappeared. This indicates that the surface is reconstructed in response to the environmental change. Especially, the surface reconstruction of poly(lsp-block-DIMA) film occurred by soaking in water only for 5 sec due to the high water solubility of PDIMA and the low T_g of PIsp (204 K).

The systematic measurement of glass transition temperature, T_g of polymeric solids at the surface region has not been done yet. T. Kajiyama, A. Takahara (3936, E1363) measured the T_g of symmetric poly(styrene-block-methyl methacrylate) diblock copolymers [P(St-b-MMA)] at the surface region on the basis of the temperature-dependent XPS (TDXPS) and the angular-dependent XPS (ADXPS). Thin films of P(St-b-MMA) diblock copolymer with thickness of ca. 50 nm were prepared by a dip-coating method. The temperature, at which the surface composition started to change, was defined as a glass transition temperature, T_g at the surface region. The surface T_g of P(St-b-MMA) diblock copolymer was determined from the change in the surface composition with temperature on the basis of the TDXPS measurement.[9] The surface T_g of P(St-b-MMA) diblock copolymer was much lower than that in the bulk sample. A decrease in T_g near the free surface could be explained by an increase in the free volume at the surface region. The depth variation of T_g for P(St-b-MMA) diblock copolymer was measured by the TDXPS measurement with the variable take-off angle of photoelectron. It was revealed that T_g of P(St-b-MMA) diblock copolymer at the surface region was more pronounced than that in the bulk sample and was expressed as

$$T_g = T_{g,\infty} - CM^{-0.45 \pm (0.02)} \tag{2}$$

where $T_{g,\infty}$ is the surface glass transition temperature at the infinite molecular weight and C is a constant. These results mentioned above apparently indicate that the larger free volume at the surface region compared with that in the bulk sample was induced by the preferential segregation of chain end groups at the surface region and/or by geometrically unsymmetrical environment at the air-polymer interface.

REFERENCES

1. J.I. Laurtitzen and J.D. Hoffman, *J. Res. Nat. Bur. St.*, **A64**, 73 (1960).
2. P.J. Flory, P.A. Orwoll and A. Vrij, *J. Am. Chem. Soc.*, **86**, 3507 (1964).
3. B. Wunderlich, *J. Phys. Chem.*, **64**, 1052 (1960).
4. A. Takahara, K. Yamada, T. Kajiyama and M. Takayanagi, *J. Appl. Polym. Sci.*, **25**, 597 (1980).
5. A. Takahara, T. Magome and T. Kajiyama, *J. Polym. Sci., Polym. Phys. Ed.*, **32**, 839 (1994).
6. Y. Takahashi, M. Iijima and E. Fukada, *Japan J. Appl. Phys.*, **28**, L2245 (1989).
7. G. Binnig, C.F. Quate and Ch. Gerber, *Phys. Rev. Lett.*, **56**, 930 (1986).
8. T. Kajiyama, K. Tanaka, I. Ohki, S.R. Ge, J.-S. Yoon and A. Takahara, *Macromolecules*, **27**, 7932 (1994).
9. T. Kajiyama, K. Tanaka and A. Takahara, *Macromolecules*, **28**, 3482 (1995).

Current Awareness

DISSERTATION ABSTRACTS

The following titles have been selected from Dissertation Abstracts International by University Microfilms International. The titles have been organized under a series of key-word headings. The full abstract may be found in the referenced issue of Dissertation Abstracts or Masters Abstracts (designated source). The last entry (DAXXXXXXX, BRDXXXX, MAXXXXX OR XX/XXXXXC) gives the identification code that may be used to order a copy of the full text when available from University Microfilms International, 300 North Zeeb Road, P.O. Box 1764, Ann Arbor, Michigan 48106, U.S.A. or by telephoning (toll-free) 800-521-3042 in the US. Dissertation titles are published with the permission of the University Microfilms International, publishers of Dissertation Abstracts International.

Title	Name	Thesis/Year/University	Source
1. POLYACETYLENES			
Cyclic homoconjugated polyacetylenes	M.J. Cooney	University of Nevada, Reno 1993	448pp. 112-B DA9411476
Toxicological evaluation of the naturally occurring polyacetylene, falcarinol	J. Avalos	University of California, Irvine 1993	136pp. 4107-B DA9402410
DC-conductivity and magneto resistance of inherently conducting polymers: polyacetylene and poly(3-alkylthiophenes)	K.A. Väkiparta	Helsinki University of Technology, Finland 1993	43pp. 55/3848c
2. NONLINEAR OPTICAL PROPERTIES			
Experimental and theoretical studies of second-order nonlinear optical properties of a chromophore-functionalized polymer	M.A. Firestone	Northwestern University 1993	388pp. 123-B DA9415724
3. OLEFIN POLYMERIZATION			
Synthesis, chemistry, and olefin polymerization reactions of d° cationic mono(pentamethylcyclopentadienyl) complexes of titanium and zirconium	D.J. Gillis	Queen's University at Kingston (Canada) 1993	232pp. 123-B DANN85188
4. ELASTOMER			
Arylsulfonylation and nitration of a poly[styrene-b-(ethylene-co-butylene)-b-styrene] thermoplastic elastomer	S.W. Jolly	The University of Akron 1993	256pp. 123-B DA9414102
5. CONDENSATION POLYMERIZATIONS			
The synthesis of carlons, polycarbosilanes, and transition-metal-containing polymers by condensation polymerizations	G.J. Kotora	Boston College 1993	299pp. 123-B DA9414145

6. AQUEOUS PHASE POLYMERIZATION

Mechanism of the aqueous phase polymerization of chlorotrifluoroethylene	D.L. Murray	The University of Akron 1992	193pp. 124-B DA9417565

7. COMPOSITES

Approaches to molecular composites from fluorine containing bismaleimides	S.M. Pendharkar	Polytechnic University 1994	185pp. 124-B DA9413149
Continuous processing of polymers and composites with microwave radiation	M. Lin	Michigan State University 1993	100pp. 1202 MA1356351
Crystallization of poly(phenylene sulfide) in fiber-reinforced composites: experimental characterization and computer simulation	N.A. Mehl	Princeton University	319pp. 6438-B DA9416135
Inclusion models of tensile fracture in fiber-reinforced brittle-matrix composites	W.B. Tsai	Northwestern University 1993	141pp. 142-B DA9415827
Micromechanical study of the interface and the elementary damage phenomena in model composites of glass/epoxy	P. Feillard	Institut National des Sciences Appliquees de Lyon (France). In French 1993	137pp. 55/2710c
Microwave processing vinyl ester and vinyl ester/glass fiber composites	R. Dhulipala	Michigan State University 1993	127pp. p.1008 MA1354863
Modeling and numerical analysis of composite manufacturing processes	W. Chang	The University of Michigan 1993	128pp. 5893-B DA9409653
Novel microporous organo-clay polymeric composite membranes	H. Lao	University of Ottawa (Canada) 1993	253pp. 5694-B DANN83821
Polymer-ceramic composite membranes in reactor applications	M.E. Rezac	The University of Texas at Austin 1993	194pp. 6347-B DA9413582
The continuous curing process for thick filament wound composite structures	H. Teng	University of Illinois at Urbana-Champaign 1993	167pp. 5758-B DA9411798
The design, fabrication and testing issues of an electrorheological fluid-filled composite helicopter rotor blade	D.M. Moorehouse	Michigan State University 1993	100pp. p.1003 MA1354903
The effect of surface sulfonation of high density polyethylene (HDPE) on the mechanical properties of HDPE/wood fiber composites	K. Haraguchi	Michigan State University 1993	115pp. 1218 MA1356331
The use of expanding monomers in carbon fibre composites	X.D. Zhang	University of Toronto (Canada) 1993	100pp. p.1014 MAMM83550
Transport and structural properties of fiber-reinforced composites and porous media of fibrous or capillary structure	M.M. Tomadakis	The University of Rochester 1993	345pp. 6350-B DA9405890
Three-dimensional, nonlinear viscoelastic analysis of laminated composites: a finite element approach	M. Wang	Oregon State University 1993	181pp. 4766-B DA9404188
The effects of jet fuel and water absorption on edge delamination of graphite-epoxy composites	R. Subramanian	Wichita State University 1993	162pp. 4794-B DA9406071
The manufacturing science of thermoplastic composites	D.G. Bland	The University of Tennessee 1993	234pp. 4805-B DA9404564
Cellulose fiber-reinforced thermoplastic composites: surface and adhesion characterization	G.B. Garnier	Virginia Polytechnic Institute and State University 1993	351pp. 4867-B DA9402329

Development of electrically/ thermally conducting polymer-matrix composites	L. Li	State University of New York at Buffalo 1993	258pp. 4868-B DA9404833
Microdeformation processes in PC/ SAN microlayer composites	K.L.K. Sung	Case Western Reserve University 1993	118pp. 4873-B DA9406231
Finite element formulation for gauge theory of brittle damage with applications to fibrous composites	C.M. Huang	Rensselaer Polytechnic Institute 1993	263pp. 4879-B DA9405677
Simulation of the thermomechanical behaviour of fiber-reinforced thermoset composites	S.C. Tseng	The University of Wisconsin-Madison 1993	245pp. 4887-B DA9330202
Carbon-plastic composite electrode for 'vanadium redox flow battery applications	S. Zhong	University of New South Wales (Australia) 1992	3230-B
Innovative design of FRP composite members combined with concrete	N. Deskovic	Massachusetts Institute of Technology 1993	3234-B
Feasible drilling conditions and hole quality criteria for PMR-15/gr composite laminates	M. Mehta	University of Cincinnati 1993	DA9329943
Crosslinking of an epoxy/amine system under a microwave field. An application to the realization of glass fiber reinforced composite materials	C. Jordan	Institut National des Sciences Appliquees de Lyon (France). In French 1992	54/5271c
Study of a glass-epoxide composite material and of its damaging: Physical, acoustic and mechanical measurements	E.L.G. Bouquerel	Institut National des Sciences Appliquees de Lyon (France). In French 1992	177pp. 54/5432c
Dimensional changes of composite materials in the space environment	R. Matthews	University of Toronto (Canada) 1992	109pp. p.1849 MAMM78485
The effects of sub-critical damage on the notched strength of carbon fibre composites	G.M. Webb	University of Toronto (Canada) 1992	132pp. p.1860 MAMM78617
The effects of Spiro ortho carbonates on properties of unsaturated polyester resin and its fibre composites	W. Zhou	University of Toronto (Canada) 1992	99pp. p.1861 MAMM78538
Investigation into toughness improvement in graphite/epoxy composites using Teflon powder	D.A. Chafey	San Jose State University 1993	134pp. p.1904 MA1352999
Application of electrochemical impedance spectroscopy as a non-destructive technique to characterize and detect degradation of carbon/nylon, 6 composites in aqueous environments	R.V. Haniyur	Florida Atlantic University 1993	152pp. p.1906 MA1351985
Laser machining of polymeric composites for joint preparation	J.D. Howard	Michigan State University 1993	64pp. p.1906 MA1352717
Interpenetrating polymer networks of polycarbonateurethane/ polymethyl methacrylate and related composite materials	P. Zhou		101pp. 3646-B DA9334472
Nondestructive evaluation of elastomer matrix composites	H.A. Huang	The Pennsylvania State University 1993	175pp. 3700-B DA9334752
Microwave properties of ferroelectric/polymer composites	F.G. Jones	The Pennsylvania State University 1993	275pp. 3767-B DA9326870
Interface characteristics and adhesion in cellulose/polypropylene composites	D.T. Quillin	The University of Wisconsin-Madison 1993	165pp. 3801-B DA9320922

Processing effects on the performance of PTC composite thermistors	R.J. Sullivan	The Pennsylvania State University 1993	163pp. 3803-B DA9334823
Preparation and properties of optically transparent, pressure-cured poly(methyl methacrylate) composite	K.D. Weaver	University of Missouri-Rolla 1993	126pp. 4181-B DA9401914
Electronically conductive polymer composites and microstructures	L.S. Van Dyke	Texas A&M University 1993	217pp. 4185-B DA9403599
Structural behaviour of plastic composite columns	Z.A. Hashem	The University of Texas at Arlington 1993	212pp. 4290-B DA9402813
Fatigue life prediction of composites based on microstress analysis	G. Shen	University of Waterloo (Canada) 1993	188pp. 4354-B DANN81217
Machining characteristics of graphite/epoxy composite	D.H. Wang	University of Washington 1993	227pp. 4357-D DA9401484
Freeze-drying method for making semi-interpenetrating polymer networks and their fiber-reinforced composites	H.J. Hsiung	The University of Tennessee 1993	347pp. 1038-B

8. DIFFUSION

Simultaneous diffusion of a disperse dye and a solvent in PET film analyzed by Rutherford backscattering spectrometry	R.M. Stinson	Cornell University 1994	104pp. 124-B DA9417426
Tracer diffusion of stars and spheres in polymer matrices	J. Won	University of Minnesota 1993	157pp. 5696-B DA9411283
Recycled cellulose fiber reinforced phenolic composites	G.F. Bixby	University of Massachusetts-Lowell 1993	87pp. p.340 MA1353263
The thermal expansion coefficient of polypropylene and related composites	Y. Okada	McGill University (Canada) 1992	137pp. p.341 MAMM80500
Deposition mechanism of coupling agents on silica and evaluation of the silica-epoxy composite using magnetic resonance techniques	R. Malik	Rensselaer Polytechnic Institute 1993	256pp. 931-B DA9420344
Intrinsic luminescence cure sensors for epoxy resins and fiber-reinforced epoxy composites	J.C. Song	The University of Connecticut 1993	250pp. 932-B DA9419419
Thermomechanical constitutive modeling of irradiated fiber-reinforced polymer composites	C.P. Chao	The Ohio State University 1994	235pp. 1058-B DA9420935
Compressive kinking failure of fibrous composites	A.M. Saleh	Rensselaer Polytechnic Institute 1993	238pp. 1068-B DA9420355
Optical detection of stress waves from glass-fiber reinforced polymer matrix composite material	R.D. Huber	The Johns Hopkins University 1994	151pp. 1123-B DA9419987
Compressive behaviour of advanced carbon fiber composite materials	Y.J. Jee	The University of Utah 1994	217pp. 1124-B DA9419711
Flange wrinkling and the deep-drawing of thermoplastic composite sheets	D.W. Coffin	University of Delaware 1993	370pp. 1138-B DA9419548
Smectite clays, muscovite mica, and hydrotalcite used as fillers and reinforcements in the formulation of epoxy-based composites	M. Wang	Michigan State University 1993	123pp. 446-B DA9418082
Wetting, spreading and interphase formation in a liquid composite molding environment	B.K. Larson	Michigan State University 1993	139pp. 509-B DA9418022

The effect of porosity on the elastic constants of thin carbon fiber composite laminates	A.M. Rubin	Washington University 1993	180pp. 552-B DA9419041
Experimental verification of the devolatilization model for AFR700 polyimide composites	J.L. Vasat	Washington University 1993	134pp. 554-B DA9419043
Acousto-ultrasonic evaluation of composite materials	M. Khazaee	New Mexico State University 1993	161pp. 559-B DA9417345
An integrated theoretical and experimental study of the fibre-matrix interface in carbon-epoxy composites	M.M. Desaeger	Katholieke Universiteit Leuven, Belgium 1993	478pp. 55/4085c
The relation between fracture toughness and microstructure in thermoplastic polymers and composites	B.E. Goffaux	Katholieke Universiteit Leuven, Belgium 1993	231pp. 55/4087c
Partitioning and tracer diffusion of polymers in porous material	Z. Zhou	University of Massachusetts 1994	122pp. 933-B DA9420701

9. POLYMER MIXTURES

Chromatographic separation of polymer mixtures using combined adsorption-size exclusion mechanisms	T.W. Wang	The University of Akron 1993	174pp. 124-B DA9414473

10. SILICONE RUBBER

Silica-silicone interactions: non-linear viscoelastic behaviour of silica-filled silicone rubber	R.L. Warley	Case Western Reserve University 1993	199pp. 125-B DA9416188

11. POLYSILOXANE

Study of the mechanical properties of nylon-polysiloxane-nylon block copolymer	T.K. Bhattacharjee	University of Massachusetts-Lowell 1993	98pp. 967p. MA1355599
Synthesis and properties of polysiloxane elastomers and composites	J. Wen	University of Cincinnati 1993	168pp. 125-B DA9416809

12. OPTICAL STORAGE

Synthesis and characterization of azo aromatic polymers for reversible optical storage	S. Xie	Queen's University at Kingston (Canada) 1993	215pp. 125-B DANN85205

13. SUSPENSIONS

Shear thickening and micro-structures of concentrated charge stabilized suspensions	M.K. Chow	University of Illinois at Urbana-Champaign 1994	281pp. 156-B DA9416346

14. POLYURETHANE

Morphology and fatigue properties of polyurethane thermoplastic elastomers	C.W.S. Myers	The University of Wisconsin-Madison 1993	199pp. 158-B DA9408575
The interactions of plasminogen with model surfaces and derivatized segmented polyurethane ureas	K.A. Woodhouse	McMaster University (Canada) 1993	DANN84961
Surface properties of polyurethanes: control of the surface free energy through the use of reactive surfactants	L. Girodet	Institut National des Sciences Appliquees de Lyon (France). In French 1993	266pp. 55/2406c 1993

Characterisation physico-chinique et etutde de la biostabilite d'une nouvelle prothese vasculaire microfibrillaire en polyurethane	Z. Zhang	Universite Laval (Canada) 1993	249pp. 4598-B DANN82335
Gas chromatography techniques for studying polycyclic aromatic hydrocarbons in aqueous solution and organic volatiles in emissions from incineration of CFC-11 containing polyurethane foam	W.F.J.C. Lien	University of Massachusetts-Lowell 1993	167pp. 4639-B DA9405122
Friction and the dynamic mechanical and thermal properties of polyurethane elastomers	K. Hanhi	Kungliga Tekniska Hogskolan, Sweden 1992	186pp. 55/806c
Buoyancy effects on smoldering of polyurethane foam	J.L. Torero	University of California, Berkeley 1992	233pp. 3308-B DA9330764
A study of the polyurethane foam extraction of complexes formed between platinum and tin halides	S.G. Schroeder	The University of Manitoba (Canada) 1990	121pp. p.1800 MAMM76805
The investigation of thin polyurethane linings as an alternative method of ground control	R.A. Mercer	Queen's University at Kingston (Canada) 1992	315pp. p.1919 MAMM76452
Accelerated aging of a radiopaque filled polyurethane	L.M. Cavallaro	University of Massachusetts-Lowell 1993	107pp. p.1926 MA1352253
Thermal stability of polyurethanes: applications to polymer processing and recycling	M.E. Hernández-Hernández	University of Minnesota 1994	129pp. 1037-B DA9420096
The role of complement activation and cellular interactions on the biocompatibility of polyurethanes	F. Lim	The University of Wisconsin-Madison 1994	231pp. 1042-B DA9410613
Photodegradation of polycarbonate, polyurethane ureas and polyureas	H.C. Shah	The University of Southern Mississippi 1993	343pp. 445-B DA9417844

15. DRUG RELEASE

Biodegradable lactide polymers: synthesis, degradation, and controlled drug release properties	X. Zhang	Queen's University at Kingston (Canada) 1993	199pp. 160-B DANN85204
Hydroxypropylmethylcellulose hydrophilic matrices: From the study of polymer and drug release mechanism to the study of the influence of different parameters on this release	V.B. Bourgeois	Universite de Dijon (France). In French 1992	140pp. 54/5112c

16. IMMISCIBLE POLYMERS

Segregation of block copolymer to the interface between strongly immiscible polymers	H.T. Dai	Cornell University 1994	180pp. 183-B DA9417414

17. MICROSCOPY

Effect of orientation on properties of reinforced polypropylene and evaluation of materials with scanning acoustic microscopy	F.J. Lisy	Case Western Reserve University 1993	247pp. 184-B DA9416203

18. NONLINEAR OPTICAL APPLICATIONS

1. Synthesis and characterization of novel polymers for potential nonlinear optical applications and 2. Exploration of poly(2-butyne-1,4-diyl)s	B. Narayanswamy	University of Cincinnati 1993	170pp. 185-B DA9416798

19. CARBON FIBER

Surface characterization of carbon fiber by infrared spectroscopy	T. Ohwaki	Case Western Reserve University 1993	291pp. 197-B DA9416189
Adhesion of novel high-performance polymers to carbon-fibers: fiber surface treatment, characterization, and microbond single fiber	C.L. Heisey	Virginia Polytechnic Institute and State University 1993	309pp. 444-B DA9419215

20. BLEND

Blend compatibilization through charge transfer interaction and investigation of the interactive group distribution by NMR	G.M. Crone	Queen's University at Kingston (Canada)	149pp. 1188 MAMM85230
Blends containing liquid crystalline polymers	D. Dutta	The University of Connecticut 1993	184pp. 6401-B DA9413111
Compatibilization of polymer blends using charge transfer interactions	C. Rondeau	Queen's University at Kingston (Canada)	140pp. 1188 MAMM85319
Study of physical and rheological properties of PC/ABS blends	C.J. Parikh	University of Massachusetts-Lowell 1993	123pp. p1059 MA1355620
Thermodynamics of mixing in model polyolefin blends (Volumes I and II)	R. Krishnamoorti	Princeton University 1994	538pp. 6234-B DA9415435
Ultrafiltration membranes from a polymer blend: hollow fiber preparation and characterization	I.M. Wienk	Universiteit Twente, The Netherlands 1993	183pp. 55/820c
Phase behaviour, crystallization kinetics and morphologies of poly(ε-caprolactone) (PCL)/ polycarbonate (PC) blends	Y.W Cheung	University of Massachusetts 1994	183pp. 929-B DA9420609
Interfacial activities of block copolymers in immiscible homopolymer blends	W. Hu	The University of Connecticut 1993	198pp. 931-B DA9419433

21. COPOLYMERS

Alpha and beta mechanical relaxations of PMMA, PS and their copolymers	E. Muzeau	Institut National des Sciences Appliquees de Lyon (France). In French 1992	221pp. 55/2725c
Characterisation of interpenetrating polymer networks based on block copolymers	J.J. Moss	University of New South Wales (Australia)	6344-B
Dielectric, piezoelectric and electroacoustic characterisation of fluorinated ferroelectric copolymers role of phase transitions	H. Lefebvre	Institut National des Sciences Appliquees de Lyon (France). In French 1993	208pp. 55/2717c
Estimation and design considerations for obtaining copolymer reactivity rations	P.J. Rossignoli	University of Waterloo (Canada) 1993	223pp. p.1012 MAMM84604
Fracture toughness of polymer/ polymer bimaterial interface reinforced with block copolymers	F. Xiao	Cornell University 1994	218pp. 6293-B DA9416684
Functionalized polyproylene and propylene graft copolymers: synthesis via the borane approach	D.C. Rhubright	The Pennsylvania State University 1993	215pp. 6236-B DA9414361
N-substituted maleimides as potential stabilizers for vinylidene chloride copolymers	J. Zhang	Central Michigan University 1994	190pp. 1188 MA1356215
Shear orientation of diblock copolymer melts	K.A. Koppi	University of Minnesota 1993	321pp. 5813-B DA94411264

Studies on the transfer and adhesion of polytetrafluoroethylene (PTFE) and PTFE-based composites during dry contact	E. Yang	Helsinki University of Technology, Finland 1993	32pp. 55/2730c
Synthesis and characterization of high performance polyimide homopolymers and copolymers	M.E. Rogers	Virginia Polytechnic Institute and State University 1993	245pp. 6236-B DA9414087
Well defined graft copolymers and end functional materials: synthesis, characterization, and adhesion studies	M.S. Sheridan	Virginia Polytechnic Institute and State University 1993	319pp. 6236-B DA9414090
SEM and FTIR investigation of bulk crystallized TPI and its epoxidized copolymers	N. Vasanthan	City University of New York 1993	158pp. 4695-B DA9405593
Solubility of homopolymer in block copolymer	K.J. Joen	University of Cincinnati 1993	150pp. 4340-B DA9329958
Synthesis and characterization of organostannane modified poly(ethylene-co-vinyl alcohol) copolymers	B.M. Pandya	Florida Atlantic University 1993	103pp. p.255 MA1353569
Study of the anionic polymerization of ε-caprolactone and related block copolymers	D. Lu	Academia Sinica, China 1989	
Synthesis and characterization of liquid-crystalline AB- and ABA-block copolymers and copolymers	K. van de Velde	Katholieke Universiteit Leuven, Belgium. In Dutch	
Characterization of styrene-butadiene-styrene block copolymers and study of their microphase	R. Yang	Academia Sinica, China 1989	

22. MICROCAPSULES

Studies on the permeability and molecular weight cut off of HEMA-MMA microcapsules	A. Kosaric Wagner	University of Toronto (Canada) 1993	142pp. 1201 MAMM83393

23. BARRIER PROPERTIES

The effect of surface sulfonation on barrier properties of polymer films	K. Wangwiwatsilp	Michigan State University 1993	101pp. 1219 MA1356390

24. POLYSILANES

The synthesis and characterization of novel polysilanes	J.P. Banovetz	Stanford University 1994	190pp. 6191-B DA9414533

25. CATALYSIS

The polymerization of formaldehyde by coordination catalysis	L.J. Starks	Georgia Institute of Technology 1993	188pp. 6213-B DA9415667

26. METHYLCELLULOSE

Effect of methyl cellulose 4000 on stability of oil-in-water emulsions	R.P. Gullapalli	The University of Tennessee Center for the Health Sciences	205pp. 6217-B DA9413186

27. COATINGS

An electrochemical impedance spectroscopic study of the hydrothermal properties of polymeric coatings	K.J. Kovaleski	Lehigh University 1994	197pp. 6224-B DA9414999
Application of polymer characterization techniques to the analysis of thin polymeric coatings	B.K. Rearick	The Pennsylvania State University 1993	144pp. 6235-B DA9414357

Inorganic-organic hybrid materials and abrasion resistant coatings based on a sol-gel approach	C.S. Betrabet	Virginia Polytechnic Institute and State University 1993	247pp. 4182-B DA9403847
Synthesis and characterization of mesogenic monomers and their applications in coatings fields	P. Xu	Eastern Michigan University 1993	166pp. p.255 MA1353985

28. PIEZOELECTRIC POLYMER

Evaluation of a piezoelectric polymer for use in display technology	S.S. Bhat	The University of Akron 1993	143pp. 6232-B DA9415530

29. METATHESIS

Synthesis of telechelic polymers through metathesis reactions and borane chemistry	M.J. Chasmawala	The Pennsylvania State University 1993	221pp. 6232-B DA9414257
Synthesis of molybdenum (VI) alkylidene complexes as catalysts for living polymerizations of terminal acetylenes and olefin metathesis	H.H. Fox	Massachusetts Institute of Technology 1993	5651-B
Rotational isomers of Mo(VI) alkylidene complexes and polymer structure: investigations in ring-opening metathesis polymerization	J.H. Oskam	Massachusetts Institute of Technology 1993	3062-B
Acyclic diene metathesis (ADMET) polymerization: the polymerization of carbonyl-containing dienes	J.T. Patton	University of Florida 1992	99pp. 3074-B DA9331198

30. DIFFUSION

Diffusion in microstructured block copolymers	C.E. Eastman	University of Minnesota 1993	117pp. 6233-B DA9412996

31. POLYIMIDE

Anisotropic structure and properties of a series of segmented rigid-rod polyimide films	F.E. Arnold, Jr.	The University of Akron 1993	274pp. 6431-B DA9415517
Growth, characterization, and kinetic studies of the low-temperature reaction, of vapor deposited polyimide films	C.D. Dimitrakopoulos	Columbia University 1993	190pp. 6401-B DA9412744
The use of polyimide for the development of micromachined materials, processes and devices	A.B. Frazier	Georgia Institute of Technology 1993	236pp. 6370-B DA9415636
Use of water sorption to probe the effect of processing on the structure of polyimide thin films	H. Han	Columbia University 1993	195pp. 6233-B DA9412761
Investigation of the chemical imidization process for polyimide	M.H.M.R. Kailani	The University of Connecticut 1993	405pp. 4699-B DA9406053
Moisture and stress effects on fretting between steel and polyimide coatings	C.C. Kang	Virginia Polytechnic Institute and State University 1993	179pp. 3287-B DA9331462

32. ANILINE

Synthesis and properties of aniline, thiophene and Schiff base copolymer	R. Hariharan	Drexel University 1993	266pp. 6233-B DA9411511

33. SEGREGATION

| Surface and interfacial segregation in end functionalized polymers | C.A. Jalbert | The University of Connecticut 1993 | 318pp. 6234-B DA9413140 |

34. MOLECULAR MODELING

| Structural and dynamical aspects of bulk amorphous polybutadienes: a molecular modeling study | E.G. Kim | The University of Akron 1993 | 157pp. 6234-B DA9415521 |

35. POLY (ORGANOPHOSPHAZENES)

| Novel approaches to functionalized poly(organophosphazenes) | C.J. Nelson | The Pennsylvania State University 1993 | 195pp. 6234-B DA9414346 |

36. POLY(ESTER IMIDES)

| Phase transitions in thermotropic liquid crystalline poly(ester imides) | R.L. Pardey Pocaterra | The University of Akron 1993 | 445pp. 6235-B DA9415518 |

37. SYNTHESIS

| The synthesis, characterization and physical properties of poly(methyl methacrylate-*b*-isobutylene-*b*-methyl methacrylate) | J.L. Price | The University of Akron 1993 | 171pp. 6235-B DA9415515 |

38. BIODEGRADABLE POLYESTERS

| Biodegradable polyesters: blocks copolymer and stereocopolymer design and characterization | M.S. Reeve | University of Massachusetts-Lowell 1993 | 161pp. 6235-B DA9411468 |

39. RAMAN SCATTERING

| Orientation and dynamics of polymeric materials studied by polarization-modulated laser Raman scattering | L.A. Archer | Stanford University 1994 | 189pp. 6332-B DA9414529 |

40. MICROPOROUS MEMBRANE

| Mechanism of microporous membrane formation by precipitation of semicrystalline polymers | L.P. Cheng | Columbia University 1993 | 353pp. 6335-B DA9412736 |

41. DISSOLUTION

| Effect of dissolution of compressed fluids on polymer properties | P.D. Condo | The University of Texas at Austin 1993 | 390pp. 6335-B DA9413469 |

42. INJECTION MOULDING

| Rheology and injection moulding of a ceramic-filled polymeric material | F. Downey | Cornell University 1994 | 211pp. 6335-B DA9416713 |

43. MODEL POLYOLEFINS

| Catalytic hydrogenation of polymers: synthesis and characterization of model polyolefins | M.D. Gehlsen | University of Minnesota 1993 | 262pp. 6336-B DA9413000 |

44. BRANCHED POLYMERS

Selected topics in randomly branched polymers	A.M. Gupta	University of Minnesota 1993	170pp. 6337-B DA9413006

45. PARTICLE INTERACTIONS

Polymer-particle interactions in aqueous solution	T.R. Ju	Stanford University 1994	117pp. 6340-B DA9414587

46. MEMBRANES

The formation and surface fluorination of gas separation membranes	J.D. Le Roux	The University of Texas at Austin 1993	288pp. 6342-B DA9413524
Synthesis and gas transport properties of aromatic polyester membrane materials	L.A. Pessan	The University of Texas at Austin 1993	206pp. 6346-B DA9413571
Membrane-based affinity chromatography: modeling and experiments of lysozyme onto Cibacron blue ligand immobilized on microporous cellulose membranes	H.C. Liu	University of Cincinnati 1993	200pp. 5815-B DA9407814
Pervaporation with latex membranes	Y. Wei	University of Waterloo (Canada) 1993	325pp. 5819-B DANN84493
Partitioning of pectinesterase on hollow fiber polysulfone ultrafiltration membranes	R. Snir	University of Georgia 1993	152pp. 4461-B DA9404684
Investigation of demixing phenomena of a polymer solution during the phase inversion process, and study of morphological and transport characteristics of the prepared membranes	M.J. Han	University of Kentucky 1993	220pp. 4808-B DA9402837
Gas separation in membranes	C.R. Gangidi	Mississippi State University 1993	202pp. p.292 MA1353877
Electrochemically controlled transport across conducting electroactive polypyrrole membranes	H. Zhao	University of Wollongong (Australia) 1993	55/3582c

47. BLENDING

Morphology development during polymer blending	U. Sundararaj	University of Minnesota 1994	252pp. 6349-B DA9415498

48. POLYMER SOLUTIONS

Flow-induced phase separation in semi dilute polymer solutions	J.W. van Egmond	Stanford University 1994	202pp. 6350-B DA9414672
Liquid-liquid equilibria of polymer solutions	D.J. Geveke	The Pennsylvania State University 1993	242pp. 3731-B DA9334738
Simulation of interdiffusion and of dilute polymer solution viscoelasticity	D.R. Whitney	University of Southern California 1993	926-B
Synthesis of mixtures of polyethylene and poly(methyl methacrylate) via structured polymer solutions	R. Janssens	Katholieke Universiteit Leuven, Belgium 1993	138pp. 55/3697c

49. SUSPENSION COPOLYMERIZATION

Synthesis and modelling of high Tg copolymers through suspension copolymerization with bifunctional initiators	M.A. Villalobos	McMaster University (Canada) 1993	414pp. 6350-B DANN84954

50. DOPED POLYMER FILMS

Ultrafast temperature jump study of energy transfer in dye doped polymer films	X. Wen	University of Illinois at Urbana-Champaign 1993	123pp. 6351-B DA9411820

51. RESIN

Selective ion-exchange in a microemulsion-based resin	J.H.K. Yang	University of Minnesota 1993	153pp. 6351-B DA9413064
Packed-bed adsorption of phenol and 2,4-dichlorophenoxyacetic acid on the macroreticular resin XAD-7	Y. Eydatoula	King Fahd University of Petroleum and Minerals, Saudi Arabia	197pp. p1008 MA1355738
Removal of organics from aqueous solutions using polymeric resins	G.M. Gusler	University of California, Los Angeles 1994	780pp. 508-B DA9418882

52. URETHANE FOAM

A chemical reaction model for polyisocyanurate urethane foam	N.E. Knox	The University of Alabama in Huntsville 1993	236pp. 6404-B DA9415113

53. ADHESIVE

Characterization of model hot melt poly(ethylene-co-vinylacetate) adhesives and their bonding to polypropylene	H.H. Shih	The University of Akron 1993	396pp. 6438-B DA9415523

54. PHASE SEPARATION

Cure shrinkage control in polymerization of multicomponent resin systems with phase separation	C.G. Becker	Rice University 1993	138pp. p.967 MA1355170
Influence of collective cohesion forces and hydrogen bonds on the phase separation of low molecular crystals and industrial semi-crystalline polymers	K.M.E. Seghers	Katholieke Universiteit Leuven, Belgium. In Dutch 1993	201pp.55/2395c

55. POLYESTERS

A Raman study of order in some complex polyesters	J.K.A. Agbenyega	University of Southampton, UK 1993	55/2402c
Dilatometric study of low-profile unsaturated polyester resins	M.R. Kinkelaar	The Ohio State University 1993	257pp. 5813-B DA9411989
Kinetic study and thermal characterization by DSC of thermosetting resins of unsaturated polyester	J.L. Martin Godoy	Universidad de Barcelona (Spain). In Spanish 1992	315pp. 54/5443c
Ultraviolet curable abrasion resistant coatings: an evaluation of the factors that influence their physical properties on thin polyester films	C.L. Allard	University of Massachusetts-Lowell 1993	123pp. p.1925 MA1352250
The effect of silane treatments on interphase properties of a glass fiber reinforced unsaturated polyester	S.M. Connelly	The University of Connecticut 1993	251pp. 3642-B DA9333197
Interfacial interactions in layered polyesters	C.B. Shah	University of Rhode Island 1993	

56. POLYPROPYLENE

Phase behaviour related to the polymerisation of gels of polypropylene	L. Bollen	Kathoilieke Universiteit Leuven, Belgium. In Dutch	138pp. 55/2404c

Polypropylene cast film extrusion using water bath cooling	P.R. Mitchell	University of Massachusetts-Lowell 1993	121pp. p1059 MA1355617
A structure-property study in realtime of the tensile deformation of oriented polypropylene thin films	R.E. Pepper	Georgia Institute of Technology 1993	452pp. 3836-B DA9400435

57. EPOXY SYSTEMS

Modification of epoxy systems with soluble and reactive rubbers: study on the chemical reaction induced phase separation	D. Chen	Institut National des Sciences Appliquees de Lyon (France). In French 1992	153pp. 55/2405c
Toughening mechanisms of an epoxy system modified with reactive liquid rubbers or preformed particles and/or glass beads	A. Maazouz	Institut National des Sciences Appliquees de Lyon (France). In French 1993	266pp. 55/2410c

58. POLY (3-METHYLTHIOPHENE)

Nucleation, growth and spectroelectro-chemical properties of thin electro-polymerized poly(3-methylthiophene) films	J.O. Lukkari	Turun Yliopisto, Finland 1993	139pp. 55/2409c
Flow injection analysis of some anionic species with electrochemical detection using conducting poly(3-methylthiophene)	G.C. Russell, III	University of Cincinnati 1993	163pp. 3032-B DA9329937

59. POLYOXYETHYLENE

The phase behaviour of polyoxyethylene surfactants, phospholipids and their mixtures	R.J. Mirkin	University of Southampton, UK 1992	55/2411c

60. ARAMIDS

Synthesis and characterization of crosslinked aramids	C. Rickert	Eidgenossische Technische Hochschule, Zurich (Switzerland). In German 1993	153pp. 55/2412c

61. POLY (DI-N-ALKYLSILOXANE)S

Cyclo- and poly(di-n-alkylsiloxane)s with ethyl, propyl, and butyl side groups	S.G. Siffrin	Universiteit Twente, The Netherlands 1993	159pp. 55/2414c

62. POLYANILINE

Characterization of the chemical properties of polypyrrole and polyaniline	P.R. Teasdale	University of Wollongong (Australia) 1993	55/2415c
Polyacetylene and polyaniline: synthesis, doping, and characterization	D.B. Swanson	University of Pennsylvania 1993	530pp. 3064-B DA9331846

63. DYNAMIC LIGHT SCATTERING

Dipolar interactions from poly ion pairs in low polar medium: study with spectroscopy and dynamic light scattering	J.J. Tossens	Katholieke Universiteit Leuven, Belgium. In Dutch 1993	173pp. 55/2416c

64. POLY (METHYLMETHACRYLATE)

Structure formation in solutions of a tactic poly(methylmethacrylate)	P.R. Vandeweerdt	Katholieke Universiteit Leuven, Belgium 1993	3pp. 55/2417c

65. POLYARYLS

Molecular mobility in parasubstituted polyaryls	L. David	Institut National des Sciences Appliquees de Lyon (France). In French 1993	1993 55/2490c

66. NONLINEAR OPTICAL EFFECTS

Second-order nonlinear optical effects in polymers: frequency doubling in poled polymers	G.E.E. S'Heeren	Katholieke Universiteit Leuven, Belgium. In Dutch 1993	180pp. 55/2503c

67. POLYACRYLAMIDE

The gelation properties of polyacrylamide gels when combined with hyaluronic acid and collagen	C.A. Nagorski	Adelphi University 1994	35pp. p.960 MA1355471
Influence of surfactants on nucleate pool boiling of aqueous polyacrylamide solutions	T.A. Wang	University of Illinois at Chicago 1993	270pp. 3823-B DA9335153

68. WATER SOLUBLE POLYMERS

Stochastic dynamics of water soluble rigid rod polymers	M.A. Jarrin	University of Massachusetts-Lowell 1993	68pp. p.967 MA1355611

69. BIOADHESIVES

Development of UV curable bioadhesives	J.J. Kao	University, of Massachusetts-Lowell 1993	77pp. p.967 MA1355612

70. ADHESION

Chemical modification of polymer surfaces using sulfonation to improve adhesion properties	H. Asthana	Michigan State University 1993	136pp. p.1007 MA1354840
Effect of interfacial bonding on the strength of adhesion	S.M.S. Lai	The University of Akron 1993	159pp. 4700-B DA9405619

71. FILMS

Diffusion and reaction of small molecules in thin polymer films	J.L. Tardiff	Michigan State University 1993	156pp. p.1013 MA1354923

72. POLYVINYL CHLORIDE

Impract modifiers for chlorinated polyvinyl chloride	K.B. Gami	University of Massachusetts-Lowell 1993	96pp. p.1059 MA1355606

73. DEGRADATION

Biodegradation of polymeric films	A.V. Yabannavar	Rutgers The State University of New Jersey-New Brunswick and U.M.D.N.J. 1993	167pp. 522-B DA9412905
Mechanistic studies of biomass-filled polyethylene films: degradation in refuse compost	C.C.W. Chen	University of Missouri-Columbia 1993	170pp. 4796-B DA9404952
Microbiological degradation of bioadditives in polyolefins	G. Grossmayer	Universitaet Wien (Austria). In German 1990	272pp. 54/5488c
Degradation and stabilisation of diisocyanage cured polybutadiene	T. Wang	Aston University, UK 1993	55/3972c

74. NON-LINEAR OPTICAL POLYMERS

Solid-state nuclear magnetic resonance of polyphosphazenes and non-linear optical polymers	S.A. Taylor Myers	Texas A&M University 1993	186pp. 5634-B DA9410918

75. POLYMERIZATION

Observation of the first titanium alkyl/alkyl complexes: the key intermediate in the Ziegler-Natta mechanism for the polymerization of alkenes	M.D. Spencer	University of Illinois at Urbana-Champaign 1993	275pp. 5659-B DA9411788
The ring-opening polymerization of ring-strained cyclic ethers	D.P.S. Riat	Aston University, UK 1993	55/814c
Surface chemistry of monolayer and thin films at metal interfaces: quantitative electron spectroscopy for chemical analysis and secondary ion mass spectrometry of adsorption and polymerization kinetics	R.W. Johnson, Jr.	State University of New York at Buffalo 1993	274pp. 3031-B DA9330081
Organoscandium complexes as mechanistic probes in the Ziegler-Natta polymerization of α-olefins and dienes	W.D. Cotter	California Institute of Technology 1993	187pp. 3057-B DA9325347
Unsaturated organosilicon polymers via acyclic diene metathesis (ADMET) condensation polymerization	D.W. Smith, Jr.	University of Florida 1992	203pp. 3077-B DA9331218

76. POLY(METHYL METHACRYLATE)

The photochemistry of various organic molecules in a poly(methyl methacrylate) matrix	M.E. O'Brien	The University of Wisconsin-Madison 1993	138pp. 5673-B DA9407384
The molecular weight dependence on the stereocomplex formation between syndiotactic and isotactic poly(methyl methacrylate)	G.R. Webster, Jr.	Temple University 1993	236pp. 3646-B DA9332867

76. POLY(VINYLSILANES)

1. Synthesis and reactivity of triphenyl lead compounds. II. Synthetic approaches to poly(vinylsilanes) and organosilicon dendrimers	D.Y. Son	Massachusetts Institute of Technology 1993	5676-B

77. NETWORK POLYMERS

Synthesis of selectively functionalized network polymers and their application as designed catalysts	J.V. Beach	University of California, Irvine 1994	174pp. 5692-B DA9412165

78. POLY(P-PHENYLENE VINYLENE)S

The effects of counter ions on the properties of poly(p-phenylene vinylene)s and their precursor polymers	A.M.P. Beerden	Limburgs Universitair Centrum (Belgium) 1993	216pp. 5692-B DA9408368

79. ELASTOMERS

Blocking of styrene and styrene derivatives from living polyisobutylene cations: synthesis, characterization and physical properties of a series of novel polyisobutylene-based thermoplastic elastomers	W.G. Hager	The University of Akron 1994	342pp. 5693-B DA9408840

Vibrational spectroscopy of elastomers	K. Jackson	University of Southampton, UK 1993	55/3696c

80. POLYPEPTIDES

Chemical sequence control of crystallization in periodic polypeptides of the sequence poly(AG)$_x$EG	M.T. Krejchi	University of Massachusetts 1993	184pp. 5693-B DA9408296
Molecular structures of elastin and elastin-based polypeptides by computation, NMR, and circular dichroism	C.H. Luan	The University of Alabama 1993	205pp. 4561-B DA9405287
Study of mobility of polypeptide and protein structures by Monte Carlo methods and under consideration of solvation effects	B.F. von Freyberg-Eisenberg	Eidgenossische Technische Hochschule Zurich (Switzerland). In German 1993	126pp. 54/4730c
Solvent effects on chain flexibility in extended polypeptides	D.J. Houpt	University Louisville 1992	205pp. 3644-B DA9332933

81. POLYETHYLENES

Effect of reactive extrusion peroxide treatments on the molecular structure and processability of linear low density polyethylenes	M.G. Lachtermacher	University of Waterloo (Canada) 1993	295pp. 5693-B DANN84453
Fatigue acceleration of crack growth in medium density polyethylene	E.A. Showaib	Case Western Reserve University 1993	204pp. 4766-B DA9406284
Creep and stress rupture of polyethylene sheet under equal biaxial tensile stresses	D.E. Duvall	Illinois Institute of Technology 1993	180pp. 4866-B DA9406290
Photocrosslinking of low density polyethylene	Q.B. Jun	Kungliga Tekniska Hogskolan, Sweden 1992	96pp. 55/809c
Kinetic and thermodynamic studies of permeation of nonpolar organic contaminants through polyethylene membrane	G.K. Manuweera	University of Missouri-Columbia 1992	160pp. 3591-B DA9327848
A study of the effects of processing parameters on the morphologies and mechanical properties of polyethylene blown films	D.M. Simpson	The Pennsylvania State University 1993	232pp. 3645-B DA9334815
The effects of polypropylene contamination on the properties of recycled high density polyethylene	M.J. Orroth	University of Massachusetts-Lowell 1993	135pp. p.255 MA1353269
Thermo-chemical recovery of low molecular weight oligomers form polyethylene waste	W. Lo	McGill University (Canada) 1992	77pp. p.293 MAMM80314

82. WATER SOLUBLE POLYMERS

Formation, structure, and stability of surfactant-free emulsions from water soluble-dispersible polymers	K. Li	The University of Akron 1993	248pp. 5694-B DA9408841

83. NETWORKS

Synthesis and characterization of novel silanol-containing polymers and copolymers, novel silanol-containing polymer blends, interpolymer complexes and semi-interpenetrating polymer networks (IPNs)	S. Lu	Polytechnic University 1994	185pp. 5695-B DA9412200

Interpenetrating polymeric networks form diacetone acrylamide oxime blocked isocyanates	S.P. Lu	University of Detroit Mercy 1993	4700-B DA9400511
Grafting of 2-(dimethylamino)ethyl methacrylate onto squalane and eicosane, a model system for grafting 2-(dimethylamino)ethyl methacrylate onto polyethylene	J.B. Wong Shing	Queen's University of Kingston (Canada) 1993	212pp. 4702-B DANN80680
Silicone networks via living ring-opening copolymerization of octamethyl-cyclotetrasiloxane and polycyclosiloxanes	P.S. Chang	Temple University 1993	210pp. 3643-B DA9332778
Hydrophilic-hydrophobic interpenetrating polymer networks	F.O. Eschbach	The University of Connecticut 1993	195pp. 3644-B DA9332884
Molecular architecture and gelation phenomena in epoxide networks	M.E. Smith	Case Western Reserve University 1993	178pp. 511-B DA9416191
The topological trapping of cyclic polymers into polymer networks	S.J. Joyce	University of York, UK 1992	213pp. 55/3698c

84. POLYSILSEQUIOXANE

Part A. An investigation of the thermally induced cluster formation in xanthan solutions by light scattering and rheological methods. Part B. The synthesis and characterization of organic bridged polysilsequioxane xerogels and applications to nonlinear optical materials by the sol-gel method	H.W. Oviatt, Jr.	University of California, Irvine 1994	245pp. 5695-B DA9412180

85. LIQUID CRYSTALLINE POLYMERS

Synthesis and associative behaviour of novel aromatic liquid crystalline polymers	Y.K. Twu	University of Massachusetts-Lowell 1993	122pp. 5695-B DA9411472
Studies of the solvation of polystyrene by binary solvents for liquid chromatography	V.L. Antonucci	Duke University 1993	276pp. 4635-B DA9404265
Liquid crystalline polymers with complex architectures	J.M. Desper	Case Western Reserve University 1993	379pp. 4675-B DA9406258
Dielectric relaxations in side-chain liquid crystalline polymers	Z. Zhong	Case Western Reserve University 1993	227pp. 4756-B DA9406253

86. MELTS

The relationship between poly-dispersity and linear viscoelasticity in entangled polymer melts	S.H. Wasserman	Princeton University 1994	377pp. 5695-B DA9410108
An experimental investigation of interfacial instabilities in superposed flow of polymer melts	G.M. Wilson	Washington University 1993	206pp. 3229-B DA9330330
Viscoelastic properties of linear long polymers in melt	P. Tong	Tulane University 1993	137pp. 932-B DA9421057
Film casting of polymer melts	V.R. Iyengar	University of Maine 1993	242pp. 509-B DA9418402

87. DRAG-REDUCING POLYMERS

Studies of the interaction between drag-reducing polymers and a turbulent flow field using PIV and FENE bead-spring model	H. Massah Bavani	University of Illinois at Urbana-Champaign 1993	374pp. 5756-B DA9411709

88. TRANSFER MOLDING

Analysis of resin transfer molding: material characterization, molding and simulation — M.J. Perry — The Ohio State University 1993 — 263pp. 5816-B DA9412039

89. PROCESS

Process defects in polymers — K. Mallick — Arizona State University 1993 — 177pp. 5906-B DA9410989

90. BIOPOLYMERS

Evaluation of starch based biopolymers for improved oil recovery — L. Ali — Texas A&M University 1993 — 167pp. 5917-B DA9410698

91. POLY(ESTERS)

Hydrolysis and controlled release characteristics of biodegradable poly(esters) — E.A. Schmitt — The University of Iowa 1993 — 241pp. 4624-B DA9404533

92. POLYPYRROLE

Polpyrrole films for the development of chemical sensors — S.J. Vigmond — University of Toronto (Canada) 1993 — 180pp. 4641-B DANN82879

93. POLYTHIOETHERS

Structural and metal-binding studies of preorganized polythioesters — J.M. Desper — The University of Wisconsin-Madison 1993 — 407pp. 4670-B DA9330166

94. POLYCARBONATES

Molecular dynamics of substituted polycarbonates and model monomers by NMR — Y.H. Lee — Purdue University 1993 — 301pp. 4691-B DA9403737

Studies of the fatigue crack initiation in polycarbonate — T.J. Chen — University of Illinois at Chicago 1993 — 134pp. 3700-B DA9335118

95. BLENDS

A study of the effects of bromination of the thermal properties of poly(2,6-dimethyl-1,4-phenylene oxide) and its blends with polystyrene — R.C. Bopp — Rensselaer Polytechnic Institute 1993 — 301pp. 4698-B DA9405657

Dynamic mechanical properties of epoxy resin/epoxidized rubber blends — P.J. Bussi — Case Western Reserve University — 172pp. 4698-B DA9406277

Complexing blends of flexible ionomers and rigid rod polymers by specific interactions — L.Y. Shao — The University of Connecticut 1993 — 232pp. 4814-B DA9405269

Rheo-optical studies on the phase transition of polymer blends — R.J. Wu — The University of Connecticut 1993 — 275pp. 4816-B DA9405273

Component contributions to the dynamics of miscible polymer blends — J.A. Zawada — Stanford University 1993 — 163pp. 4816-B DA9404047

Probing the limits of toughening by diluent-induced localized plasticization in polymer blends — S.H. Spiegelberg — Massachusetts Institute of Technology 1993 — 3226-B

The morphology of liquid crystalline polymers and their blends — K.R. Schaffer — University of Massachusetts 1993 — 178pp. 3325-B DA9329664

Unsaturated polyester-poly(vinyl acetate) blends: Study of morphology and shrinkage compensation — M. Ruffier — Institut National des Sciences Appliquees de Lyon (France). In — 167pp. 54/5456c

Preparation and study of polymer blends of recycled poly(ethylene terephthalate) and virgin poly(butylene terephthalate)	D.H. Mehta	French 1993 University of Massachusetts-Lowell 1993	163pp. p.1927 MA1352262
Structure and deformation behaviour of amorphous ionomers, their blends and plasticized systems	X.Ma. Rutgers	The State University of New Jersey-New Brunswick 1993	179pp. 3644-B DA9333425
Ternary polymer blends: miscibility, crystallization and melting behaviour and semi-crystalline morphology	M.J.M. Vanneste	Katholieke Universiteit Leuven, Belgium 1993	130pp. 55/3703c
Poly(styrene-co-N-maleimides): preparation by reactive extrusion, molecular characterization and miscibility behaviour in blends	I. Vermeesch	Katholieke Universiteit Leuven, Belgium 1993	117pp. 55/3705c

96. NONLINEAR OPTICAL POLYMERIC

Relaxations in second order non-linear optical polymeric systems	J.I. Chen	University of Massachusetts-Lowell 1993	152pp. 4698-B DA9405119

97. LATEX PARTICLES

An investigation into the structure and breakup of aggregated latex particles	M. Durali	Lehigh University 1993	230pp. 4699-B DA9406006

98. POLYMETHYLENE

A study of the atomistic details of structure and dynamics of polymethylene crystals via molecular dynamics simulations	G. Liang	The University of Tennessee 1993	200pp. 4700-B DA9404590

99. BIOMOLECULES

Design of novel intelligent material by incorporating biomolecules into electroactive polymeric thin film systems	J.O. Lim	University of Massachusetts-Lowell 1993	128pp. 4700-B DA9405123

100. POLYPYRROLE

Investigation of the electropoly-merization kinetics and deposition processes during the formation of electronically conducting polypyrrole	D.E. Raymond	University of Alberta (Canada) 1993	240pp. 4701-B DANN81985

101. POLY(METHYL METHACRYLATE)

In situ laser interferometry and fluororescence quenching measurements of poly(methyl methacrylate) thin film dissolution	F. Wang	University of Toronto (Canada) 1993	231pp. 4702-B DANN82785

102. INTERFACES

Thermodynamics and kinetics of polymer adsorption at interfaces	Z.M. Gao	Lehigh University 1993	150pp. 4722-B DA9406013
Chemical reactions at the polymer/ substrate interfaces as determined using X-ray photoelectron spectroscopy (XPS) and infrared spectroscopy techniques	R. Horner	University of Cincinnati 1993	200pp. 3286-B DA9329956

103. POLYOLEFIN

Steric stabilization in polyolefin asphalt emulsions	S.A.M. Hesp	University of Toronto (Canada) 1991	183pp. 4809-B DANN82769

104. RUBBER

N-nitrosamines in rubber	E. Annys	Rijksuniversiteit te Gent, Belgium 1993	146pp. 55/665c
Thermo-oxidative degradation and stabilization of rubber materials	B. Mattson	Kungliga Tekniska Hogskolan, Sweden 1993	55/810c
Anisotropy in rubber materials induced by processing or fatigue	H. Lavebratt	Kungliga Tekniska Hogskolan, Sweden 1992	166pp. 55/1334c
Impact modification of polybutylene terephthalate (PBT) by blending with ethylene propylene diene rubber (EPDM)	C.J. Parikh	University of Massachusetts-Lowell 1993	112pp. p.1927 MA1352270

105. RAMAN SPECTROSCOPY

Analytical Fourier transform Raman spectroscopy: development, and applications to polymeric systems	G.J. Ellis	University of Southampton, UK	55/673c
Fourier transform Raman spectroscopy of some polymeric materials	C.H. Jones	University of Southampton, UK 1992	55/808c

106. BIODEGRADABLE

Biodegradable ion-exchange microspheres for site-specific delivery of adriamycin	CH.F.M. Cremers	Universiteit Twente, The Netherlands 1993	156pp. 55/772c

107. MATRIX

The degradation of gel-spun poly(β-hydroxybutyrate) fibrous matrix	L.J.R. Foster	Aston University, UK 1992	55/803c

108. POLYVINYLCHLORIDE

Ferrocene containing smoke suppressants and flame retardants for semi-rigid polyvinyl chloride	J. Grant	Council for National Academic Awards, UK 1992	332pp. 55/805c

109. POLYMER SYNTHESIS

The reaction of isocyanates with carboxylic acids and its application to polymer synthesis	A.H.M. Schotman	Technische Universiteit te Delft (The Netherlands) 1993	136pp. 55/815c

110. HYDROGEL POLYMERS

Synthesis of novel cyclic monomers for hydrogel polymers	C.B. St. Pourcain	Aston University, UK 1992	55/816c

111. POLYTETRAHYDRO-FURANEN

Synthesis of new telechelic polytetrahydrofuranen	G.G. Trossaert	Rijksuniversiteit te Gent, Belgium. In Dutch 1993	140pp. 55/818c

112. ROTATIONAL MOULDING

Rotational moulding of reactive plastics	E.M.A. Harkin-Jones	Queen's University of Belfast, Northern Ireland 1992	424pp. 55/1333c

113. POLY (VINYLFERROCENE)

Part I. Voltammetric studies of potassium iodide at gold and platinum electrodes. Part II. Electrode position and characterization of poly(vinylferrocene) films — R.J. Holt — State University of New York at Buffalo 1993 — 215pp. 3029-B DA9330077

114. EMULSION POLYMERIZATION

Tubular reactors for emulsion polymerization — D.A. Paquet, Jr. — The University of Wisconsin-Madison 1993 — 292pp. 3222-B DA9315001

115. CONDUCTIVE POLYMER

Conductive polymer fibers and low dimensional organic crystals — S. Li — University of Minnesota 1993 — 129pp. 3288-B DA9331922

116. BIODEGRADATION

Biodegradation, biocompatibility, strength retention and fixation properties of polylactic acid rods and screws in bone tissue: An experimental study — A.E. Majola — Helsingin Yliopisto (Finland) — 117pp. 54/4944c

117. INJECTION MOLDING

Calculation and control of heat transfer in injection molding — K.M.B. Jansen — Technische Universiteit te Delft (The Netherlands) 1993 — 145pp. 54/5270c

Development of a cavity pressure based production monitoring system for injection molding — A.P. Chinnaswamy — University of Massachusetts-Lowell 1993 — 67pp. p.1926 MA1352880

Injection molding process monitoring based on cavity pressure — W.J. Qiu — University of Massachusetts-Lowell — 83pp. p.1927 MA1352652

Fiber orientation during injection molding of glass-fiber-reinforced thermoplastics: an experimental and numerical investigation — J.P. Greene — The University of Michigan 1993 — 443pp. 3732-B DA9332073

Sensor concepts for polymer injection molding — C.L. Thomas — Drexel University 1993 — 129pp. 3822-B DA9333519

118. PVC MEMBRANE

Ion and water transfer in a potassium-selective PVC membrane — K. Syverud — Norges Tekniske Hogskole (Norway) 1992 — 109pp. 54/5268c

119. THERMAL DEGRADATION

Volatile compounds from large-scale thermal degradation of polystyrene, poly(acrylonitrile-styrene) and poly(acrylonitrile-butadiene-styrene) plastics — M.M. Shapi — Helsingin Yliopisto (Finland) 1992 — 111pp. 54/5187c

Permeation of isomeric diethers through oriented polypropylene film and the thermal degradation of vinylidene chloride/butyl acrylate and vinylidene chloride/masked 3,4-dihydroxystyrene copolymers — Z. Ahmed — Central Michigan University 1993 — 254pp. p.255 MA1353693

120. POLYTETRAHYDRO-FURANS

Evaluation of selected aliphatic acids and polytetrahydrofurans for use as calibration agents for thermospray HPLC/MS	T.R. Dombrowski	University of Nevada, Las Vegas 1993	66pp. p.1791 MA1352612

121. BIOMATERIAL

Surface modification of a biomedically important polymer, poly(ethylene terephthalate): an initial step towards the creation of a permanent non-thrombogenic biomaterial	L.N. Bui	University of Toronto (Canada) 1992	103pp. p.1806 MAMM78598

122. VISCOELASTIC PROPERTIES

The structure and viscoelastic properties of oriented polypropylene	Y. Jian	University of Toronto (Canada) 1992	114pp. p.1857 MAMM78544

123. CARBON FIBRE

Mechanical properties of aligned short carbon fibre reinforced epoxy	M.M.C. Ko	University of Toronto (Canada) 1992	113pp. p.1857 MAMM78638

124. PVC COMPOUND

The effect of PHR variations on the physical properties of a flexible PVC compound	J.E. Macaluso	University of Massachusetts-Lowell 1993	114pp. p.1927 MA1352261

125. THERMOPLASTICS

Production of thermoplastic elastomers based on an ethylene-co-vinyl acetate soft phase by blending with commercial thermoplastics	F. Niedermeier	University of Massachusetts-Lowell 1993	144pp. p.1927 MA1352646
Synthesis and characterization of co-polymers containing sequential thioethylene units: potential engineering thermoplastics	J. Muthiah	The University of Southern Mississippi 1993	141pp. 4183-B DA9402542

126. URETHANES

Synthesis and characterization of polyol prepolymers from lignin and their use in urethanes	F.F. Chang	University of Detroit Mercy 1993	221pp. 3643-B DA9321931

127. CONDUCTING POLYMERS

Electrochemical studies of heterocyclic conducting polymers	R. John	University of Wollongong (Australia) 1992	3644-B
Synthesis and characterization of organic and organometallic conducting polymers and oligomers	M.O. Wolf	Massachusetts Institute of Technology 1994	894-B
Charge localization and delocalization phenomena in conducting polymers	J. Joo	The Ohio State University 1994	196pp. 961-B DA9420970

128. POLYIMIDES

Synthesis and characterization of soluble, high temperature aromatic polyimides	T.M. Moy	Virginia Polytechnic Institute and State University 1993	265pp. 3645-B DA9400146

129. CARBOSILANE POLYMERS

The synthesis and characterization of unsaturated carbosilane polymers — S.J. Sargeant — University of Southern California 1993

130. COLLOIDS

Plasma protein interactions with co-polymer-stabilized colloids — J.T. Li — The University of Utah 1993 — 335pp. 3735-B DA9332576

131. STRUCTURE FACTOR

Light scattering study of the structure factor of dilute polymer coils in solution — L. Sung — University of California, Santa Barbara 1993 — 166pp. 3681-B DA9334888

132. POLY(PROPYLENE)

Gas vapor transport in rolltruded polymers: isotactic poly(propylene) and poly(aryl-ether-ether-ketone) — R.J. Ciora, Jr. — University of Pittsburgh 1993 — 182pp. 3730-B DA9333150

133. POLY(2-VINYL PYRIDINE)

Template effect on the copolymerization of styrene and methacrylic acid in the presence of poly(2-vinyl pyridine) — Q. Xu — State University of New York at Albany 1993 — 62pp. 3646-B DA9333234

134. POLYAMIDE

The importance of cavitation process in the toughening of epoxy and polyamide — D. Li — The University of Michigan 1993 — 225pp. 3799-B DA9332119

135. POLYSTYRENE

The effect of stress rate on crack damage evolution in polystyrene and PEEK — B.L. Gregory — University of Illinois at Chicago 1993 — 159pp. 3798-B DA9335124

136. SUSPENSION POLYMERIZATION

Suspension polymerization of methyl methacrylate initiated by benzoyl peroxide: kinetics of initiation and modelling of polymerization reactions — S.S. Shetty — University of South Florida 1993 — 218pp. 3739-B DA9323701

137. FLUOROCARBON COMPOUNDS

Gas permeation through composite membranes prepared by plasma polymerization of fluorocarbon compounds — S.J. Oh — Polytechnic University 1994 — 149pp. 3737-B DA9332722

138. NONLINEAR OPTICAL MATERIALS

Computer-aided selection of polymeric nonlinear optical materials — S.E. Zutaut — The University of Alabama in Huntsville 1993 — 107pp. 3805-B DA9333526

139. DRUG DELIVERY

Characterization of factors affecting *in vitro* degradation of poly(D,L-lactide-co-glycolide) for drug delivery	A.C.G. Hausberger	University of Kentucky 1993	283pp. 4096-B DA9402838

140. ELECTROACTIVE POLYMERS

Syntheses and properties of poly([6.2]cyclophane-1,5-diene)s: electroactive polymers	D.J. Guerrero	The University of Oklahoma 1993	127pp. 4159-B DA9400077

141. POLYARYLENES

Structural effects on the electronic and electrochemical properties of polyarylenes	A.D. Child	The University of Texas at Arlington 1993	184pp. 4182-B DA9402807

142. DRAG REDUCTION

Investigations of polymeric drag reduction: a proposed model and experimental evidence	J.P. Dickerson	The University of Southern Mississippi 1993	204pp. 4183-B DA9402524

143. WATER-SOLUBLE POLYMERS

The synthesis and characterization of hydrophobically associating water-soluble polymers	F.S.Y. Hwang	University of Southern California 1993	4183-B

144. POLYOLEFINS

Ethylene-poly(alkene oxy) block cooligomers as reagents for surface modification of polyolefins	B. Srinival	Texas A&M University 1993	191pp. 4184-B DA9403589
Time controlled photo-oxidation of polyolefins by polymer bound additives	S. Islam	Aston University, UK 1992	55/3695c

145. POLY(AMIDE)S

2,5-Thiophene-based liquid crystalline poly(amide)s, poly(arylene ether ketone)s, and poly(benzimidazole)s	S. Stompel	The University of North Carolina at Chapel Hill 1993	176pp. 4184-B DA9402187

146. PHOTOPOLYMERS

Holographic studies of liquid crystalline photopolymers	J. Zhang	Syracuse University 1993	153pp. 4185-B DA9401722

147. ELECTRON BEAM RESIST

The role of residual casting solvent in determining the lithographic and dis-solution behaviour of poly(methyl methacrylate), a positive electron beam resist	R.R. Criss, Jr.	The University of Texas at Dallas 1993	170pp. 4224-B DA9334688

148. POLY(VINYL CHLORIDE)

Effects of plasticizers on the behaviour of highly practicized poly(vinyl chloride) ion-selective electrode membranes	M.A. Simon	The University of North Carolina at Chapel Hill 1993	116pp. 4271-B DA9402184

149. ACOUSTIC MISSION

Failure mechanism characterization in graphite/epoxy using acoustic emission data	T.M. Ely	Embry-Riddle Aeronautical University 1993	103pp. p.260 MA1353831

150. EPOXY

The design, construction, analysis, and testing of an elastically tailored carbon fiber/epoxy wing box structure	J.D. Baldwin	Mississippi State University 1992	59pp. p.286 MA1353312
Ultrasonic imaging of ply orientation in graphite epoxy laminates using oblique incidence techniques	R.W. Sullivan	Mississippi State University 1993	136pp. p.332 MA1353907

151. POLYETHYLENE TEREPHTHALATE

Sorption of carbon dioxide on annealed polyethylene terephthalate	Z.J. Sethna	University of Louisville 1993	234pp. p.295 MA1353551

152. METALLOCENE POLYMERS

Synthesis and characterization of metallocene polymers	K. Jang	Brandeis University 1994	245pp. 902-B DA9417720

153. POLY(1,4-PHENYLENE VINYLENE)

The development of versatile substituent placement in the polyelectrolyte soluble precursor synthesis of poly(1,4-phenylene vinylene)s and their analogues	A.M. Sarker	University of Massachusetts 1994	188pp. 907-B DA9420685

154. POLYHETEROCYCLIC

Electrically conductive polyheterocyclic compounds: a mechanistic study of the oxidative polymerization of thiophenes and pyrroles (Volumes I and II)	J. Tian	Drexel University 1994	513pp. 932-B DA9421438

155. VISCOELASTIC

Dynamic viscoelastic studies of sol-gel transitions	Q. Yu	University of Southern California 1993	933-B

156. ADHESIVES

Organosolv lignin characterization, modification and application in phenol-formaldehyde adhesives	S.G. Allen	University of Toronto (Canada) 1993	181pp. 1032-B DANN86267

157. HYDROPHILIC POLYMERS

Transport and interaction of ionizable drugs and proteins in hydrophilic polymers	M.T. am Ende	Purdue University 1993	426pp. 1033-B DA9420779

158. POLY-β-HYDROXYBUTYRATE

Multivariable control in the continuous production of poly-β-hydroxybutyrate	P.A. Gostomski	Rensselaer Polytechnic Institute 1993	160pp. 1036-B DA9420323

159. POLY(ETHYLENE TEREPHTHALATE)

The thermal and mechanical behavior of poly(ethylene terephthalate) fibers incorporating novel thermotropic liquid crystalline polymers	S.L. Joslin	University of Massachusetts 1994	164pp. 1124-B DA9420637

160. MICROSPHERES

Characterization of metered dose inhaler formulations of poly(D,L-lactide-co-glycolide) microspheres	M.G. Kulkarni	University of Kentucky 1994	233pp. 381-B DA9418507

161. POLYORGANOSILICON

Synthesis and chemistry of polyorganosilicon compounds. Mechanistic and structure-property relationships by spectral and thermal techniques, and positive-ion and collisionally activated dissociation mass spectral studies of organosilanes	K.W. Beyene	Cleveland State University 1993	340pp. 442-B DA9418109

162. POLYISOBUTYLENE

Synthesis and characterization of novel, well-defined, polyisobutylene-based polymers	B.J. Chisholm	The University of Southern Mississippi 1993	264pp. 443-B DA9417826

163. POLY(ARYLENE VINYLENES)

Synthesis and doping of poly(arylene vinylenes)	W.A. Eevers	Universitaire Instelling Antwerpen (Belgium) 1993	177pp. 444-B DA9419250

164. POLYBENZOXAZOLES

Synthesis and characterization of high performance polybenzoxazoles	W.D. Joseph	Virginia Polytechnic Institute and State University 1993	365pp. 444-B DA9419216

165. POLY(AMIDE-IMIDES)

Synthesis and characterization of soluble, melt-processible aromatic poly(amide-imides)	V.N. Sekhari-puram	Virginia Polytechnic Institute and State University 1994	277pp. 445-B DA9419231

166. DECOMPOSITION

Perfluoropolyalkylether decomposition on catalytic aluminas	W. Morales	Cleveland State University 1994	117pp. 510-B DA9416880

167. POLY(3-OCTYLTHIOPHENE)

Electrochemical characteristics of poly(3-octylthiophene) film electrodes	J.B.M. Bobacka	Abo Akademi, Finland 1993	122pp. 55/3568c

168. THERMOSENSITIVE POLYMERS

Thermosensitive polymers and hydrogels based on N-isopropyl-acrylamide	H. Feil	Universiteit Twente, The Netherlands 1994	153pp. 55/3689c

169. POLY[(R)-3-
HYDROXYALKANOATES]
Prospects of bacterial poly[(R)-3-
hydroxyalkanoates]   G.J.M. deKoning   Technische Universiteit   132pp. 55/3687c
                                                           Eindhoven, The
170. POLY(ETHER ETHER   Netherlands 1993
KETONE)S
The preparation and
characterisation of   J.E. Denness   University of York, YK   199pp. 55/3688c
trifluoromethylated poly(ether   1992
sulphone)s and poly(ether ether
ketone)s

INDEX

339

Printed and bound by CPI Group (UK) Ltd, Croydon, CR0 4YY

24/10/2024

01778291-0007